IMAGING IN HIGH ENERGY ASTRONOMY

IMAGING IN HIGH ENERGY ASTRONOMY

Proceedings of the International Workshop held in Anacapri
(Capri–Italy), 26–30 September 1994

Edited by

L. BASSANI

and

G. DI COCCO

Istituto Tecnologie e Studio Radiazioni Extraterrestri, CNR,
Bologna, Italy

Partly reprinted from *Experimental Astronomy*, Vol. 6, No. 4, 1995

SPRINGER SCIENCE+BUSINESS MEDIA, B.V.

Library of Congress Cataloging-in-Publication Data

```
Imaging in high energy astronomy   proceedings of the international
  workshop held in Anacapri (Capri-Italy) 26-30 September 1994 /
  edited by L. Bassani, G. Di Cocco.
       p.   cm.
  "The International Workshop on Imaging in High Energy Astronomy
was held on the Island of Capri (Italy) between the 26th-30th
September 1994"--Pref.
  Includes index.
  ISBN 978-0-7923-3788-1     ISBN 978-94-011-0407-4 (eBook)
  DOI 10.1007/978-94-011-0407-4
  1. Imaging systems in astronomy--Congresses.   I. Bassani, L.
II. Di Cocco, G.  III. International Workshop on Imaging in High
Energy Astronomy (1994   Capri, Italy)
QB51.3.I45I474  1995
522'.2--dc20                                        95-25632
```

ISBN 978-0-7923-3788-1

Printed on acid-free paper

CONTENTS

Contributed Papers

Modulation Techniques

New Detectors and Imaging Methods

PREFACE

The International Workshop on Imaging in High Energy Astronomy was held on the Island of Capri (Italy) between the 26th-30th September 1994. Eleven years had passed since the first and only previous conference on this subject,which was held in Southampton,U.K, in July 1983. During this period we have witnessed many developments in the field of astronomical x and gamma-ray Imaging techniques and we have all become acquainted with images of the sky at high energies.However there was never the opportunity for all interested people to meet and discuss the new techniques developed and the problems encountered,or to consider new ideas and prospects for the future of Imaging in high energy astronomy.The Capri Workshop was organized precisely to give the scientific community this opportunity and we hope that it fulfilled this expectation.

On behalf of all those involved in the organization of the meeting, we also express the hope that these Proceedings, which contain an almost complete collection of all the papers given at the meeting, will be useful to scientists involved in the field.

As a general consideration it seems to us that the content of this volume represents the state of art in the field of Imaging in High Energy Astronomy at the present date. The high quality of the papers was assured by a refereeing process made possible by a group of collegues who took upon themselves the not easy task to provide authors with their comments and evaluation. To all of them our deepest thanks.

The attendence at the meeting of approximately 100 active researchers in the field can be directly attributed to the entusiasm of both the Scientific Committee (G.Di Cocco, A.J.Dean, G.J. Fishman, J. Paul, T. Prince, J.P. Roques, V. Schönfelder, G. Skinner, P. Ubertini) e Local Organizing Committee (E. Caroli, G. Malaguti, F. Schiavone, A. Spizzichino, J.B. Stephen); we wish to thank them all for their effort.

We also like to acknowledge the work of all session Chairmen,the financial help of the CNR, Laben and CREASO and the help of the local staff at the Europa Palce Hotel in Anacapri, where the meeting took place.

As a final remark we like to say that we left Capri with a sense of optimism which we hope all partecipants shared with us: after successfull missions like Granat and CGRO, the future looks extremely promising in the field of Imaging in High Energy Astronomy, naturally due to the next gamma-ray mission, INTEGRAL, but also because of the many new ideas and prospects that lie ahead of it.

L. Bassani
G. Di Cocco

CODING (AND DECODING) CODED MASK TELESCOPES

G. K. SKINNER

University of Birmingham, Edgbaston, Birmingham B15 2TT, UK

June 7, 1995

Abstract. Many designs of masks for coded aperture telescopes have been proposed and a number of different configurations for instruments considered. Their advantages and disadvantages and some of the considerations involved in designing an instrument and in choosing a mask are reviewed. The methods of image reconstruction, which strongly influence the choice of design, are discussed and a way of quantifying the effectiveness of a mask pattern when used with a detector of finite resolution is presented.

Key words: X-rays – Gamma-rays – Imaging – Coded-masks

1. Introduction

Coded mask telescopes are increasingly used for hard X-ray and gamma-ray observations – for example all three of the X/γ-ray instruments on the INTEGRAL mission (Winkler *et al.*, 1994) are expected to use this technique. With such instruments an image is not recorded directly but is recovered only after the recorded shadow of the mask is processed by a 'reconstruction' algorithm. This post-processing operation can be regarded as an intrinsic part of the instrument, even though it is usually carried out retrospectively on the ground. Some aspects of the choice of a mask pattern and of image reconstruction algorithm are reviewed here.

Fig. 1 shows an image from a simulation of a γ-ray observation of the central part of the galaxy with a coded mask telescope. The source distribution which was assumed contained 8 point sources. Most are discernable, but there is a halo around each and also ghost peaks. The imperfect reconstruction may be understandable given that the mask pattern used was that in Fig. 2 – far from the traditional mask designs which are well known and discussed below! In fact a wide variety of mask patterns can be used (as illustrated in a companion paper, Skinner and Rideout, 1995). Some, however, are more satisfactory than others; §3 will discuss why.

2. Reconstruction algorithms

The image in Fig. 1 was obtained by correlating the (simulated) recorded data pattern with the mask pattern. One way of justifying the use of this procedure is as follows. Consider the case where there is only one source, in position j. There is then a function P_{ij} which describes, for every pixel i in the detector, whether the source illuminates that pixel or not. Ideally P_{ij}

Experimental Astronomy **6**: 1–7, 1995.

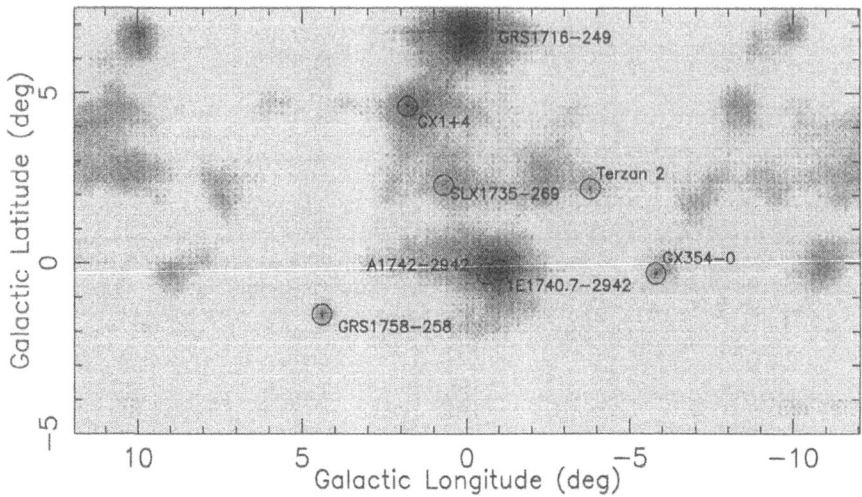

Fig. 1.   Reconstruction of a simulated coded mask observation of the galactic centre region

will be 0 or 1 depending on whether the line joining the source to a given pixel passes through a transparent or opaque part of the mask. In real cases, because of blurring in the detector, or because of the properties of practical materials we may wish to attribute intermediate values to the P_{ij}.

Fig. 3 illustrates how the signal D_i detected in each pixel might vary with P_{ij}. If the source is weak or absent we will just see detector background; if it is strong there will be a linear dependence of the detected signal on P_{ij}. The gradient of a line through the points gives an estimate of the strength of the source at the assumed position. Using standard formulae for the best fit line, the best estimate of the source strength is:

$$\hat{S}_j = \frac{4}{n\Delta^2}\sum_i D_i P_{ij} - \frac{4}{n^2\Delta^2}\sum_i D_i \sum_i P_{ij} \pm \frac{2}{\Delta}\sigma_D \qquad (1)$$

$$= \frac{4}{n\Delta^2}\sum_i D_i(P_{ij} - \bar{P}_j) \pm \frac{2}{\Delta}\sigma_D \qquad (2)$$

where there are n detector pixels and where $\Delta^2/4 = \frac{1}{n}\sum_i P_{ij}^2 - \left(\frac{1}{n}\sum_i P_{ij}\right)^2$. It is here assumed that the uncertainty in each data point, σ_D, is the same, which will be the case if the background is uniform and dominates over detected photons.

If the mask pattern is used cyclically, or if the entire shadow of the mask is always recorded (*i.e.* in what is sometimes known as the fully coded field of view – FCFOV), Δ will be a constant which is characteristic of the mask pattern. For reasons which become apparent in §4, we will refer to it as the 'coding power' of the mask pattern. Δ is actually twice the *rms* deviation of

Fig. 2. Pattern used as a mask in the simulation in Fig. 1

the mask transmission, as seen by the detector, from its mean. Similarly, in the cyclic case or within the FCFOV the $\sum_i P_{ij}$ term is a constant. So, for a given dataset, is $\sum_i D_i$. Thus the second term in (1) is a constant offset in the image, while the first is proportional to the cross-correlation between the observations D_i and the P_{ij} values which describe the mask pattern. In the case where there is assumed to be only one source, the correlation with the mask pattern (suitably scaled) is thus shown to provide the best estimate of the source intensity. So we see there is a sound theoretical basis for using correlation maps, though the map heights at positions away from a single known source must not necessarily be regarded as representing source intensity distribution.

In the presence of (potentially) more than one source, more sophisticated analysis techniques can be considered. In the limit it is possible, in principle, to solve for n source intensities simultaneously (or $n - 1$ plus a background term). This amounts to inverting the $n \times n$ matrix of P_{ij} values (in which the row of values P_{in} corresponding to the background 'pseudo-source' is all 1s). For n up to a few hundred the matrix inversion is perfectly practicable, but of course, as is usual with inverse problems, difficulties can arise from the amplification of noise.

Multiplying the vector of observations by the inverse matrix corresponds to using for the estimate of a given S_j a linear combination of the D_i in which the weights are in general different from those in (1). As (1) gives the best estimate in the presence of random noise, the susceptibility to noise will in general be increased by the removal of cross-talk between source pixels.

An alternative way to consider the problem is in the Fourier domain. This is particularly natural for cyclic systems completely sampled on a reg-

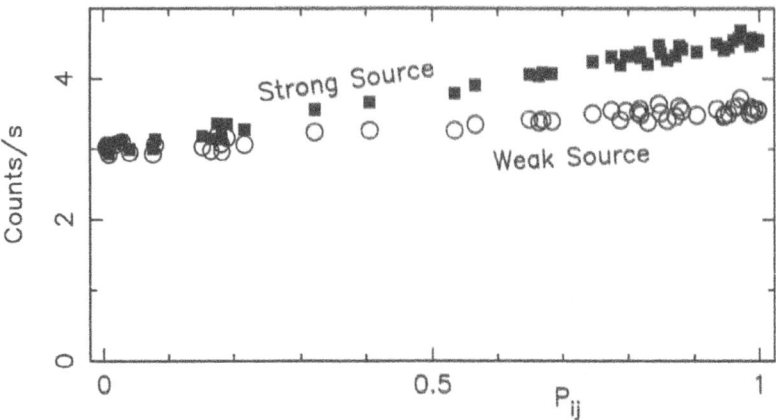

Fig. 3. Illustration of a possible scattergram of count rate in a pixel against P_{ij} (see text).

ular grid, which have a position independent point source response function (PSF). The PSF of such a system is the autocorrelation function (ACF) of the mask pattern (*e.g.* Fenimore and Cannon, 1978). Ideally the PSF (and hence the ACF) should be a δ-function and so would have all spatial frequencies present equally. The Fourier transform of the ACF of a function is the power spectrum of that function, so to achieve a δ-function PSF the mask pattern should have a flat power spectrum. Thus the sidelobes and ghosts apparent in Fig. 1 result from spatial frequencies which are under- or over-represented in the mask pattern. These spatial frequency components can be re-emphasised or attenuated in the recorded data as part of the reconstruction process, but again at the expense of amplification of random noise. The Fourier Transforms and the filtering are linear operations, so again the estimate of a given S_j is a linear combination of the D_i, which in general is different from that given by (1), and so not optimum from the point of view of signal to noise ratio.

3. Mask Patterns

For some mask patterns, it may be that the estimate $\hat{S}$ provided by (1) is in practice independent, or almost independent, of the presence of another source in the field of view. If we consider only the first term in (1), the independence will be achieved if

$$\sum_i P_{ij} P_{ik} = 0 \qquad (j \neq k). \qquad (3)$$

As the P_i are necessarily positive, this cannot be achieved, but it is possible to find mask patterns for which

$$\sum_i P_{ij}P_{ik} = K \qquad\qquad (j \neq k), \qquad\qquad (4)$$

where K is a constant. Skinner and Ponman (1994) have discussed the degree of residual correlation that remains between $\hat{S}$ values in this case, which can be very small.

It is the mask patterns which fulfil (4) that are described as 'optimum coded'. An important class of such patterns are closely related to cyclic differences sets (CDSs). In fact any rectangular mask pattern for which (4) is true and in which the numbers of elements in the two orthogonal directions, N_x, N_y, have no common factors can be used by the reverse of the process described by Proctor *et al.* (1979) to generate a CDS. Such patterns are often referred to as 'URAs', though strictly this term refers only to those related to CDSs of the twin prime class and with $(N_x N_y + 1)/2$ open elements (Fenimore and Cannon, 1978).

There are also some rectangular arrays which are optimum coded, and in which N_x, N_y do have common factors, that do not correspond to conventional 1-d cyclic difference sets. The relevence of these to coded mask telescope design has only recently been appreciated (Kopilovich and Sodin, 1994); they include square configurations which may simplify instument design.

If condition (4) is obeyed, then the ACF of the mask pattern must be constant for all non-zero offsets and the PSF is thus a delta-function super-imposed on a flat background. The inverse matrix corresponding to coding by such arrays is, within a multiplying factor and an offset, just the transpose of the forward matrix and so no undue noise amplification takes place. Furthermore, suitably sampled, the power spectrum is flat.

Thus for masks with this property, correlation with the mask pattern, multiplication by the inverse matrix, and inverse filtering are (with suitable scale factors and offsets) the same thing – all of the advantages of the various reconstruction techniques can be had simultaneously.

Gottesman and Fenimore (1989) introduced the concept of MURAs (Modified URAs). For these the inverse matrix is very similar to the transpose of the forward matrix, differing from it (again apart from a factor and an offset) only in one element per row. Like that for 'optimum coded' masks, it is skew-symmetric (so multiplying by it can be achieved by cross-correlation with a single row) and is bi-valued (so that noise propogation is minimised). The ACF is not precisely constant for all non-zero offsets, but is nearly so (see *e.g.* Skinner and Rideout, 1995).

4.   Coding Power of Mask

The uncertainty term in (1) allows us to consider how to choose a mask pattern for best sensitivity. For a given detector noise, the uncertainty is inversely proportional to the coding power, Δ, of the mask pattern, which is simply proportional to the *rms* deviation of P. As P necessarily lies in the range 0–1, the best one can do is $P = 0$ and $P = 1$ occuring equally frequently, with no intermediate values. This is the well known result that in the case we are considering (detector background dominates) the mask should be 50% open, 50% closed, which gives $\Delta = 1$. The concept of coding power tells us nothing about the way to lay out the mask as Δ depends only on the *proportions* of different transmissions which occur. It can be used however to assess proposed coding schemes, in particular ones in which intermediate transmissions occur. For example a similar concept has been used by Skinner and Grindlay (1993) to consider the effectiveness of masks with some elements whose transmission varies with photon energy.

5.   Real detectors

It has so far been assumed that the detector is perfect, but in practice the spatial resolution is likely to be limited. Caroli *et al.* (1987) have modelled the loss of sensitivity resulting from recording the shadowgram with a detector whose spatial response can be described by a Gaussian function. The coding power approach can be used as a way of quantifying the loss in this and other cases (In't Zand *et al.*, 1994, have used an approach to this problem very similar to that given here).

Use of a detector of limited spatial resolution is equivalent to recording with an ideal detector the shadow cast by a blurred version of the mask. The P values for the blurred mask will include intermediate values as well as 0s and 1s, so the coding power will be reduced. In fact for the case considered by Caroli *et al.*, one needs to evaluate Δ for a function which is the mask pattern convolved with a Gaussian of a width corresponding to the detector resolution.

Another case which is of increasing importance is where the detector is divided into a large number of well defined pixels of finite size (*e.g.* CCD pixels, discrete scintillator elements, or Germanium detectors). If we assume that all relative positions of the boundaries of the detector pixels and the mask element shadows are equally likely to occur, then a typical value for the sensitivity can be found by considering the coding power of the mask convolved with the detector pixel shape (*e.g.* a square or circular patch).

Fig. 4 shows this for a URA mask with square elements when the shadow is recorded by a pixellated detector with square pixels of various sizes. For this particular case it is easy to show that

$$\Delta = \begin{cases} (1 - \frac{d}{3m}) & \text{if } m > d \\ \frac{m}{d}(1 - \frac{m}{3d}) & \text{if } m < d \end{cases}$$

The relationship for elements which are hexagonal or other shapes is extremely close to that for square ones. The figure also shows the relationship for a

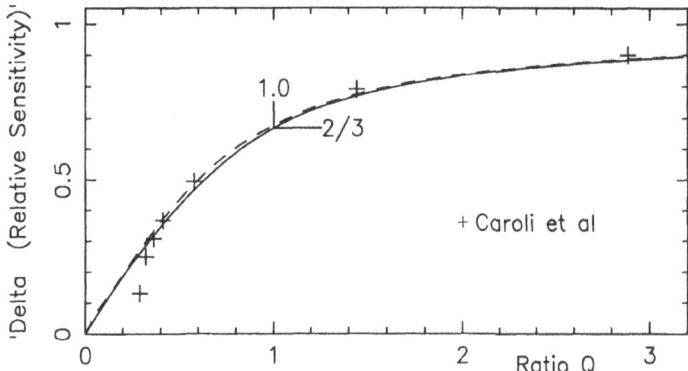

Fig. 4. Coding power for a URA-based mask with square elements of size m viewed by a detector with square pixels of side d as a function of the ratio $Q = m/d$. The corresponding curve for a continuous detector is plotted as a dotted line, using $Q = m/(2\sqrt{3}w)$ where w is the σ of the Gaussian function describing the detector resolution. The numerical factor $1/(2\sqrt{3})$ is the *rms* uncertainty in a variable which is known to plus or minus half a pixel.

continuous detector with Gaussian resolution function.

6. Conclusions

The theory behind the selection of mask patterns for coded mask telescopes in the idealised case is relatively well understood, as are the techniques available to analyse the data. The effect of finite detector resolution is, however, only one of many complicating effects which are encountered in practice. Design choices for new generations of coded mask telescope will involve the exploration of these as well as the considerations discussed above.

References

Caroli *et al.*: 1987, *Space Sci. Rev.* **45**, 349.
Fenimore, E.E., and Cannon, T.M.: 1978, *Appl. Opt.* **17**, 337.
Gottesman, S.R. and Fenimore, E.E.: 1989, *Appl. Opt.* **28**, 4344. MURAs
In't Zand, J.J.M., Heise, J., and Jager, R.: 1994, *Astron. Astrophys.* **288**, 665.
Kopilovich, L.E., and Sodin, L.G.: 1994, *Mon. Not. R. Astr. Soc.* **266**, 357.
Proctor, R.J., Skinner, G.K., and Willmore, A.P.: 1979, *Mon. Not. R. Astr. Soc.* **187**, 633.
Skinner, G.K., and Grindlay, J.E.: 1993, *Astron. Astrophys.* **276**, 673.
Skinner, G.K., and Ponman, T.J: 1994, *Mon. Not. R. Astr. Soc.* **267**, 518.
Skinner, G.K., and Rideout, R.M:. 1995, This volume.
Winkler, C. *et al.*: 1994, *Ap. J. Suppl. Ser.* **92**, 327.

IMAGING TECHNIQUES APPLIED TO THE CODED MASK
SIGMA TELESCOPE

A. GOLDWURM

Service d'Astrophysique /DAPNIA/CEA, Centre d'Etudes de Saclay,
91191 Gif sur Yvette Cedex - France

Abstract. After more than four and a half years of successful operation aboard the Russian GRANAT space observatory, the French soft gamma-ray telescope SIGMA can be considered a milestone in the application of the coded mask aperture technique to high energy astronomy. The unprecedented imaging performance attained by SIGMA, coupled to the long observation time have yielded impressive results. Here I briefly describe the SIGMA imaging system and review the standard image reconstruction techniques and analysis procedures applied to the SIGMA data.

Key words: Coded masks – SIGMA

1. Introduction

Coded aperture imaging systems nowadays find their major application in high energy astronomy, and in particular in the hard X-ray (10-30 keV) and soft gamma-ray (30-2000 keV) domains where conventional focussing techniques are still very difficult to implement and where the high and variable background limits the perfomance of standard on/off monitoring techniques (for a review see Caroli et al. 1987). In coded aperture telescopes source radiation is spatially modulated by a mask of opaque and transparent elements before being recorded by a position sensitive detector, allowing simultaneous measurement of the background flux. Reconstruction of the sky image is generally based on a correlation procedure between the recorded image and a decoding array derived from the mask pattern. Special mask patterns, including those called uniformly redundant arrays (URA) allow the reconstructed image to be free of secondary lobes (Fenimore & Cannon 1978). The angular resolution of such a system is then defined by the angle subtended by one hole at the detector but the sensitive area depends on the number of all transparent elements. In conditions of high background (typical of the gamma-ray domain) the maximum sensitivity is obtained when this number is half the total number of elements in the basic pattern. To have a sidelobe-free response a source must be able to cast on the detector a whole basic pattern (fully coded source). To make use of all the detector area and to allow more than one source to be fully coded, avoiding ambiguities in source position, the basic pattern is normally taken of the same size and shape of the detector and the total mask made, in the case of rectangular geometry, of a mosaic of nearly 2×2 cycles of the basic pattern. Such a system is called an *optimum coded aperture system* (OCAS) (Skinner & Ponman 1994). The

Experimental Astronomy 6: 9–18, 1995.

TABLE I

Imaging Properties of SIGMA

FCFOV	4.8° × 4.3°
EXFOV (0.5 sensitivity)	11.4° × 10.6°
EXFOV (0.0 sensitivity)	18.1° × 16.8°
Angular Resolution	≈ 15'-21' (FWHM)
Localization Accuracy	≈ 0.5'-5'
Sensitivity (at 2 σ for 20 h obs.)	≈ 26 mCrab
Angular Mask El. Size	12.9'
Angular Pixel Size	1.62'

dimensions of the mask, of the detector and their separation define the fully coded field of view (FCFOV) of the telescope. A source outside the FCFOV may project a part of the mask pattern and it is said to belong to the partially coded field of view (PCFOV). As the projected pattern is incomplete the contribution of such a source to the reconstructed FCFOV image cannot be apriori subtracted and it produces secondary lobes (coding noise). On the other hands the modulated radiation from PC sources can be reconstructed by extending with a proper normalization the correlation procedure to the PCFOV (sect. 6). The extended FOV of the telescope (EXFOV) is therefore composed by the central FCFOV of constant sensitivity surrounded by the PCFOV of decreasing sensitivity. A source outside the EXFOV simply contributes to the background level.

The launch at the end of 1989, of the SIGMA telescope aboard the GRANAT space observatory, the first soft gamma-ray (30-1300 keV) optimum coded aperture telescope on a satellite, can be considered as a milestone in the application of coded masks to high energy astronomy. At present (October 1994), after more than 4 and a half years of operation, SIGMA has collected ≈ 9200 hours of effective time of high quality data comprising about 800 observations. Its imaging performance (Table 1), never attained before in this energy domain, coupled to the long observation time have yielded impressive results (e.g. Mandrou et al. 1994, Goldwurm et al. 1994a).

2.   The SIGMA Coded Aperture Imaging System

The SIGMA telescope (Paul et al. 1991, Mandrou et al. 1992) is composed of a position sensitive Anger type gamma camera (GCA), sensitive to photons between 30 and 1300 keV, and a 2.5 cm thick coded mask of square tungsten elements of 9.4 mm size placed 2.5 m from the detector plane. The GCA is a single circular 1.25 cm thick NaI(Tl) crystal of 57 cm diameter, viewed by 61 photomultipliers (PM) in a honeycomb structure. Event positions, reconstructed using the response of all PMs, are coded in fine square pixels of 1.175 mm size and images of the central detector useful zone of 248×232 pixels ($= 794$ cm^2), which corresponds to the mask basic pattern, are stored in the memory. Positions are estimated with a finite spatial resolution which ranges (σ) between 3 mm (200 keV) and 8 mm (30 keV). The mask is a repetition of a URA basic pattern of 31×29 elements, half of which transparent. Each square mask element corresponds to 8×8 detector pixels. To reduce the strength of the ghosts in the PCFOV the total mask is formed by 53×49 elements instead of the maximum allowed 61×57. The mask support is opaque and a passive shielding tube is placed between the mask support and the base of the GCA which is surrounded at the bottom and on the sides by an active anticoincidence shield of CsI(Tl). This geometry (Fig. 1) and the GCA performances provide the imaging properties reported in Table 1. During the 4 day orbit of the GRANAT spacecraft, SIGMA typically performs 3 observations of ≈ 20 hours each during which images are collected at intervals of a few hours in different energy bands. GRANAT is stabilized on 3 axes, one of them constantly pointed towards the sun. However the stabilization precision is only $\approx \pm 20'$ and the satellite drifts would lead to the smearing of the projected mask pattern, if no correction was applied. Two optical star trackers follow the movement of 2 preselected stars and give estimates of the 3-axis components of the spacecraft attitude drifts every 4 seconds, which are also stored in the memory. The on-board computer then recalculates the event positions to correct for the estimated drifts along the 2 axes of the detector. At the edges of the useful zone the events which due to the correction would be moved outside the image are actually replaced on the other side of the detector array. In this way, because of the cyclic structure of the mask, the flux of a fully coded source is reconstructed correctly.

3. Uniformity Correction of Detector Images

One major problem in coded aperture systems is the non-uniform distribution of the background along the detector plane. The SIGMA telescope works in conditions of a low number of sources and a high background level, the latter being therefore the dominant component of the noise. Even a small fraction of systematic modulation of the background level produces strong noise in the deconvolved images (Laudet & Roques 1988). In the SIGMA telescope there are 2 main sources of non-uniformities. First of all the inherent defects of the GCA (non-linearity and varying efficiency) will give rise to constant structures on a short spatial scale corresponding to the distance between the GCA PMs ($\approx$ few mask elements.) The modulation induced by the external background is instead a large scale structure of conical shape peaked about the center of the detector (Fig. 1) and variable on timescales from hours/days (activation during Van Allen belt passage, solar flares) to weeks/months (orbit parameters, solar cycle CR modulation) (Cordier 1991). The constant orbit circularization and the approaching of solar minimum make this component vary mainly on the longer time scales. We have measured these components by performing observations of source empty fields during the entire SIGMA mission ($\approx$ each 3 months). During these observations the star sensors were off so that no attitude correction was applied to the recorded images. If U is the GCA non-uniformity and B the background structure, then the recorded detector image during an observation is

$$D = (S * M)U^d + (BU)^d$$

where the superscript "d" means convolution with the histogram of the on-board drift corrections H_c (e.g. $U^d = U * H_c$). Indeed the term B is independent of small direction changes and is modulated by U before the on-board drift correction is applied. The source term S on the other hand is modulated by the mask M and then replaced by the drift correction algorithm after being modulated by the corresponding U^d. The empty field image D_{EF} is instead given by $D_{EF} = B_{EF}U$ and a correction of the systematic noise can be obtained with

$$\frac{D - D_{EF}^d}{U'^d} = (S * M)\frac{U^d}{U'^d} + \frac{((B - B_{EF})U)^d}{U'^d}$$

if from D_{EF} it is possible to estimate the contribution (U') of U. If U' is a good estimate of U and B_{EF} is close to B we obtain the needed correction. The empty-field images are grouped together to reduce their statistical noise and separation of the U component is performed by high-pass filtering (B contributes mainly at low frequencies). The main problem with this procedure is the variability of B. Also U is not completely constant and the array U' derived from the empty fields is not always a good estimate of the GCA non-uniformity. Because these residual effects are dominated by the background term a simple division by D_{EF}^d often gives very similar corrections. Residual systematic noise is then reduced by different techniques. One standard way is to perform a local background subtraction after deconvolution, where the local background is measured in a circular corona far enough to avoid source pixels (flat-field filter). Other methods employ subtraction of a fitted parabolic function or filtering before deconvolution (Schmitz-Fraysse et al. 1994). In general these techniques, not based on a model of the background distribution, prevent the study of extended diffuse emission, but they are quite effective to allow the search and analysis of discrete sources. Finally an estimate of the goodness of the correction is given by the ratio of the measured variance of the deconvolved image and its expected value for pure

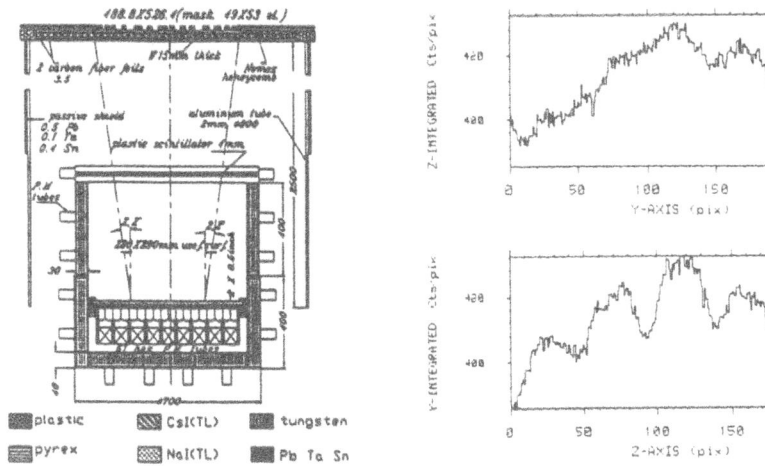

Fig. 1. The SIGMA imaging system (left). Integration on the two axes of a SIGMA
35-150 keV empty field image (the 2 components U and B are clearly visible) (right).

statistical noise (which in FCFOV of OCAS is simply the total number of
counts in the detector array). This value is used to choose the best set of
empty fields to correct the given observing periods. Its square root is also
used to increase error bars before proceeding to the scientific analysis.

4. Standard Deconvolutions in the FCFOV

For an OCAS with a URA mask a simple cross-correlation of the recorded
image and a deconvolution array G derived from the mask by substituting 1 for transparent and -1 for opaque elements ($G = 2M - 1$) provides a
sidelobe-free image of the source flux distribution in the FCFOV, apart from
a flat background level (Caroli et al. 1987). Other values than $+1$ an -1 can
be used leading to different normalizations (Skinner & Ponman 1994), but
in any case the background level must be estimated to evaluate the source
fluxes. To increase the telescope sensitivity, detector pixels are usually of
smaller size than the mask elements, and the correlation can take the form
of *fine cross-correlation* (FC), where the array G is itself divided in finer
elements and then correlated. In the *delta-decoding* (DD) (Fenimore & Cannon 1981) form, instead, the fine decoding array contains only one pixel per
mask element with a value ($+1$ or -1) different from 0. DD is a more general
form because a simple convolution of a DD image with a block function of
one mask element dimension provides a FC decoded image. Delta-decoded
images also conserve the important statistical property that image pixels are
statistically quasi-independent (they are actually independent only within

a mask element). It can be shown indeed that the correlation coefficient between two reconstructed sky mask-elements is given by $-\frac{1}{n}$ where n is the total number of basic pattern elements (e.g. Skinner & Ponman 1994). This is of course a very important property because the standard least squares method can be applyed to a DD image, and indeed fine analysis of source parameters is best performed using this technique (sect. 5). Fine correlation on the other hand provides a representation of the source distribution which is convenient for the search of significant excesses. When the detector spatial resolution is finite and is not negligible with respect to the mask element size, an even more sensitive reconstruction is what we call *PSF deconvolution* (PSFD). A DD image is, in this case, convolved with the DD point spread function (PSF) (sect. 5) rather than with a block function. It can be shown that, in this way the statistical high frequency noise is suppressed and the signal to noise ratio (S/N) of point-like sources optimized. Clearly in FC and PSFD images pixels are highly correlated. In the standard analysis of SIGMA images, after uniformity correction we apply the PSFD and perform a flat-field filter (FF) before running the search algorithm for significant excesses. However in PSFD images, sources are spread out and enlarged because of the PSF convolution, and it is not the best procedure to use when we look for close sources. Moreover the FF filter also modifies the final image creating a corona of negative values around each source (Fig. 2). The final data analysis procedure is an iterative process for which first a PSFD plus a FF filter is performed and then the image is analyzed to search for signals at known source positions or significant excesses (see Caroli et al. 1987 for a discussion on significant detection in OCAS). The stronger detected source is then subtracted and the procedure re-starts. The reconstructed source count rates (in FC units) are then converted into photon fluxes with the telescope energy response matrix (Barret & Laurent 1991) and used to perform standard timing/spectral analysis.

5. Delta-Decoding Point Spread Function and Fine Analysis

We use delta-decoded images to evaluate source parameters and their errors by comparing DD image sectors with the telescope point spread function (PSF) by means of the chi-square fitting technique. For PSF we mean the final response of the imaging system to a point-like source after deconvolution. In the FCFOV of an optimum coded aperture system the PSF is given by the convolution between a function describing the blurring of the detector finite spatial resolution (typically a bidimensional Gaussian function) and a function describing the discrete deconvolution process. In the case of DD the latter is a one-mask-element wide block function convolved with a one-pixel wide block function. In the case of fine cross-correlation this function is instead a square-pyramidal function of width (FWHM) equal to one mask

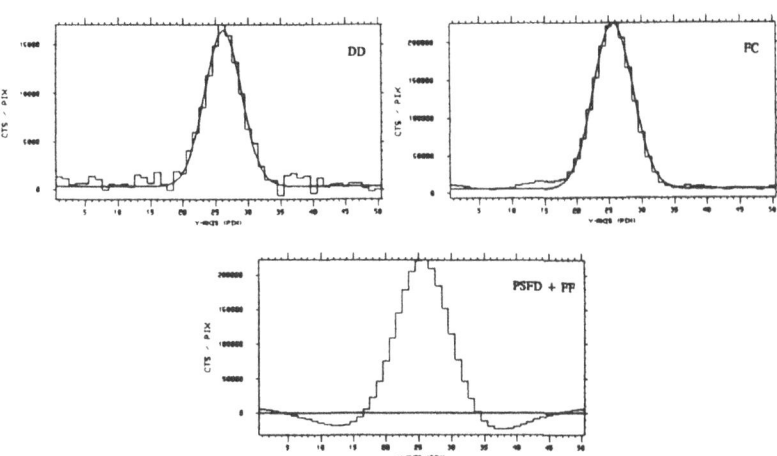

Fig. 2. Delta-decoding, fine correlation, and PSF deconvolution plus FF filter, for a 35-75 keV image of Cyg X-1. Solid lines in DD and FC histograms show the relative PSFs.

element (Fenimore & Cannon, 1981). The PSF for DD can be conveniently described by an analytical function (Fig. 2) which depends, in addition to the source flux and position and background level, on the detector spatial resolution. This is a key parameter of the PSF because it influences its width and the total imaging sensitivity of the telescope. Derivatives of the PSF can be computed and the covariance matrix of the chi-square fit (Press et al. 1986) between the PSF and a DD image sector calculated. The procedure and in particular the error estimation have been checked by numerical simulations and it was found acceptable down to S/N ratios of $\approx$ 4-5. This technique has been applied to the data collected during the mission from strong and well identified sources (Crab, Cyg X-1, X-Ray Nova Persei 92) to perform in-flight calibrations of the SIGMA imaging system and in particular to measure the spatial resolution and the systematic position offset, their dependence with energy and their variability. Once these key parameters are obtained and the model of the telescope PSF fixed, the technique can be applied to obtain the error boxes of the detected sources and to study images with close sources (for which correlation techniques cannot be applied). A clear example of the power (and the limits) of the SIGMA imaging system is the detection of A 1742-294. In September 1992 the shape of the Galactic Center source 1E 1740.7-2942 appeared clearly elongated. By fitting the PSF for 2 close sources we could derive their positions and errors and identify the second 5.3 σ excess with the X-ray burster A 1742-294 (Goldwurm et al. 1994b), some 30' away from the main source. Another relevant application was in the analysis of the Galactic Center images performed to set upper limits to the > 30 keV emission from Sgr A* (Goldwurm et al. 1994a).

6. EXFOV Deconvolution and Analysis

The standard deconvolution in FCFOV can be extended in the PCFOV by extending the correlation of the decoding array G with the detector array D in a non-cyclic form (i.e. padding G with 0 elements). The advantage with respect to cyclic correlation is of course that only the detector section modulated by the PC source is used to reconstruct the signal, reducing the statistical error at the correct source position and reducing the significance of the ghost peaks. However to ensure a flat image in the absence of sources, detector pixels which for a given sky position correspond to mask opaque elements must be balanced, before subtraction, with the factor $b = \frac{n_+ - 1}{n_-}$, where n_+ is the number of pixels corresponding to transparent elements and n_- to opaque ones for that given sky position (assuming 1 pixel per mask element). We can write this operation in one-dimensional notation as

$$S_i = \sum_k M^+_{i+k} D_k - b_i \sum_k M^-_{i+k} D_k$$

where the decoding arrays are obtained from the mask M by $M^+ = M$ and $M^- = 1 - M$ and then padded with 0's, and where the sum is performed over all detector elements. In the FCFOV we obtain the same result of the standard FCFOV correlation. However we have to consider the onboard drift corrections. These corrections replace part of the modulated counts of a PC source on the other side of the detector image and, using the previous formula, they are not summed at the source position but rather at the FCFOV ghost peak. To consider this effect a weighting array W is built by the convolution of the drifts histogram and a unity array of the dimension of the detector image. W has dimensions greater than the detector array, which is then enlarged to these dimensions by cyclic repetition of its elements and multiplied by the W array before being decoded. We have

$$S_i = \sum_k M^+_{i+k} W_k D^E_k - b_i \sum_k M^-_{i+k} W_k D^E_k$$

where now $b_i = \frac{\sum_k M^+_{i+k} W_k - 1}{\sum_k M^-_{i+k} W_k}$ and where the sums are extended to the whole enlarged detector array D^E. For the variance however the weights W_k which refer to the same pixel in D must be summed before squaring and the formula for the variance image (which is not constant outside the FCFOV) is

$$V_i = \sum_k^{d_D} D^E_k \left(\sum_{h=0}^{1} M^+_{i+k+hd_D} W_{k+hd_D} \right)^2 + b_i^2 \sum_k^{d_D} D^E_k \left(\sum_{h=0}^{1} M^-_{i+k+hd_D} W_{k+hd_D} \right)^2$$

where now the first sums are over the dimension d_D of the original detector array D, and the array W is padded with 0's. The varying effective area can be calculated by similar formula (the first sum of the 2nd term of this equation without the value D_k^E) and a normalization is included to obtain an intensity image in equivalent counts for a FC deconvolution in FCFOV (our standard units). All this can be performed in DD, FC or PSFD form, where for PSF-deconvolution in EXFOV we mean a DD deconvolution followed by a convolution with the PSF for DD in FCFOV. Naturally in this case also the variance image must be convolved and all corresponding normalizations performed. This procedure can be performed with a fast algorithm which is an extension of the algorithm proposed by Roques (1987) for the FCFOV and which exploits the symmetry of URA masks. Alternatively it is possible to reduce some of the previous formula to a set of correlations which can be computed by means of the FFT. In the EXFOV standard analysis, images are deconvolved in PSFD form, FF filtered, divided by the image of the standard deviations ($\sigma = \sqrt{V}$) (increased by a factor which accounts for residual systematic noise (sect. 3)) and values > 5-6 (significant excesses) are then searched for throughout the image along with significant signals (> 3 σ) at known X-ray source positions.

7.   Model of the PCFOV Response and Sum of Images

SIGMA was not conceived to observe in the PCFOV and a model of the PCFOV imaging response is needed to a improve the analysis and perform image cleaning. Because of the partial modulation the true PSF in the PCFOV is not the same as in the FCFOV, it varies with the position and it contains sidelobes. The reasoning adopted to show that in DD images, pixels are statistically independent is not valid anymore. However pixels in PCFOV DD images are still weakly correlated locally and the PSF in the central peak is not too different from the FCFOV PSF. Therefore PSFD using the FCFOV PSF is still very useful to increase the S/N ratio for point-like PC sources and the least squares technique (with the FCFOV PSF) can be applied to PCFOV DD image sectors to determine source position. The bias introduced by using the incorrect PSF can be evaluated by modelling the response for a given sky position and running the algorithm on the modelled image. Such a response model is necessary also to obtain the correct fluxes. It must take into account the mask element thickness, the effects at the mask border, the different opacity of the mask support and of the passive shield, and the transparent zones in or between these elements. This model is also used to perform the image cleaning from coding noise. Each source in the EXFOV creates in the decoded image a noise composed of diffuse modulation and 8 main source-like ghosts, located at distances multiple of the basic pattern from the source peak. Because the mask is 9 elements smaller in each direction than a 2×2 pattern, ghosts will be less significant than the source peak and in general no ambiguity arises. However, when a source

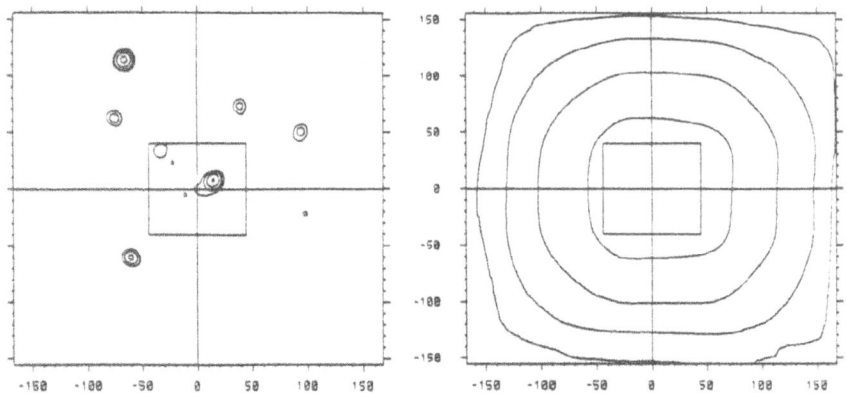

Fig. 3. Sum of 109 SIGMA EXFOV images of the Galactic Center in the 35-75 keV band plotted in the EXFOV frame of the 1st observation (the central FCFOV is also shown). Contours are in units of standard deviations (from 4.5 σ to 40 σ) (left). Image of the sensitivity ($\frac{1}{\sigma}$) for the same field (contours from 30 % to 90 % of maximum) (right).

is detected, the cleaning of the coding noise is essential before searching for other excesses or adding several decoded images. An iterative procedure which performs uniform correction, deconvolves the intensity and variance EXFOV images, cleans the coding noise of detected sources, applies a FF filter and then performs a weighted sum at each pixel (using the variance array and exposure time) for a set of images of superposed sky regions has been developed and tested. It was used to add 109 images of the Galactic Bulge region, for a total of $\approx$ 1800 hours of effective exposure time, and to build the most precise images (1 σ errors < 2-3 mCrab) of this sky region (Fig. 3) ever obtained in this energy domain (Goldwurm et al. 1994a).

References

Barret, D., & Laurent, P.: 1991, *N.I.M.P.R.* **A307**, 512
Caroli, E., et al.: 1987, *Space Sci. Rev.* **45**, 349
Cordier, B.: 1991, *Thèse doctorale*, Université de Paris VII
Fenimore, E.E., & Cannon, T.M.: 1978, *Appl. Opt.* **17(3)**, 337
Fenimore, E.E., & Cannon, T.M.: 1981, *Appl. Opt.* **20(10)**, 1858
Goldwurm, A., et al.: 1994a, *Nature* **371**, 589
Goldwurm, A., et al.: 1994b, *AIP Conf. Proc.* **304**, 421
Laudet, P., & Roques, J.P.: 1988, *N.I.M.P.R.* **A267**, 212
Mandrou, P., et al.: 1992, *AIP Conf. Proc.* **232**, 492
Mandrou, P., et al.: 1994, *Ap. J. Suppl.* **92**, 343
Paul, J., et al.: 1991, *Adv. Space Res.* **11(8)**, 289
Press, W.H., et al.: 1986, *Numerical Recipes*, Cambridge University Press
Roques, J.P.: 1987, *App. Opt.* **26(18)**, 3862
Schmitz-Fraysse, M.C., et al.: 1994, *these proceedings*, Kluwer Academic Publishers
Skinner, G.K., & Ponman, T.J.: 1994, *M.N.R.A.S.* **267**, 518

WIDE FIELD MONITORING OF THE X-RAY SKY
USING ROTATION MODULATION COLLIMATORS

NIELS LUND and SØREN BRANDT

Danish Space Research Institute
Gl. Lundtoftevej 7
DK 2800 Lyngby, Denmark

October 6, 1994

Abstract.
Wide field monitoring is of particular interest in X-ray astronomy due to the strong time-variability of most X-ray sources. Not only does the time-profiles of the persistent sources contain characteristic signatures of the underlying physical systems, but, additionally, some of the most intriguing sources have long periods of quiesense in which they are almost undetectable as X-ray sources, interspersed with relatively brief periods of intense outbursts, where we have unique opportunities of studying dynamical effects, in, for instance, the evolution of accretion discs. Another question for which wide field monitors may provide key information, is the origin and nature of the cosmic gamma ray bursts.

Rotation Modulation Collimators (RMC's) were originally introduced in X-ray astronomy to provide accurate source localizations over extended fields. This role has since been taken over by the grazing incidence telescope systems. The potential of the RMC's as wide field monitors have recently been demonstrated by the WATCH instruments on GRANAT and EURECA. It now appears likely, that for use on large, 3-axis stabilized spacecraft, a pinhole camera system may provide better sensitivity than an RMC-system of corresponding physical dimensions. But due to its simplicity, low data rate, and ability to work on spin stabilized (micro)satellites, the RMC wide field monitor may still have a role to play in the X-ray astronomy of the future.

Key words: X-Ray Astronomy – All-Sky Monitors

1. Introduction

The Rotation Modulation Collimator was originally conceived by Mertz [9] as a technique to encode an image formed in the focal plane of an optical Schmidt-telescope to allow electronic read-out by a photomultiplier. The technique was soon adapted to image forming in X-ray astronomy by Schnopper et al. [11], and was successfully used in a number of rocket flights to localize X-ray sources with good precision [12,14]. In 1974 the UK ARIEL-V and in 1975 the NASA SAS-3 satellite was launched both equipped with RMC systems. The source localizations provided by ARIEL-V and SAS-3 led to the discovery of many new X-ray sources and the identification of the optical counterparts of several sources [13,15,16].

The development of the grazing-incidence X-ray telescope made RMC-systems obsolete for precision localization of X-ray sources, but the wide-field capabilities of the RMC could not easily be matched by systems based

Experimental Astronomy 6: 19–24, 1995.
© 1995 *Kluwer Academic Publishers.*

Fig. 1. The typical WATCH-EURECA cross-correlation map with the ghost image suppressed by the proper choice of the grid shift parameter, δ, (see text). This image was derived from real data with Sco X-1 in the field.

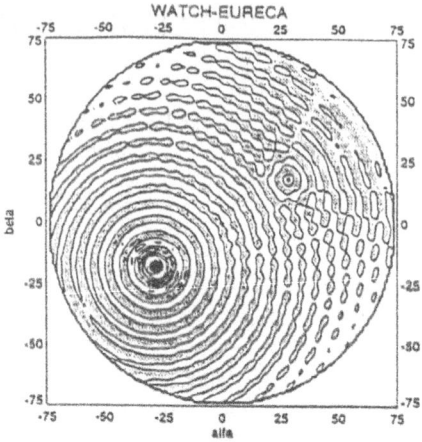

on reflecting optics. Nishimura pioneered the use of very wide-field RMC-instruments for localization of cosmic gamma-ray bursts [10]. A further development was made by Lund [8], who introduced the "striped detector" (see below) which obviated the need for the monitor counter used by the Nishimura group, even when observing the rapidly flickering gamma bursts. Detectors based on this principle, have recently been flown on both the russian GRANAT mission and on the ESA EURECA mission.

2. The WATCH Detector

The basic idea behind the RMC-system is to use a rotating modulator with two parallel grids in front of a non-imaging detector. In the WATCH detector the rear modulation grid and the monolithic detector of the classical RMC-system is replaced by a detector plane with interleaved stripes of two separate detectors - in WATCH the two detector systems are realized as alternating strips of NaI(Tl)- and CsI(Na)-scintillators. The period of the scintillator strip pattern is equal to the period in the modulation grid.

WATCH is designed specifically for wide-field monitoring, and, therefore, the field of view is very large, with an opening angle of more than 130°. The large opening angle implies large parallax effects, and, consequently, the detector stripes must be thin compared to the strip width. This requirement is difficult to fulfill for gas detectors, which is the reason for using scintillators, despite their high threshold energy and limited energy resolution. The energy range for EURECA-WATCH is from 8 keV to 180 keV in NaI and from 15 keV to 180 keV in CsI.

RMC-systems are ideally suited for spinning spacecraft, but, actually, the first two flight opportunities for WATCH have both been on 3-axis stabilized satellites, GRANAT and EURECA. Therefore the current WATCH instruments includes a motor to drive the modulation unit. The detector plane and the modulation grid rotates as a unit, and the light flashes from the two scintillators are recorded in a common (stationary) photomultiplier and separated electronically afterwards.

The instrument data system collects count rate data and modulation patterns in several energy bands. A burst trigger system is included to allow data to be collected at high rates following a trigger. A serious limititation, particularly for the instruments on GRANAT, has been the very limited data rate, about 35 bits/s per instrument (GRANAT carries 4 WATCH units, of which, unfortunately, one was damaged during launch). The EURECA instrument data rate was higher, about 120 bits/s.

3. Properties of Wide-Field RMC's

The instantaneous signal from an RMC-detector depends on the fraction of the detector illuminated through the modulation grid. As the modulator rotates, this fraction will, to first order, vary as:

$$\Delta_I = \text{saw}(L/p \tan(\theta) \sin(x - \phi) + \delta)$$

where saw() is the sawtooth function normalized between 0 and 1 and having a period of 1. $\text{saw}(0) \equiv 0$. L is the separation between the grid and the scintillator and p is the grid period. x is the rotation phase of the modulator, θ is the off axis angle and ϕ the azimuthal angle for the source. The parameter δ is the shift of the scintillator grid with respect to the modulation grid. $\delta = 0$ means that the modulation grid, when projected along the spin axis, covers one of the two scintillator systems completely.

The above expression is only a first approximation to the real modulation function. Particularly for large off-axis angles, effects arizing from the finite thickness of the modulation grid, becomes important. Some of the higher order effects also depend on the energy of the observed photons.

Analysis of RMC-data usually is carried out through a cross-correlation technique. The observed modulation pattern is cross correlated with model patterns calculated for trial positions in the sky. Sources are identified through high values of the cross-correlation function around their position. Figure 1 shows a cross-correlation map based on six hours of observation with WATCH-EURECA. The bright source in the image is Sco X-1. The very strong sidelobes in the form of concentric circles around the source is the major drawback of the RMC-technique. It is the price we pay for reducing the two dimensional sky-image to a one dimensional modulation pattern.

Obviously, the sidelobes from a strong source can easily overwhelm signals of weaker sources in the same field. However, if the signal from the strong source can be well modelled, it can be subtracted from the input signal, and with the signal removed, the sidelobes disappear as well. Then it frequently becomes possible to detect weaker sources in the field.

Fig. 2. The sidelobes in the combined image resulting from adding the cross-correlation maps of many pointings. The solid line show a section through the combined image of Sco X-1 based on data with many different pointings. The dotted line shows a section through the cross-correlation map based on a single pointing. The distant sidelobes are strongly reduced in the combined image.

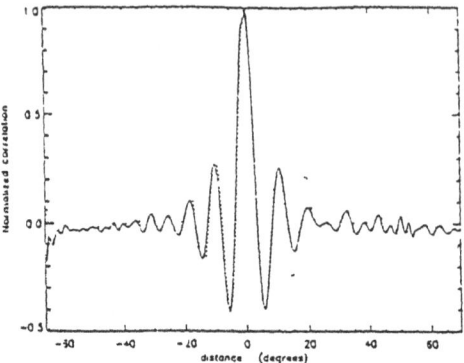

The GRANAT- and EURECA-WATCH detectors differs in the shift of the scintillator grid with respect to the modulation grid (the value of the δ-parameter). The GRANAT instruments uses a value of 0.25, resulting patterns which are antisymmetric around $(x - \phi) = 0$. This is an advantage for the onboard source localization algorithm, but produces a negative ghost image in the correlation map, of the same intensity as the true image, but symmetrically placed with respect to the spin axis. The EURECA-WATCH employs a scintillator with a shift-parameter close to 0.125. This entails difficulties for the on-board source finding algorithms, but, as illustrated in figure 1, the ghost image and its sidelobes now are much weaker than the real image.

The sidelobes shown in figure 1 appears as equidistant circles only because we plot the correlation map on a plane perpendicular to the RMC spin axis. When projected on the celestial sphere the sidelobes are neither circular nor equidistant. This means, that if we overlay the correlation maps for many observations, with different orientations of the spin axis, then only the central peaks will add constructively from all exposures, while the sidelobes will have differing phases, and therefore diminish in relative importance in the combined image. This is illustrated in figure 2.

The limiting sensitivity of WATCH for an integration time of $5 * 10^4$ s is about 100 mCrab. This allows to monitor about 40 of the brightest persistent X-ray sources. The limiting sensitivity is very dependent on the lower energy threshold of the detector. We feel confident that we, with an improved design, may push down this threshold from the 8-9 keV effectively achieved in EURECA and GRANAT to 5-6 keV.

4.   Monitoring the X-ray Sky

A number of persistent, bright X-ray sources have been monitored over long periods during the GRANAT and EURECA missions [1,2]. The period development have been followed for the pulsars Vela X-1 [7], GX 301 − 2

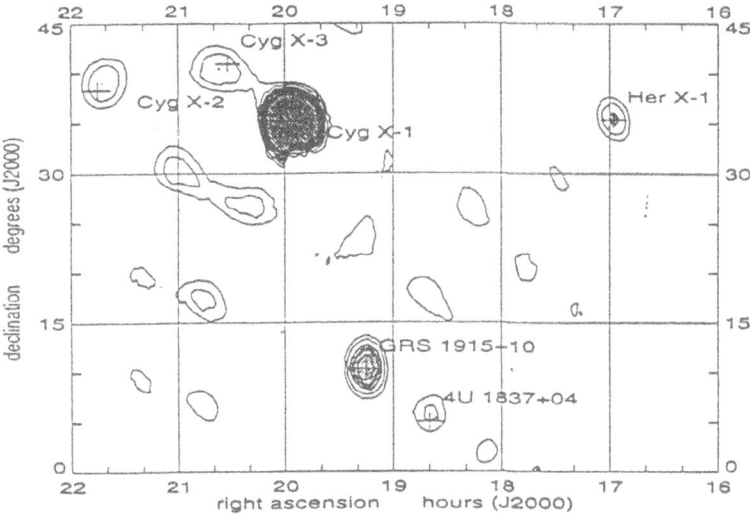

Fig. 3. WATCH-skymap during the EURECA offset pointing on 28-29 April 1993. GRS 1915 + 105 is seen to be active. Her X-1 is in the main-on state of its 35 day cycle.

[3] and A 1118 − 61 [1]. Several very interesting transient sources have been discovered during the GRANAT and EURECA missions, in particular Nova Musca 1991 [1], GRS0834-43 [6] and GRS 1915+105 [2].

In order to display our results, and search for faint sources in the presence of stronger ones, we have developed a technique to remove the sidelobes of the strong sources from our correlation maps. After identifying a source, we fit its intensity and then subtract the source signal from the modulation pattern. We repeat this procedure for all identifiable sources. We keep the information about the subtracted sources (positions and intensities) together with the final residual pattern. We then compute the correlation maps based on the residual patterns from many observations, and combine those into a global map. This brings out the weak sources, not identified from the individual observations. Finally we add to the combined map all the central correlation peaks corresponding to the identified, strong sources. The result of such a procedure is shown in figure 3. Note that the sidelobes are present only for the weakest sources in this map. Using this technique we have f.i. measured the quiescent flux from the X-ray burst source 4U 0614+09 at a level of 25 mCrab [1].

5. Gamma Burst Localizations

The WATCH detector was originally designed primarily to provide rapid localizations of gamma-ray bursts. Until now more than 70 bursts have been identified in the WATCH data. Of these, about 40 have been localized to better than 1 degree [1,2]. A major contribution to the current uncertainty in our burst positions is the lack of knowledge of the GRANAT attitude at the time of the bursts. To remedy this, we are now preparing to include the data from the SIGMA star tracker in our WATCH analysis software.

6.   Competing Techniques

The development of imaging proportional counters with large sensitive areas, and the parallel evolution in the data storage and processing techniques has revived the simple pinhole camera as a powerful wide-field X-ray monitor [5]. There can be no doubt that such systems can be made more sensitive than a WATCH-type RMC, particularly in the very important energy range around 2 keV. It remains to be seen how the pinhole camera will perform as a gamma-burst detector. Another interesting technique also discussed at this meeting is the Lobster Eye Camera [4]. However, this camera type tends to become very bulky in all-sky configurations.

7.   Future Applications of the RMC-technique

In the future, RMC-instruments may find applications as low cost, wide-field monitors for spin stabilized microsatellites. The alternative techniques requires three-axis stabilized platforms to perform well, and much higher data rates as well. The WATCH experience have shown, that the performance of an RMC is adequate for providing the first alert for, and the initial positions of, transient sources like X-ray novae. Plans are also underway in several laboratories to establish ground-based optical telescopes with possibility for rapid follow up of satellite gamma-burst localizations. A spaceborne RMC detector would complement such concepts very well.

References

1. Brandt, S.: 1994, Ph.D. Dissertation, Danish Space Research Institute
2. Castro-Tirado, A.J.: 1994, Ph.D. Dissertation, Danish Space Research Institute
3. Castro-Tirado, A.J., Brandt, s: et al.: 1993, A&A Suppl., 97, 257
4. Gorenstein, P.: 1994, these proceedings
5. in 't Zand, J.J.M., Priedhorsky, W.C et al.: 1994, these proceedings
6. Lapshov, I. Yu., Dremin, V.V. et al.: 1992, Sov. Astr. Lett., 18(1), 12
7. Lapshov, I. Yu., Sunyaev, R.A. et al.: 1992, Sov. Astr. Lett., 18(1), 16
8. Lund, N.: 1981, Astrophysics and Space Science, 75, 145
9. Mertz, L.: 1968, in *Proc. Symp. Mod. Opt.*, ed. J. Fox, Polyt. Inst. Brooklyn, 17, 787
10. Nishimura, J., Fuji, M. et al.: 1978, Nature, 272, 337
11. Schnopper, H.W., Thompson, R.I. and Watt, S.: 1968, Space Science Reviews, 8, 534
12. Schnopper, H.W., Bradt, H.V. et al.: 1970, ApJ, 161, L161
13. Schnopper, H.W., Delvaille, J.P. et al.: 1976, ApJ, 210, L75
14. Cruise, A. and Willmore: 1975, MNRAS, 170, 165
15. Eyles, X. et al.: 1975, Nature, 254, 577
16. Wilson, X. et al.: 1975, ApJ, 215, L111

HARD X–RAY IMAGING VIA FOCUSING OPTICS WITH MOSAIC CRYSTALS

F. FRONTERA

Dipartimento di Fisica, Università di Ferrara and Istituto TESRE, CNR, Bologna, Italy

and

G. PARESCHI

Dipartimento di Fisica, Università di Ferrara, Italy

Abstract. In spite of the tremendous potential of hard X-ray astronomy (>10 keV) for studying high energy phenomena in celestial objects, the current generation of direct-viewing telescopes is heavily noise limited. It can accurately study only the strongest sources. Thus focusing of hard X-rays is mandatory in order to overcome these sensitivity limitations. Several focusing techniques of hard X-rays (>10 keV) are under study. We will discuss the Bragg diffraction technique and the imaging performance of a concentrator configuration based on this technique. Apart from its unprecedented flux sensitivity, the Bragg concentrators show intrinsic capabilities as polarimeters.

Key words: Mosaic crystals – X-Ray imaging – Concentrators – X-ray Astronomy

1. Introduction

Hard X-ray (>10 keV) celestial observations performed with balloon experiments and satellites, like those recently performed with the SIGMA/GRANAT and CGRO missions have confirmed that hard X-ray astronomy can provide key relevance information for the high energy astrophysics. However the current generation of hard X-ray telescopes, that is based on direct-viewing detectors (with or without mask) has a limited sensitivity mainly due to the Poissonian variance of the detector background level that increases with the detection area. The only practical way to overcome the above sensitivity limitations is to focus hard X-rays collected over a large passive area onto a small area detector. Indeed in this case the telescope sensitivity is proportional to S^{-1}, where S is the photon collecting area, rather than to $S^{-0.5}$, as in the case of direct-viewing detectors. Hard X-ray concentrators should provide not only a much higher flux sensitivity but also a higher imaging performance with respect to the mask imaging systems.

Studies of hard X-ray concentration techniques are currently under way by various groups. They include the use of multiple small angle reflections in the interior of glass microcapillaries (Gorenstein, 1991) or micro-channel plates (Chapman et al., 1993), the reflection from multilayer structures with graded d-spacing (Joensen et al., 1993), the Bragg diffraction technique from mosaic crystals in reflection configuration ('Bragg geometry') (De Chiara and Frontera, 1992). The latter technique has the unique capability to focus

Experimental Astronomy 6: 25–31, 1995.

the most energetic hard X-rays (up to about 200 keV), and, in transmission configuration ('Laue geometry'), can efficiently focus higher energy gamma rays (Melone et al., 1993). Here we will present the main imaging features of the Bragg concentrators in reflection configuration.

2.   Reflection of hard X-rays with mosaic crystals

In a previous paper (De Chiara and Frontera, 1992) we derived the expression of the reflectivity of a mosaic crystal in Bragg geometry in the general case of linearly polarized X-rays and we discussed the optimization criteria of the hard X-ray reflectivity of mosaic crystals.

A material that satisfies our concentrator requirements resulted to be graphite (002) with the following mosaic parameter values: crystal thickness 2 mm, the microblock thickness about 100 times the lattice spacing d_{002} of graphite, mosaic spread $\beta \approx 0.2°$ (fwhm).

The hard X-ray reflectivity of flat samples of pyrolytic graphite (002), that were provided according to our requirements by Advanced Ceramics (Cleveland, OH, USA), was measured in our X-ray facility (Frontera et al., 1993). Results of these reflectivity measurements performed at different grazing angles and thus at different Bragg energies were already reported (Frontera et al., 1991). The measured integrated reflectivity at different Bragg energies resulted to be in good agreement with that expected.

3.   Bragg concentrators and their imaging capabilities

Thanks to the larger reflecting angles of the hard X-ray radiation from mosaic crystals with respect to the other reflecting techniques mentioned, paraboloidal mirrors do not imply prohibitive focal lengths, even though they show worse imaging capabilities than Wolter I configurations. In spite of that, in order to optimize mirror effective area, we assumed paraboloidal mirrors. We studied concentrators consisting of a set of confocal paraboloidal mirrors with same height. Each mirror is made of many small pieces of mosaic crystal with the average reflecting planes parallel to the crystal surfaces. The concentrator effective area, for linearly polarized radiation with polarization angle ϕ, is a function of the photon energy and ϕ (De Chiara and Frontera, 1992). From the dependence of the effective area on angle ϕ derives the intrinsic capability of these concentrators as polarimeters that we recently reported (Frontera et al., 1994).

In order to derive the imaging properties of Bragg concentrators we developed a Monte Carlo code that simulates a parallel X-ray beam incident on the concentrator top. In the code the actual position of the mirror crystals, including possible misalignments of the single pieces with respect to the perfect parabolic geometry, is taken into account ((Pasqualini et al., 1992)). It is also described the spread of the reflected photons around the geometrical focus due to the angular distribution of the perfect crystallites of the mosaic crystals around their average lattice plane. The description of the normals of the crystallites planes around the average normal was made in a similar fashion to that followed by Sanchez del Rio et al. (1992). As crystal material we assumed graphite (002) with the mosaic parameters given above.

Here we show results of performance of a medium size concentrator configuration, with the parameters given in Table 1. As focal length we intend the distance mirror top-focal plane. This configuration, named HAXTEL (Hard X-ray Telescope), can easily be accommodated aboard a free-flying satellite.

Table 1. HAXTEL configuration

Number of concentrator mirrors	28
Mirror height h (cm)	60
Top radius of the outer mirror (cm)	134
Top radius of the inner mirror (cm)	10
Focal length D (cm)	379
Mirror mechanical support material	nickel
Mirror support thickness (mm)	1
Nominal operation band (keV)	10-140
Effective area @ 15 keV (cm^2)	1000
Effective area @ 60 keV (cm^2)	100
Effective area @ 100 keV (cm^2)	35
Sensitivity @15 keV (10^{-6} Crab)	8
Sensitivity @60 keV (10^{-6} Crab)	190
Sensitivity @100 keV (10^{-6} Crab)	670

Effective area and energy range of operation can be increased with respect to HAXTEL by increasing the focal length and the concentrator diameter, as can be seen from fig. 1. The diameter is constrained by the lower energy threshold of operation. In the case of HAXTEL, an energy threshold of 40 keV implies a concentrator diameter of only 70 cm.

The light distribution on the focal plane for centrally incident photons of 15 keV and 80 keV is shown in fig. 2. As can be seen, the concentrator image shows a remarkable peak whose full width at half maximum (fwhm) depends on energy. This spread corresponds, in the case of HAXTEL, to an angular resolution of about 4 arcmin at 15 keV and to 1.5 arcmin at 80 keV. The resolution capability can be also seen in fig. 3, where the light distributions due to two X-ray sources separated by 10 arcmin are superposed. As can be seen, the sources are clearly resolved.

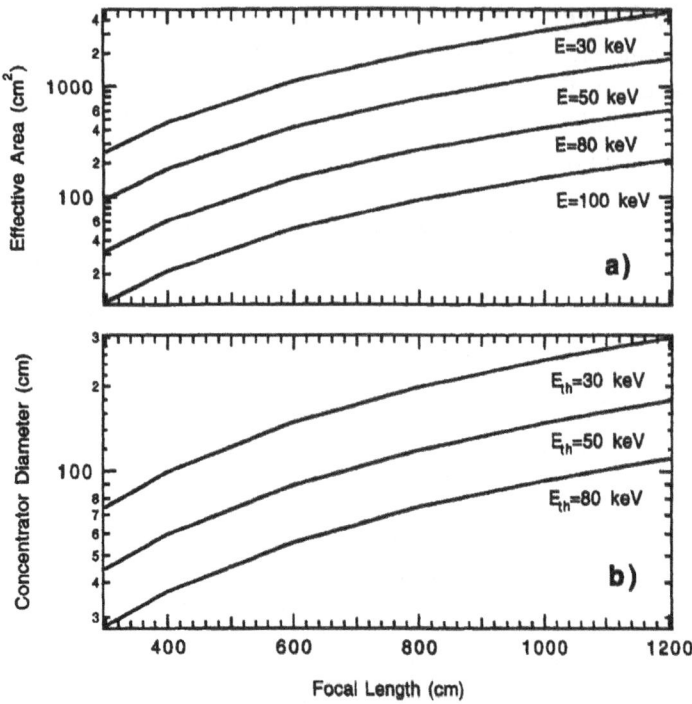

Fig. 1.   (a) Effective area as a function of the focal length at different energies. (b) Concentrator diameter as a function of focal length for different threshold energies.

The field of view of the instrument was also investigated. It results to be about 40 arcmin. The off-axis images show aberration defects (coma).
In order to evaluate the flux sensitivity of HAXTEL, a high efficiency and position sensitive focal plane detector was assumed. For the present evaluation, we assumed detection efficiency and energy resolution of a high purity Germanium detector 1 cm thick and a background level of 1.0×10^{-4} counts/(cm^2 s keV) through the 10–40 keV energy range, 4×10^{-5} counts/(cm^2 s keV) through the 40–100 keV energy range and 2×10^{-5} counts/(cm^2 s keV) above 100 keV (Vallerga et al., 1983). Some figures of the resulting 3σ sensitivity for an observation time of 10^5 s are given in Table 1. As discussed above, these sensitivity figures could be increased by increasing the focal length of the telescope. Also the use of mosaic crystals of germanium (111) as reflectors for the inner mirrors is expected to increase the effective area at high energies (>80 keV) by a factor up to two. However this evaluation, based on theoretical calculations, requires to be confirmed with measurements.

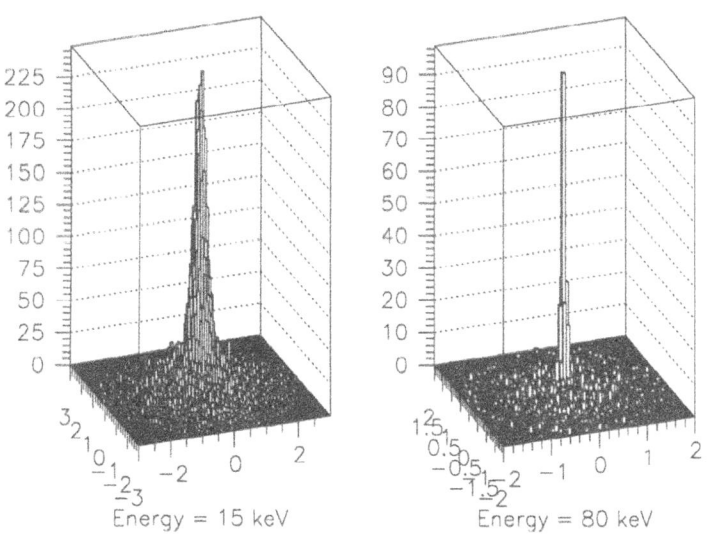

Fig. 2.   (a) Point spread function of HAXTEL at 15 keV (b) Point spread function of HAXTEL at 80 keV. Units for axes x and y in cm.

4.   Discussion and Conclusion

As a result of our investigation it is clear that Bragg reflection from mosaic crystals is good candidate technique for building a hard X-ray concentrator with operative range up to 100 keV and beyond. The only material tested thus far is pyrolytic graphite (002). However other mosaic materials could be used. One of them is Germanium (111). We intend in the future to test other materials, as soon as they will be available by suppliers with proper specifications. The graphite (002) with the required mosaic spread has been successfully tested and, important thing, can be found on the market, even though the availability of such well-ordered material is, for time being, the exception and not the rule. However it is possible to produce such a material in large quantities via an initial developing program. Also the production cost, now expensive, after a developing phase, is expected to sensitively decrease.

The Bragg concentrator configuration we considered for performance evaluation has been a set of confocal paraboloidal mirrors. The paraboloids do not offer the best imaging performance, even though they allow to achieve angular resolutions of a few arcmin, that represent a remarkable improve-

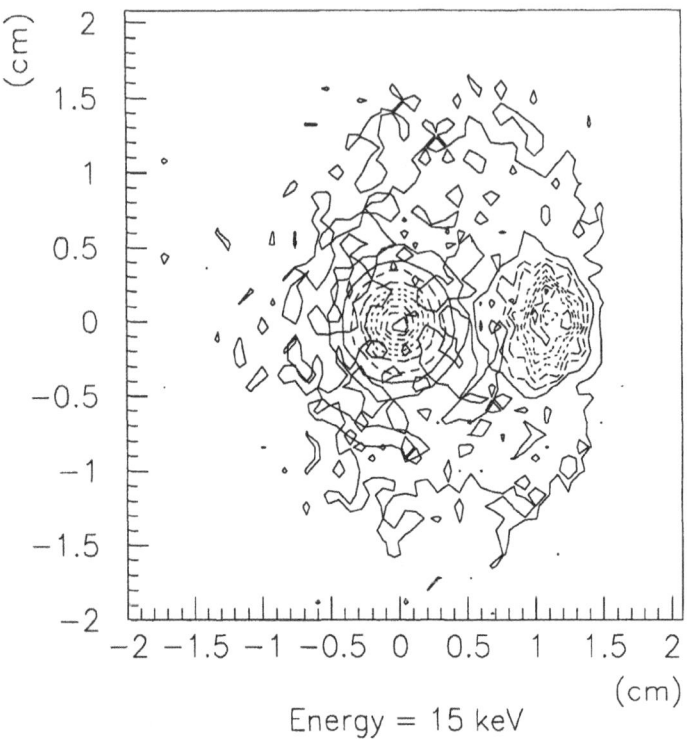

Energy = 15 keV

Fig. 3.   Contour plot of the 15 keV light distribution on the focal plane due to two X-ray sources 10 arcmin apart.

ment with respect to the current generation of high energy imaging instrumentation. A drawback of the Bragg concentrators is their large diameters. These strongly depend on the minimum reflecting energy. As an example, in the case of the HAXTEL configuration, a minimum reflecting energy of about 10 keV requires a 268 cm diameter, but a minimum energy of 40 keV requires a telescope diameter of only 70 cm.

The joint use of a Wolter I concentrator based on multilayers to focus low energy (<40 keV) X-rays and a Bragg concentrator like HAXTEL to focus higher energy X-rays appears a possible telescope configuration that efficiently focuses X-rays in a broad energy band (from 1 keV to about 200 keV). A replication technique to fabricate multilayer mirrors is currently investigated, that would render feasible a Wolter I multilayer concentrator

(Citterio et al., 1994) operating up to 40-60 keV.

As a conclusion, hard X–ray concentration is by now within reach for astrophysical applications and future hard X-ray missions cannot neglect the concentration technique that opens exciting perspectives to hard X–ray astronomy.

Acknowledgements

This research is supported by the Italian Space Agency ASI, Consiglio Nazionale Ricerche and Ministero Università e Ricerca Scientifica e Tecnologica of Italy. We want to thank Advanced Ceramics Corp., Cleveland, OH (USA) for supplying us samples of pyrolytic graphite for reflectivity tests.

References

Chapman, H. N., Nugent, K. A., and Wilkins, S. W.: 1993, 'X-ray focusing using cylindrical-channel capillary arrays', *Applied Optics* **32**, 6316.

Citterio, O., Frontera, F., Gorenstein, P., and Pareschi, G.: 1994 'New Conception Multilayer Hard X-ray Telescope for X-ray Astronomy', Proc. of the Capri Workshop on *Imaging in High Energy Astronomy*, in press.

De Chiara, P. and Frontera, F.: 1992, 'Bragg diffraction technique for concentration of hard X-rays for space astronomy' *Applied Optics* **31**, 1361.

Frontera, F. De Chiara, P., Gambaccini, M., Landini, G. and Pasqualini, G.: 1991, 'Hard X-ray Imaging via Crystal Diffraction: First Results of Reflectivity Measurements', *Proc. SPIE* **1549**, 113.

Frontera, F. Del Guerra, A., Gambaccini, M., Marziani, M., Pasqualini, G., Franceschini, T., Landini, G. and Silvestri, S.: 1993, 'Hard X-ray (>15 keV) Facility for Calibration of Space Astronomy Experiments', *IEEE Trans. on Nucl. Sci.*, **40**, 874.

Frontera, F. De Chiara, P., Pareschi, G., and Pasqualini, G.: 1994, 'Polarimetric performance of a Bragg hard X-ray (>10 keV) Concentrator', *Proc. SPIE* **2283**, 85.

Gorenstein, P.: 1991, 'Modelling of Capillary Optics as a Focussing Hard X-ray Concentrator', *Proc. SPIE* **1546**, 91.

Joensen, K. D., Hoghoj, P., Christensen, F., Gorenstein, P., Susuni, J., Ziegler, E., Freund, A., and Wood, J.: 1993, 'Multilayered supermirror structures for hard X-ray Synchrotron and astrophysics instrumentation', *Proc. SPIE* **2011**, 360.

Melone, S., Francescangeli, O., and Caciuffo, R.: 'Gamma-ray focusing concentrators for astrophysical observations by crystal diffraction in Laue geometry', Rev. Sci. Instrum., **64**, 3467.

Pasqualini, G., De Chiara, P. and Frontera, F.: 1992, 'Development status of Bragg hard X-ray Concentrators for Space Astronomy', *Il Nuovo Cimento*, **15C**, 879.

Sanchez del Rio, M., Bernstorff, S., Savoia, A., and Cerrina, F.: 1992, 'A Conceptual Model for Ray Tracing Calculations with Mosaic Crystals', *Rev. Sci. Instrum.*, **63**, 932.

J. V. Vallerga, R. K. Vanderspeck and G. R. Ricker, *Nucl. Instr. and Meth.*, 213, 145 (1983).

A HARD X-RAY TELESCOPE/CONCENTRATOR DESIGN BASED ON GRADED PERIOD MULTILAYER COATINGS

F. E. CHRISTENSEN, K. D. JOENSEN[*], P. GORENSTEIN[*], W. C.
PRIEDHORSKY[**], N. J. WESTERGAARD, H. W. SCHNOPPER

Danish Space Research Institute
Gl. Lundtoftevej 7, DK-2800 Lyngby, Denmark

Abstract. It is shown that compact designs of multifocus, conical approximations to highly nested Wolter I telescopes, as well as single reflection concentrators, employing realistic graded period W/Si or Ni/C multilayer coatings, allow one to obtain more than 1000 cm^2 of on-axis effective area at 40 keV and up to 200 cm^2 at 100 keV. The degree of concentration is defined by a focusing factor i.e., the effective area divided by the half power focal area. For the cases studied, this is 400 at 40 keV and 200 at 100 keV for a 2 arcmin imaging resolution. This result is quite insensitive to the specifics of the telescope configuration provided that mirrors can be coated to an inner radius of 3 cm. Specifically we find that a change of focal length from 5 to 12 m affects the effective area by less than 10%. In addition the result is insensitive to the thickness of the individual mirror shell provided that it is smaller than roughly 1 mm. The design can be realized with foils as thin ($\leq$0.4 mm) as used for ASCA and SODART or with closed, slightly thicker (~1.0 mm) mirror shells as used for JET-X and XMM. The effect of an increase of the inner radius is quantified on the effective area for multilayered mirrors up to 9 cm. The calculated Field of View (full width at half maximum), ranges from 9 arcmin at 1 keV to $\geq$ 5 arcmin at 60 keV. Finally, the continuum sensitivity of the design assuming a signal to noise ratio of 5 and a 10% energy bandwidth has been calculated. For a balloon flight observation of 10^4 sec. with a telescope having 2 arcmin imaging resolution the point source sensitivity is ~3 · 10^{-6} photons/cm^2/s/keV up to 70 keV for a W/Si coated telescope and up to ~100 keV for a Ni/C coated telescope. For a satellite observation time of 10^5 sec and an imaging resolution of 1 arcmin the sensitivity is ~10^{-7} photons/cm^2/s/keV which demonstrates the great potential of this hard X-ray imaging telescope in the energy range up to 100 keV.

1. Introduction

Instruments with focusing optics in the 10-100 keV energy band have not been flown previously and many interesting problems in high energy astrophysics await the launch of such a mission (Gorenstein et al. 1994). Recently, there have been several activities aimed at the development of focusing optics in this energy band. Among them are highly nested Si-wafers in a multifocus Kirkpatrick-Baez geometry (Elvis et al. 1988), slumped microchannel plate arrays as proposed by Fraser et al. (1993), microcapillary arrays as described by Gorenstein (1991) and

[*] Center for Astrophysics, Harvard Smithsonian Astrophys. Obs., USA.
[**] On leave from Los Alamos National Laboratory, USA.

Experimental Astronomy 6: 33–46, 1995.
© 1995 *Kluwer Academic Publishers.*

the use of mosaic Graphite crystals for a hard X-ray concentrator as proposed by Schnopper (1981) and Frontera et al. (1991). Christensen et al (1991) proposed a hard X-ray telescope design that combines grazing incidence Wolter I geometry with Bragg reflection in an in-depth, graded period, multilayer coating. This technique combines the desirable imaging properties of a Wolter I telescope and the high reflectivity in the hard X-ray band provided by the graded period multilayer mirrors - also known as X-ray supermirrors. Christensen et al (1991) aimed at generating a peaked effective area near 40 keV. These mirror designs have recently been extended and applied to multifocus Kirkpatrick-Baez designs that provide a significant effective area up to 100 keV as described in a series of papers by Joensen et al. (1992, 1993, 1994). A design for an all sky, hard X-ray, survey instrument based on a one dimensional Lobster eye geometry and X-ray supermirrors is described by Priedhorsky and Christensen (1994). Non focusing, imaging optics above 10 keV that are based on coded aperture technology have been successfully flown on the GRANAT mission. For pointed observations, the inherent sensitivity of coded aperture telescopes is limited by the large detector background and it is clear that an order of magnitude or more gain in sensitivity can only be achieved by using focusing optics. In addition, the imaging resolution of focusing optics is required to provide an improvement in the imaging resolution to 1 arcmin Half Power Diameter (HPD) or less.

This paper concentrates on the study of what can be obtained with compact, multifocus, supermirror designs of conical approximations to Wolter I geometries that have focal lengths from 5 to 12 meters and a geometrical aperture of ~1.5 m^2. The next section gives the baseline design of the telescope/concentrator system. This is followed by a short description of the supermirror design and, finally, the calculated performance of the proposed design including on-axis effective area, focusing ability, field of view (FOV), continuum sensitivity and a comparison to a non focusing system of the same size.

2. Telescope/concentrator baseline design

The baseline design that was studied is shown in Figure 1. A 1.3 by 1.3 m aperture, optimally filled with a multiplicity of focusing units is assumed. Each unit is either a double reflection conical approximation to a Wolter I telescope or a single reflection conical approximation to a parabolic concentrator. The focal length is FL and the focal plane is an array of hard X-ray detectors that, for the present study, is assumed to have a quantum efficiency of 1 in the whole energy band. Calculations of effective area etc. for the 4 focal lengths listed in Table 1 have been made. The number of modules that effectively fills the aperture for each focal length and the associated outer radius, R_o, of each focusing unit, assuming a conical Wolter I telescope, is also listed. The outer radius is found by calculating the radius for each focal length that gives an on-axis angle of incidence of ~0.25 deg. Above this angle, the X-ray reflectivity is no longer sufficient at energies ≥50 keV to make an efficient use of the double reflection mirror combination. R_o is determined in a similar way for the concentrator case.

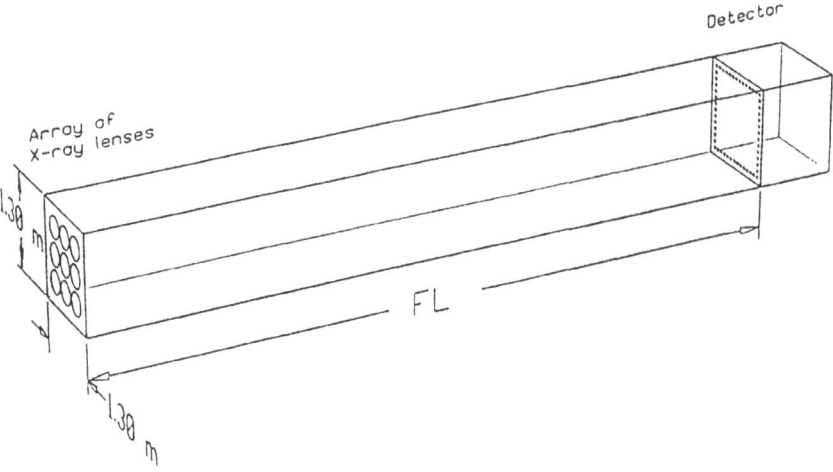

Figure 1. Baseline configuration. An aperture of 1.3 x 1.3 m is optimally filled with focusing units. A detector array is placed in the focal plane at a distance FL behind the telescope.

Table 1.

Focal length m	Number of modules	Outer radius of telescope unit cm
5	49	8.5
6	36	10.5
8	16	14.0
12	9	19.0

The other parameters that are required to fully specify the telescope/concentrator unit are the inner radius, R_i, (the smallest radius for which the supermirror coating can be applied) the mirror thickness, t, the minimum spacing between each mirror shell, s, and the length of each mirror in the direction of the optical axis, l (thus, for the telescope case, the total length of the 2 mirror combination is 2·l). For all calculations $R_i = 3$ cm is assumed unless otherwise stated. With this choice of R_i, and with the values of R_o listed in Table 1, the number of mirror shells per unit, N_M, is a function of t, s, l and FL as listed in Table 2 for the two extreme focal lengths of 5 and 12 m. Two cases are given for each focal length. One case is a thin foil case where t = 0.2 mm, s = 0.5 mm and l = 15 cm. The other case is a "thick" foil case where t = 1.0 mm, s = 1.0 mm and l = 30 cm. Clearly the "thick" foil case reduces significantly the number of mirrors to be coated.

Table 2.

Focal length m	Foil thickness mm	Min. foil separation mm	Mirror length cm	Number of mirror-shells/unit
5	0.2	0.5	15	89
5	1.0	1.0	30	30
12	0.2	0.5	15	280
12	1.0	1.0	30	98

Substrates that may be useful for thin foil cases (t ≤ 0.3 mm) are thin glass sheets, thin polished Si-wafers and electroformed Ni foils. All of these can be obtained in a realistic size assuming that l ~15 cm. They all have demonstrated microroughnesses below 5 Å and, for the thin glass and the Si-wafers, the deposition of state-of-the-art multilayers has been demonstrated. For the "thick" foil cases (t ≥ 0.5 mm) and l ~30 cm the obvious candidates are replicated mirror shells similar to those produced within the SAX (G. Conti et al. 1993), the JET-X (A. Wells et al. 1991) and the XMM (Gondoin et al. 1994) telescope programmes; all based on the replication of Ni shells.

3. Supermirror design.

For the present study the supermirror design uses W/Si or Ni/C multilayer coatings and the d-spacing progression is based on the work of Joensen et al., 1992. It prescribes a power law progression in the in-depth graded period variation of the form:

$$d_i = a \ (b-i)^{-c} \qquad\qquad (1)$$

where d_i is the spacing of the i'th period and a, b, c are constants. A thorough discussion of eq. 1 is given by Joensen et al. (1992). A number of physical parameters set the boundary conditions for eq. 1. These are the minimum d-spacing, d_{min}, the maximum d-spacing, d_{max}, the number of periods, N, the ratio of the heavy element to the light element, Γ, and the interfacial roughness, σ. An optimum value of c = 0.27 (Joensen et al. 1992) has been used and Table 3 lists the values of d_{min}, d_{max}, Γ, σ and N used for the two material combinations as well as the resulting values of a & b.

Table 3.

Material combination	d_{min} Å	d_{max} Å	Γ	σ Å	N	a	b
W/Si	25	200	0.35	4	300	116.526	300.135
Ni/C	25	200	0.5	5	600	140.572	600.271

The value of σ = 4 Å for W/Si and 5 Å for Ni/C is based on recent test coating results by Joensen et al. (1992, 1993). The use of W as the heavy material limits the useful range of W/Si coatings to energies below the K absorption edge of W (69.525 keV) whereas the Ni/C combination is useful up to at least 100 keV. The optical constants used to calculate reflectivities of the coatings are obtained from the work of Cromer and Lieberman (1970).

4. Calculated properties

4.1 EFFECTIVE AREA

Using the baseline design of section 2 and the supermirror design of section 3 for a telescope array with W/Si multilayers, the on-axis effective area shown in Figure 2a for the four focal lengths listed in Table 1 is obtained. This calculation assumes that t = 0.2 mm, s = 0.5 mm and l = 15 cm. A similar calculation with t = 0.4 mm, s = 0.5 mm and l = 30 cm as well as a calculation with t = 1.0 mm, s = 1.0 mm and l = 30 cm shows that the loss of effective area relative to what is shown in Figure 2a is <20% for all four focal lengths. Figure 2a shows that more than 1000 cm^2 can be obtained up to the cut-off energy of 70 keV. The focusing factor defined as the effective area divided by the half power focal area for the 5 m focal length case is shown in Figure 2b and is in excess of 200 up to 70 keV.

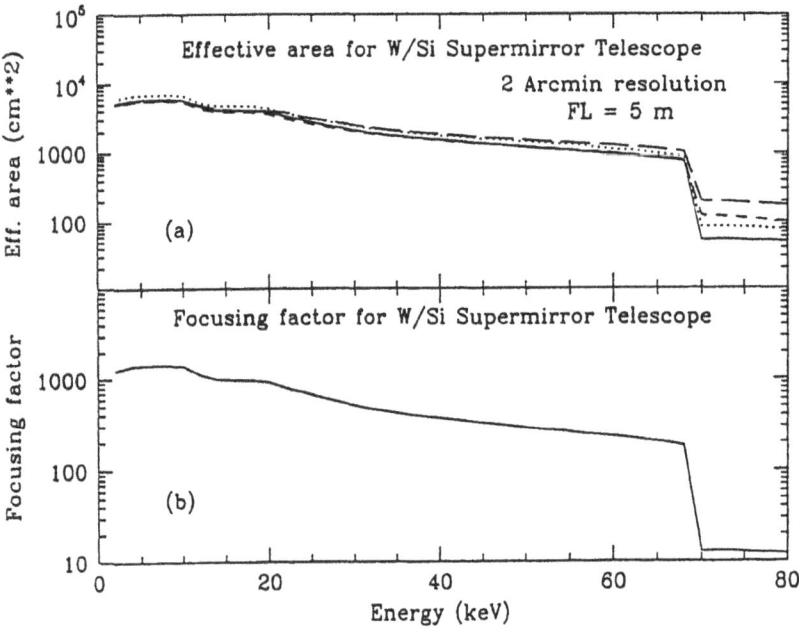

Figure 2. On axis effective area (a) and focusing factor (b) for a W/Si supermirror telescope. In Figure 2a the full line is for a 5 m focal length case, the dotted line is for a 6 m case, the short dashed line is for a 8 m case and the long dashed line is for a 12 m case.

Figure 3a and 3b shows the comparison between the effective areas obtainable with a Ni/C coating, a W/Si coating and a conventional Au coating for 8 m focal length telescope array (3a) and concentrator array (3b), respectively. The values of t, s and l are the same as the ones used in Figure 2a. Obviously, the Ni/C coating is preferable to the W/Si coating for energies above 70 keV and the Au coating lies well below the two supermirror coatings above 40 keV for the telescope case and above 30 keV for the concentrator case. An increase of the interfacial microroughness for the multilayers will decrease the reflectivity (Spiller et al, 1985) and thus decrease the effective area. If σ is increased from 4 to 5 Å for the W/Si case the reduction of the effective area is less than 10% up to 70 keV. An increase of σ from 5 to 6 Å for the Ni/C coating decreases the effective area by slightly more than 10% up to 100 keV. This indicates that the microroughness is a more critical parameter for the Ni/C coating than for the W/Si coating. This is because a larger number of periods are used for the Ni/C case, necessitated by the smaller reflectivity from each Ni/C interface, compared with the W/Si interface.

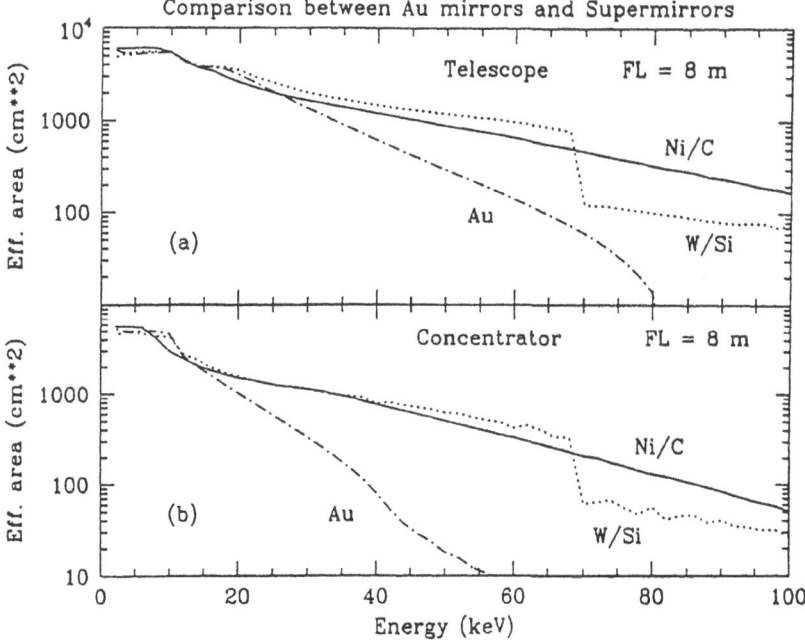

Figure 3. Comparison of the on axis effective area for a W/Si supermirror, a Ni/C supermirror and a Au coated telescope case (a) and concentrator case (b).

The calculations presented thus far have been based on the assumption that the supermirror coating can be applied to mirrors having a radius of 3 cm. Figure 4 shows how the on-axis effective area varies as this minimum multilayered radius is increased to 9 cm. The focal length for the W/Si calculation was 8 m. For the Ni/C calculation it was set to 10 m. In both cases it is assumed, that mirrors with radii from 3 cm up to the minimum multilayered radius were coated with Au.

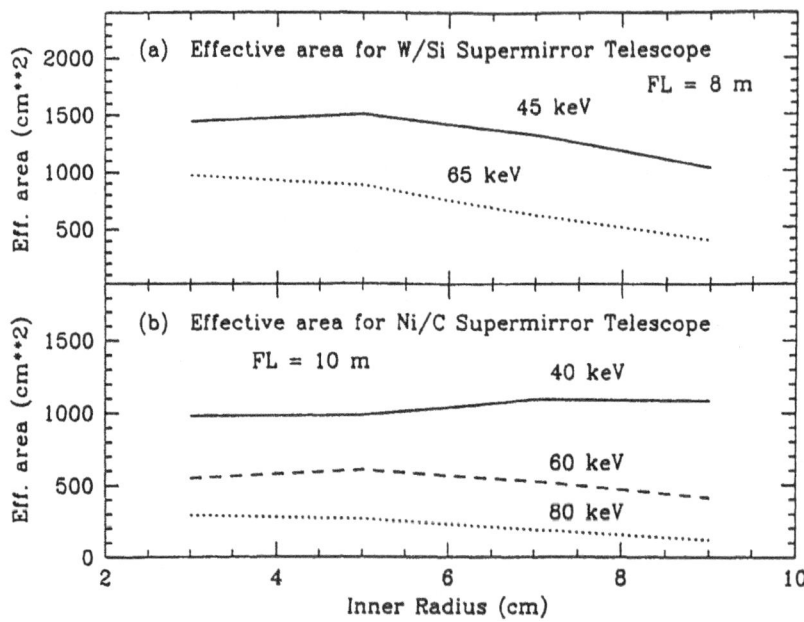

Figure 4. On axis effective area versus the minimum multilayered radius. Figure 4a is for a W/Si supermirror telescope. Figure 4b is for a Ni/C supermirror telescope. It is, in both cases, assumed that mirrors having a radius from 3 cm to the specific radius in question is coated with Au.

Figure 4a shows the result for the W/Si telescope at 45 and 65 keV. Figure 4b shows the result for a Ni/C telescope at 40, 60 and 80 keV. A loss of effective area clearly results from increasing the minimum multilayered radius. The decrease is, however, slow and no significant loss would result from going to a minimum multilayered radius of 6 cm. There is even a slight increase of effective area for the 40 keV case in figure 4 b. This is due to the fact that for the "lower" energies and the innermost mirrors a Au coating is a slightly more efficient reflector than the multilayer coated mirrors. In addition the design parameters of the telescope array (same as the ones used in figure 2) does not represent the most efficient packing of the central aperture. Ideally one would reduce the spacing between reflectors to a value smaller than the minimum spacing used in this calculation (0.5 mm). This features generally reduces the effect of the minimum multilayered radius for the 8 m and the 10 m focal length cases shown in figure 4. The plot, however, shows that the reduction of effective area versus minimum multilayered radius is larger for the 8 m case than for the 10 m case. To evaluate the loss in effective area for the smallest focal length studied, namely 5 m, Figure 5 shows the on-axis effective area for the W/Si telescope case for minimum multilayered radii of 3 and 5 cm, respectively. A more significant loss is observed than for the 8 m and 10 m cases of Figure 4 but a quite satisfactory effective area still results.

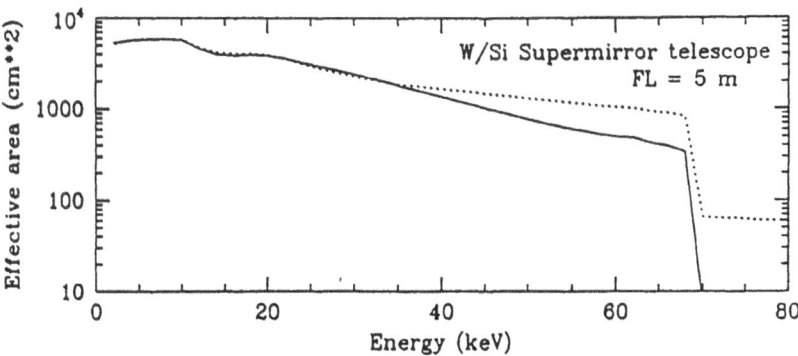

Figure 5. On axis effective area for a 5 m focal length case with W/Si supermirrors. The full line is for a case where the minimum multilayered radius is 5 cm. The dotted line is for a case where the minimum multilayered radius is 3 cm.

4.2 FIELD OF VIEW.

By using ray-tracing, the effective area versus off-axis angles for the baseline design has been calculated. Figure 6 shows the normalized effective area versus off-axis angle for a W/Si supermirror telescope with a focal length of 8 m. Curves are shown for 1, 10 and 60 keV.

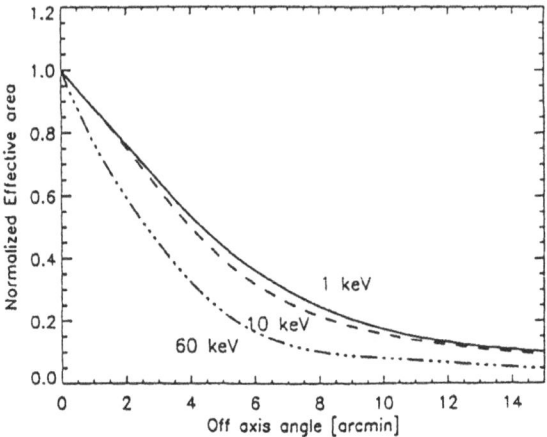

Figure 6. The normalized effective area verus off axis angle (radius) for a W/Si supermirror telescope with 8 m focal length.

The field of view as defined by the full width (diameter) at half maximum is 5.2, 8 and 8.8 arcmin at 60, 10 and 1 keV, respectively.

4.3 SENSITIVITY.

The continuum sensitivity of the baseline design is calculated under the assumption that we require a signal to noise ratio (S/N) of 5 where:

$$\frac{S}{N}(E) = \frac{S(E) \cdot A_{eff} \cdot \Delta E \cdot T}{\sqrt{S(E) \cdot A_{eff} \cdot \Delta E \cdot T + B \cdot A_{FOC} \cdot \Delta E \cdot T}} \qquad (2)$$

Here E is the energy in keV, S(E) is the source spectrum in photons/s/cm^2/keV, ΔE is the energy bandwidth in keV which is set to 10% of E, A_{eff} is the effective area in cm^2, T is observation time in seconds and B is the detector background which is assumed to be 10^{-4} photons/s/cm^2/keV. The multiple foci focal area, A_{FOC}, has conservatively been set to 4 times the multiple foci half power focal area assuming that only a small fraction of source counts falls outside these areas on the detector, the diffuse background has been neglected since it is small compared to the detector background. Figure 7 shows the resulting continuum sensitivity for a balloon observation time of 10^4 sec., an imaging resolution of

2 arcmin and 3 g/cm² residual atmosphere. Figure 8 shows the continuum sensitivity for a satellite observation time of 10^5 sec., an imaging resolution of 1 arcmin and no residual atmosphere.

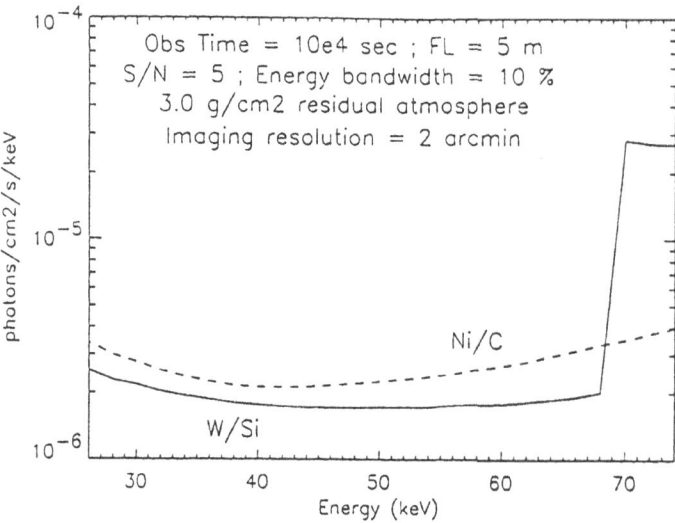

Figure 7. Continuum sensitivity for a balloon relevant observation including atmospheric absorption. Parameters are given in the plot and in the text.

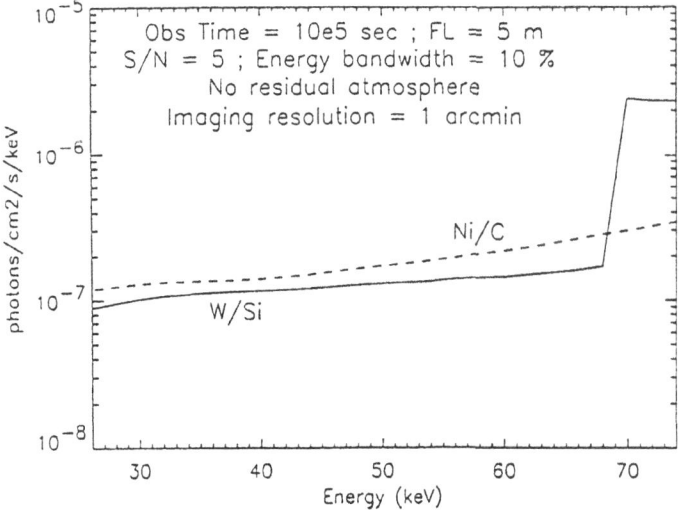

Figure 8. Continuum sensitivity for a satellite relevant observation. Parameters are given in the plot and in the text.

4.4 COMPARISON TO A NON FOCUSING TELESCOPE.

For a background limited observation the figure of merit for comparing a focusing with a non focusing imaging telescope, is (see equation 2):

$$FOM = \frac{A_{eff}}{\sqrt{A_{FOC}}} \tag{3}$$

For a non focusing telescope based on a coded aperture, A_{FOC} is the complete detector area and A_{eff} is half of the detector area. As in the previous section, for the focusing telescope it is assumed that A_{FOC} is 4 times the half power focal area and A_{eff} is taken from the calculations corresponding to the 8 m focal length case shown in Figure 3a. Figure 9 shows FOM for the W/Si supermirror telescope, the Ni/C supermirror telescope and for a non focusing system of the same size. For the supermirror telescopes, an imaging resolution of 1 arcmin is assumed.

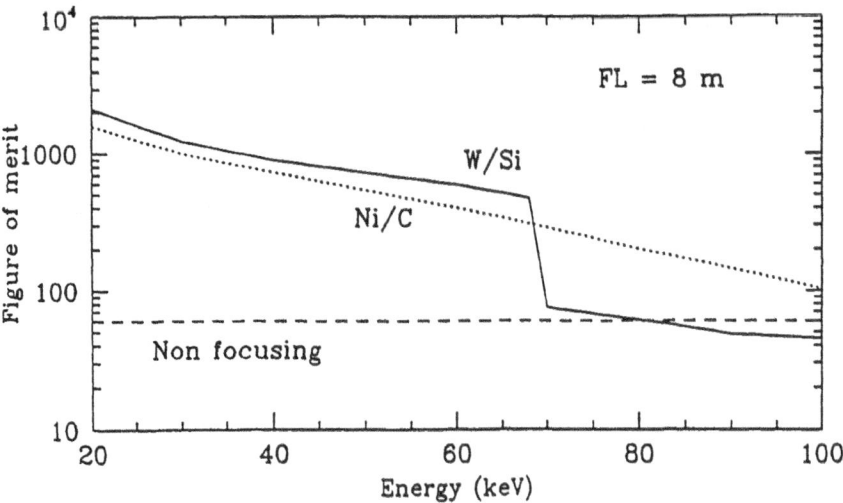

Figure 9. Figure of merit for comparing focusing W/Si and Ni/C supermirror telescopes with a non focusing system of the same size.

Figure 9 clearly demonstrates that more than an order of magnitude incrase in sensitivity can be obtained with the focusing telescope for energies up to ~80 keV.

5. Future developments

The test coating results of Joensen et al (1992, 1993) provide the basis for future hardware developments of conical Wolter I telescopes. There are three key issues that must to be addressed. These are the technique for depositing supermirror coatings either on highly curved substrates or on the inside of highly curved closed conical shells; the subsequent establishment of a mass production facility as initially described (for flat mirrors) by Joensen et al (1994); and, finally, the development of stress-free coatings and/or mounting techniques that allow stressed mirrors to be mounted in a structure that allows 1 to 2 arcmin imaging resolution to be obtained.

6. References

Christensen, F.E., Hornstrup, A., Westergaard, N.J., Schnopper, H.W.: 1991, *Proc. SPIE*, **1546**, 160.

Conti, G., Mattaini, E., Santambrogio, E., Sacco, B., Cusumano, G., Citterio, O., Brauninger, H., Burkert, W.: 1993, *Proc. SPIE*, **2011**, 118.

Cromer, D.T., Lieberman, D.: 1970, *Report LA-4403*, Los Alamos Scientific Laboratory, New Mexico, USA.

Elvis, M., Fabricant, D., Gorenstein, P.: 1988, *App. Optics*, **27**, 1481.

Fraser, G.W., Brunton, A.N., Lees, J.E., Emberson, D.L.:1993, *Nucl. Inst. Meth. A*, **334**, 404.

Frontera, F., De Chiara, P., Gambaccini, M., Landini, G., and Pasqualini, G.: 1991, *Proc. SPIE*, **1549**, 113.

Gondoin, P., De Chambure, D., van Katwijk, K., Kletzkine, P., Laine, R., Stramaccioni, D., Aschenback, B., Citterio, O., Willingale, R.: *Proc. SPIE*, **2279**, To be published.

Gorenstein, P.: 1991, *Proc. SPIE*, **1546**, 91.

Gorenstein, P.: 1994, To be published in Exp. Ast.

Joensen, K.D., Christensen, F.E., Schnopper, H.W., Gorenstein, P., Susini, J., Høghøj, P., Hustache, R., Wood, J., Parker, K.: 1992, *Proc. SPIE*, **1736**, 239.

Joensen, K.D., Høghøj, P., Christensen, F.E., Gorenstein, P., Susini, J., Ziegler, E., Wood, J.: 1993, *Proc. SPIE*, **2011**.

Joensen, K.D., Gorenstein, P., Wood, J., Christensen, F.E., Høghøj, P.: 1994, *Proc. SPIE*, **2279**, To be published.

Priedhorsky, W.C., Christensen, F.E.: 1994, Submitted to Exp. Ast..

Schnopper, H.W.: 1981, *App. Optics*, **20**, 1089.

Spiller, E., Rosenbluth, A.E.: 1985, *Proc. SPIE*, **563**, 221.
Wells, A., Stewart, G.C., Turner, M.J.L., Watson, D.J., Whitford, C.H., Antonello, E., Citterio, O., Braüninger, H., Cropper, M.S., Curtis, W.J., Peskett, S. Eyles, C.J., Goodall, C.V., Mineo, T., Sacco, B., Terekhov, O.: 1991, *Proc. SPIE*, **1546**, 205.

Review of Crystal Diffraction and its Application to Focusing Energetic Gamma Rays

Robert K. Smither, Patricia B. Fernandez, Timothy Graber
Advanced Photon Source, Argonne National Laboratory,
9700 S. Cass Avenue, Argonne, IL 60439 USA

and
Peter von Ballmoos, Juan Naya, Francis Albernhe, G. Vedrenne
Centre d Etude Spatiale des Rayonnements, 9, du Colonel-Roche, 31029 Toulouse,
FRANCE

and
Mohamed Faiz
Physics Department, KFUPM, Dhahran 31261, SAUDI ARABIA

Abstract. The basic features of crystal diffraction and their application to the construction of a crystal diffraction lens for focusing energetic gamma rays are described using examples from the work performed at the Argonne National Laboratory. Both on-axis and off-axis performance are discussed. The review includes the use of normal crystals, bent crystals, and crystals with variable crystal-plane spacing to develop both condenser-type lenses and point-to-point imaging lenses.

1. Introduction

The basic design for a crystal lens that focuses high-energy gamma rays[1-3] is shown in figure 1. One uses crystal diffraction from rings of crystals to collect gamma rays incident on a large area and concentrates them in a small focal spot for detection with a relatively small detector. Each ring uses a different set of crystalline planes and/or a different crystalline material.

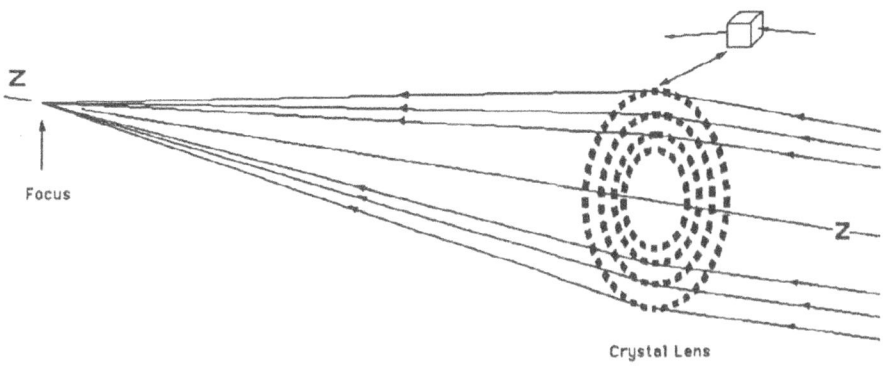

Figure 1. Crystal lens made of rings of cubes of single crystals.

Experimental Astronomy **6**: 47–56, 1995.
© 1995 *Kluwer Academic Publishers.*

The radius of each ring is adjusted so that the Bragg angle, θ, is given by $\sin\theta=\lambda/2d$, where λ is the wavelength of the gamma ray and d is the spaceing between crystalline planes. This approach can give one a large increase of signal in the detector without increasing the background in the detector and make a major improvement in the sensitivity of the instrument. Crystal-diffraction lenses divide into two basic types, Bragg lenses (surface-diffraction) and Laue lenses (volume-diffraction), which are commonly used for lower-energy gamma rays and high-energy gamma rays, respectively. Figure 2 compares these two cases. For energetic gamma rays the diffraction angle (Bragg angle) is quite small and Laue geometry makes much more efficient use of the crystalline material.

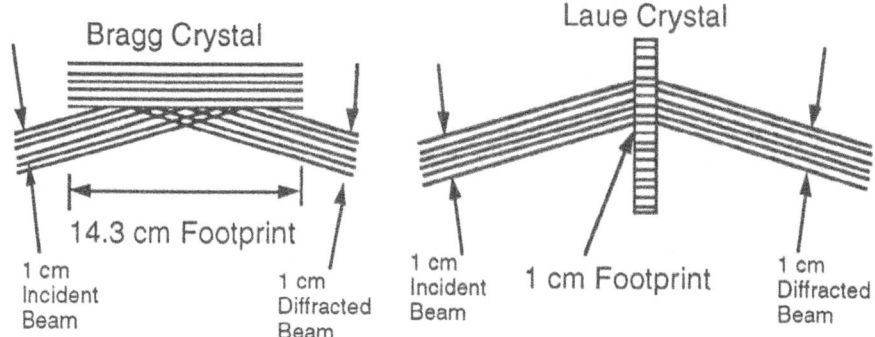

Figure 2. Comparison of Bragg and Laue Diffraction Crystals

Crystal lenses are further subdivided into condenser type lenses[1-6] that concentrate a large amount of radiation on a small focal spot and whose spatial resolution in the image plane is about the size of the individual diffraction crystals, and the point-to-point focusing type lenses[1,2,3,5,7,8] that produce real images of the source as well as a high concentration of photons at the focal point. These focusing point-to-point lenses come in two varieties. They both use crystals that are bent on a radius centered on the axis of the lens. One uses normal crystals and gives a focal spot with a diameter equal to the radial width of the crystal element.

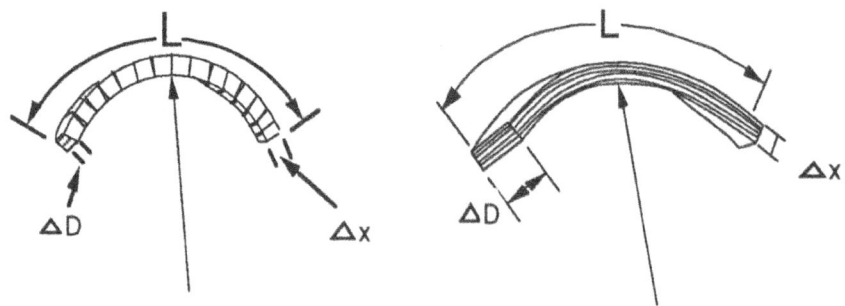

Figure 3. Comparison of the unbent and bent crystal lens elements.

The other uses a bent variable metric crystal and has the potential of producing a focal spot of 0.1 mm or less for a point source and could generate an image with sub-millimeter spatial resolution. Argonne has been the leader in developing the Laue type lens for high-energy gamma rays and has experimented with all three different types mentioned above. Figure 3. compares the individual crystal lens elements for the unbent and bent cases. Bending the crystals eliminates the spreading of the diffracted beam in the plane perpendicular to the diffraction plane (sagittal focusing). To focus the diffracted beam in the diffraction plane one must change the lattice spacing between the crystal planes as a function of radius. The spacing needs to decrease with increasing radius so that the Bragg angle can increase with radius and focus radiation diffracted from the whole crystal to a point focus.[1] These bent crystal geometry's are complicated and required large lens structures (1-m dia.) so only partial lens sections were built to demonstrate the principals involved.

2. The Argonne Crystal Lens

The first complete lens constructed at Argonne was of the condenser type that used small cubes of normal (unbent) germanium crystals. The lens has a 45-cm diameter with 600 germanium crystals arranged in 8 rings. Figure 4 shows schematic drawing of this lens with its 8 rings of crystal cubes.

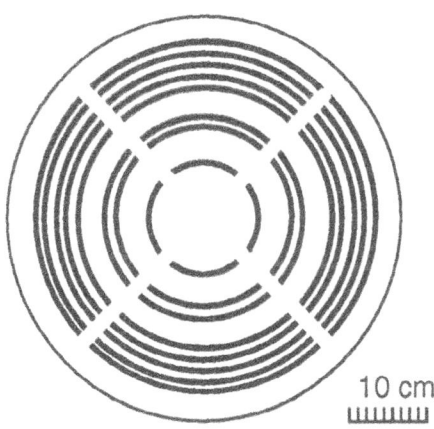

Figure 4.   Schematic drawing of the crystal lens with outer dia. of 45 cm.

Each crystal cube was mounted on an aluminum plate, one end of which was bolted to the lens frame and the other end was free to move. The diffraction angle of each crystal was adjusted separately by applying a light force on the free end of the aluminum plate through a soft spring. This resulted in a slight bending of the plate and allowed one to adjust each crystal so that the appropriate crystalline planes make the right Bragg angle (+/- 5 arc sec) with the incident radiation to satisfy the required relation, $\lambda = 2d \sin\theta$. A ^{137}Cs source of 661.65 keV was used to align the individual crystals, one at a time. Seven different gamma-ray energies have been used in the tests with this lens system. They are listed in Table I, with their source, full lens focal lengths, and distances from source to lens and lens to detector. All of these measurement were made without retuning the lens.

Table I Gamma-ray sources tested with ANL crystal lens.

Energy (keV)	Source	Focal Length (meters)	Source-Lens (meters)	Lens-detector (meters)
611.65	^{137}Cs	10.92	24.75	19.54
511.00	^{22}Na	8.43	19.11	15.09
413.7	^{239}Pu	6.83	15.47	12.22
383.71	^{133}Ba	6.33	14.35	11.33
375.2	^{239}Pu	6.19	14.03	11.08
355.94	^{133}Ba	5.87	13.31	10.51
302.83	^{133}Ba	5.00	11.33	9.70

All that was needed to focus a new energy was to change the distance from the source to the lens and place the detector at the new focal point as explained above.

3. Crystal Diffraction Formula

The crystal diffraction angle θ_B is defined by equations:

$$\sin\theta_B = n\lambda / 2d \tag{1}$$

and

$$\lambda(\text{Å}) = 12.397 / E_\gamma(\text{keV}) \tag{2}$$

where λ is the wavelength of the gamma ray (in Angstrom units in eq. 2), d is the crystalline spacing, n is he order of the diffraction, and E_γ is the energy of the gamma-ray, in keV.   For a 100-keV gamma ray, diffracted by the [111] planes of germanium, θ_B is about 1.0895 degrees, 0.01902 radians. Thus the surface of a Bragg diffraction crystal will need to be 52.6 times longer than the height of a Laue crystal used to diffract the same size beam. (see figure 2)

Each ring uses a different set of crystalline planes and can be of different crystalline material as well. The focal length (F.L.) for each ring (the distance from the ring to the focus for a distant source) is given by equation 3,

$$\text{F.L.} = R / \tan 2\theta \tag{3}$$

where R is the radius of the ring and θ is the Bragg diffraction angle defined above. By adjusting R one can make the focal lengths of all the rings the same and obtain a single focus for the full lens. Substituting equations 1 & 2 in 3

$$\text{F.L.} = R / \tan (2 \arcsin [12.397 \, n/(2d \, E\gamma)]) \tag{4}$$

For small Bragg angles

$$F.L. = [R\ d(\text{Å})\ /(\ 12.397\ n)]\ E\gamma(keV) \tag{5}$$

so for any ring with a given R and d,

$$F.L. = \text{constant} \times E\gamma \tag{6}$$

If all of the rings have the same focal length, then eq. 6 is true for the full lens. For a source at a finite distance from the lens, eq. 7 and eq. 8 must be satisfied along with eq. 1 & 2.

$$L_S = R\ /\ \tan(\theta - \alpha) \tag{7}$$

$$L_D = R\ /\ \tan(\theta + \alpha) \tag{8}$$

where L_S is the distance from the source to the lens, L_D is the distance from the lens to the detector, and α is the angle between the crystalline planes and the axis of the lens. For the case of a distant source $\alpha = \theta$. The small angle approach gives:

$$1\ /\ F.L. = \ 1\ /\ F_S\ + \ 1\ /\ L_D \tag{10}$$

which is the simple lens formula for a simple convex lens with visible light.

A typical scan over a small source (^{137}Cs) is shown in Figure 3. The background count rate is very small, less than 0.1 counts per sec., so the wings are a real part of the scan. These wings come about because the view of a single crystal is not a small circular (angular) area of space but rather a strip across the sky. This strip is narrow in the diffraction plane but can be quite wide in the plane perpendicular to the diffraction plane. This wide angular width depends on, the width of the individual

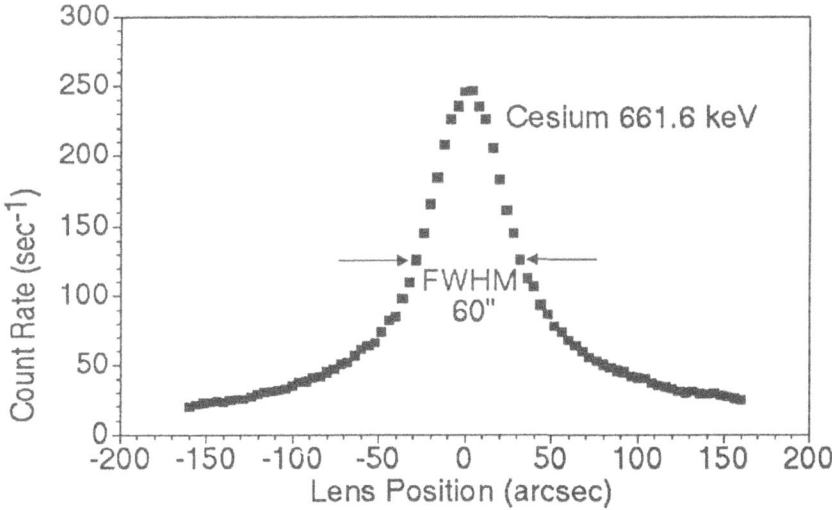

Figure 5. Vertical scan of crystal lens over a 3-mm dia. 137 Cs source.

crystals, the width of the detector and the distance from the lens to the detector. If the crystal is 2-cm wide; the detector 6-cm wide; and the distance from lens to detector is 800 cm, then the angular width of the strip of the sky viewed by the crystal is 0.01 radians, which is 2063 arc seconds or 34.37 arc min. The wide wings seen in figure 5 would extend out either side of the peak by 1032 arc sec.

4. Off-Axis Response

The wide wings in the field of view give the lens considerable off-axis sensitivity and make it much easier to locate a source. They also contribute significantly to the sensitivity (count rate) of the lens when the source has a finite angular size. For an extended source the wings can contribute more to the count rate in the detector than the central region of the field of view. This is because, although the sensitivity of the lens to off-axis sources decreases with the angular distance off-axis, the area being viewed increases at about the same rate. Thus each angular ring of the source contributes the same amount of focused flux to the diffracted beam. If the source exhibits a continuum energy spectrum, then a more complicated picture emerges. The corresponding view for an individual crystal for wavelengths other than that for which the lens focus was adjusted, will still be a strip of the sky but now it will be displaced from the center of the field of view. The strength of this response is strongest from a ring surrounding the center of the field of view. If the source is an extended source, then this wavelength will be focused on the detector as a ring surrounding the center of the detector and a series of wavelengths will be focused on the detector as a series of rings. If the source has an irregular shape, this shape will be imaged on the detector. If a monochromatic source is off center to the left, its image at the focal point will be off center to the right. Thus, the crystal lens has many of the features of a simple convex lens for visible light. This discussion suggests that a multi-element detector would be quite useful at the focal point of the lens, both for locating the source and for determining its size and shape. Just such an experiment was performed at Argonne using a multi-element Ge detector supplied by the astrophysics group from Toulouse. This collaborative (Argonne + Toulouse) experiment is described in an other paper presented at this meeting. [9]

5. Angular Resolution And Diffraction Efficiency

The angular resolution and the diffraction efficiency both depend on the physical properties of the crystals, alignment errors of the crystals, and on the design of the lens. These components are strongly coupled. Lenses that have good angular resolution tend to have good energy resolution, high diffraction efficiencies, large effective areas and low backgrounds for a narrow band width of gamma-ray energies. While lenses that are designed to focus large energy bandwidths tend to have poor angular resolution, smaller effective areas, and lower diffraction efficiencies for gamma-ray sources with narrow bandwidths.

The full Argonne lens contains 600 germanium crystals in the form of 1 cm cubes, mounted in 8 rings. The diffraction planes used and the number of crystals in each ring are, respectively: [111], 28; [220], 52; [311], 60; [400], 76; [331], 84; [422], 92; [333], 100; [440], 108. They are all used in Laue (transmission) diffraction and have mosaic structures that are much larger than their Darwin widths. For a simple unbent crystal cube, the diffraction efficiency (the number of gamma rays focused on the detector divided by the number of gamma rays incident on the crystal) is the product of the transmission of the gamma rays through the crystal times the diffraction coefficient for the crystalline planes and is given by equation 11,

$$\text{Diff. Eff.} = (e^{-\mu x}) [0.5 (1 - e^{\sigma x})] \qquad (11)$$

where μ is linear absorption coefficient; σ is the linear diffraction coefficient; and x is the thickness of the Laue crystal. For the 661.65-keV line from [137] Cs, the transmission is 70% for a 1 cm germanium crystal and the maximum diffraction coefficient is 50%, giving 35% for the maximum efficiency. The individual crystals have diffraction efficiencies for the Cs line that range from 4% to 15%, depending on their mosaic structure and the crystalline planes used. These low efficiencies result from the use of crystals that are too perfect. The average acceptance angle of these crystals is 2 arc sec. The 3-mm dia. Cs source intercepts an angle of 25 arc sec as seen from the lens and the 1-cm crystals intercept an angle of 82 arc sec as seen from the source.   Both of these subtended angles are much larger than the acceptance angle of the crystals so only a small region of the crystal can diffract gamma rays from a point in the source, and only a fraction of the crystal can diffract gamma rays from any part of the source. If these geometric effects are combined with the equation 11, they would predict that the diffraction efficiency of the lens would be less than 1 percent and its performance, very poor. The increase in the efficiency was obtained by cutting slots in the back of each crystal and wedging the three sectors apart so they made a 27-arc-sec angle with their adjacent sections. This reduced the miss match between the acceptance angles and the size of the crystals and the size of the source. Further improvements in the efficiency of the crystals was made by roughing the surfaces of the crystal. This increased the mosaic structure of small regions of the crystals near the surfaces. The net effect was to improve the efficiency from less than 1 percent to the 4 to 15 percent mentioned above.

Table II gives the diffraction efficiencies with the present lens for the measured sources, normalized to a constant distance from the source to the lens of 24.75 m and a constant diameter source, 3-mm dia., (25 arc sec). The last column gives the effective area of the full lens at that energy.

Table II Diffraction efficiencies and effective areas for the 45 cm Lens

Energy (keV)	Element	Diffraction Efficiency	Eff. Area, cm^2
661.65	[137] Cs	0.070	39.
511.00	[22] Na	0.14	78.
413.7	[239] Pu	0.22	123.
383.7	[133] Ba	0.25	140.
355.9	[133] Ba	0.27	151
302.8	[133] Ba	0.24	134.

If the present crystals are replaced with crystals that have the optimum mosaic structure widths and the optimum thickness, the projected diffraction efficiency will be 25 to 40 % for these gamma-ray energies. This corresponds to an effective area of 150 to 240 cm^2. A lens with an effective area of this size should be quite acceptable for a balloon experiment. In all of these experiments, the direct beam from the source to the detector is blocked. If the central region of the lens is opened up so that the detector can see the source, then the effective areas will increase by the area of the detector (28 cm^2).

The width of the scan over the Cs source (60 arc sec), as shown in figure 3, is due to the width of the source (25 arc sec), the size of the 3 subsections (25 arc sec) of each crystal and the 1.6 size increase in width that comes from the circular geometry of the lens. The calculated width, 57 arc sec, is close to the measured value of 60 arc sec. Relatively little of the width is due to the misalignment of the crystals.

6. Standard And Multi-Element Detectors

The Argonne lens system was tested with a standard intrinsic-germanium gamma-ray detector and also with a 3 x 3 matrix of germanium detectors brought to Argonne by the astrophysics group from Toulouse.[8] More detail on the test results of this lens / matrix-detector combination can be found in the paper presented by Juan Naya at this conference . The use of the matrix detector reduces the background count rate under the peak by a factor of 8, while maintaining a detector efficiency of the detector the same size as the full matrix. This improves the signal to noise by a factor of 2.8 .

7. Tunable Crystal Lens

The major advantage of the Argonne-type lens that adjusts all of the crystals to focus a narrow band of wavelengths and thus maximize its sensitivity for that wavelength, is also a major drawback for some experiments. Concentrating on one gamma-ray energy is all right for a balloon experiment, where the observation time is limited but it would not be acceptable for a satellite experiment, where one would like to view many different wavelengths. This means that one must be able to retune the lens to a new wavelength. The Toulouse - Argonne collaboration has devised at least two different ways of accomplishing this. Details on how this can be accomplished can be found in a paper presented at this conference by Peter von Ballmoos.[10] Experiments performed with single crystal elements have achieved short term reproducibility of one arc sec for the diffraction angle of the crystal for these systems.

8. Calculated Performance For A Balloon Flight Experiment

The expected performance of the lens during a balloon flight was calculated for a lens with the same structure as the Argonne lens but with germanium crystals with a mosaic structure width of 10 arc sec, when viewing a monochromatic, distant, point source. The calculation assumes a total diffraction crystal area of 600 cm^2 and an effective area (diffraction efficiency x transmission x total area) of 150 cm^2. The detector system was a 3 by 3 matrix of germanium gamma-ray detectors with a full-energy-peak efficiency of 54 percent. A 3σ experiment for 511-keV gamma rays, with a counting time of 20 hours will have a sensitivity of 2.3 x 10^{-5} photons cm^{-2} sec^{-1}. This calculation follows the calculations in references 4 and 10. The authors in reference 10 calculate a sensitivity for 511-keV gamma rays in a 3σ satellite experiment lasting 10^6 seconds for a lens with a diameter of 90 cm and a more efficient lens geometry, of 1×10^{-6} photons cm^{-2} sec^{-1}. If one scales this design down to the 45 cm dia. of the Argonne lens, the sensitivity becomes 4×10^{-6} photons cm^{-2} sec^{-1}.

Acknowledgment

This work was supported in part by the DOE contract No. W31-109-38-ENG. The germanium crystal lens was designed and built with funds supplied by the DOE Office of Nonproliferation and National Security for the work performed at ANL for the project entitled "Crystal Diffraction Lens For Long-Range Passive Detection Of Fissile Material".

References:

[1]   "New Method for Focusing X-Rays and Gamma-Rays", R.K. Smither, Rev. Sci. Instrum., 44, 131-141 (Feb. 1982).

[2]   "New Method for Focusing and Imaging X-Rays and Gamma-Rays with Diffraction Crystals", R. K. Smither, Sym. on Future X-Ray Experiments, X-Rays in the 80's", GSFC, Oct. 1981, NASA Tech. Mem. No. 83848 (Nov. 1981)

[3]   "A Positron Annihilation Radiation Telescope Using Laue Diffraction in a Crystal Lens", R. K. Smither and Peter von Ballmoos, INTEGRAL Workshop, Feb. 2-5 at Les Diablerets, Switzerland. AP supp., Vol. 92, 1994 June, 663

[4]   "A Bragg Crystal Flux Concentrator for Annihilation Radiation", R.K. Smither and N. Lund, 16th Inter. Cosmic Ray Conference, India (1983). Conference proceedings, supplement issue of AP.

[5]   "Crystal Diffraction Telescope for Discrete line sources", R.K. Smither, et. al., GRO Science Workshop, GSFC, April 1989, NASA Report, Ed, W. Neil Johnson.

[6]   "A Study of Focusing Telescopes for Soft Gamma Rays", Niels Lund, Experimental Astronomy Vol 2, (1992) 259

[7]   "Gamma-Ray Telescopes Using Variable-Metric Diffraction Crystals", R.K. Smither, 11th Texas Symposium on Relativistic Astrophysics", Austin, Texas, Dec. 1981, Annal. of New York Acad. Sci., 422 (1983) 384

[8]   "Crystal Diffraction Lenses for Imaging Gamma-Ray Telescope", R.K. Smither, "13th Texas Symposium on Relativistic Astrophysics," p 55-59, Ed. MAP. Ulmer, Northwestern Un., World Scientific Publishing Co., Singapore

[9]   "Experimental Results Obtained with the Positron-Annihilation Radiation Telescope of the Toulouse-Argonne Collaboration", Juan Naya, et. al., contributed paper to this conference.

[10] "A Space Borne Crystal Diffraction telescope for the Energy range of Nuclear Transitions", Peter von Ballmoos, et. al., contributed paper to this conference.

HARD X-RAY ALL-SKY IMAGING WITH BATSE/CGRO*

The Earth Occultation Transform Imaging

S. N. ZHANG,** B. A. HARMON and G. J. FISHMAN

NASA/Marshall Space Flight Center
Huntsville, AL 35812 USA

and

W. S. PACIESAS

University of Alabama in Huntsville
Huntsville, AL 35899 USA

Abstract. The large Area Detectors (LADs) of the BATSE experiment aboard the Compton Gamma-ray Observatory have been used recently as the first hard X-ray all-sky imager at energies between 20 keV and 300 keV. The Earth occultation process is formulated in terms of a curved Radon transform convoluted by a step transform in a selected field of view (FOV) ranging from $5°\times5°$ to $40°\times40°$. The Maximum Entropy Method is then used to reconstruct an image in the hard X-ray sky. Multiple images of different regions of the sky can be produced simultaneously. A source location accuracy of $0.1°$ for strong sources and a sensitivity limit of 100 mCrab have been achieved in an one-day integration period.

Key words: γ-rays: Imaging – Earth Occultation, Radon Transform, Maximum Entropy

1. Introduction

The scientific capabilities of the BATSE experiment aboard the Compton Gamma Ray Observatory have been enhanced substantially by the development of a new all-sky imaging method[1, 2]. Traditionally, imaging at the hard X-ray energies has only been made by employing modulators or transformers carried aboard together with hard X-ray detectors. Scientifically it is very important to have a sensitive all-sky monitor with imaging capability operating at hard X-ray energies constantly. This is demonstrated by the recent discoveries of several hard X-ray transients by the BATSE team [3], especially the X-ray Nova Scorpii (GRO J1655-40) using the new imaging technique described in this paper. BATSE's discovery and quick location of the source [4, 6] have led to an optical identification[5], radio[7, 8] and soft x-ray observations of this source[9, 10].

2. Formulation of the Earth Occultation Process

The BATSE's eight LADs are non-collimated and cover the full sky with almost uniform sensitivity[12, 13]. The Earth blocks $\sim30\%$ of the total solid

* Invited Paper on "Imaging in High Energy Astronomy", Sept., 1994, Capri, Italy
** also Universities Space Research Association

Experimental Astronomy 6: 57–62, 1995.
© 1995 *Kluwer Academic Publishers.*

angle at any time. When the Earth occults a source in a sky region observable by BATSE/CGRO, an occultation step is produced on top of the time variable background as detected by the LADs. The step has a finite width ranging from several seconds to tens of seconds depending on the angle of the source from the orbital plane and on the energy of the X-ray photons. Two steps, a rising and a setting step, may be generated within each orbital period of ∼90 minutes. The size of the steps and thus the source flux can be calculated by fitting the data with a pre-calculated occultation profile and a background model. This is the so called Standard Earth Occultation Analysis, which has been applied to the BATSE data analysis since the launch of the CGRO in April 1991[14].

Mathematically, the Earth occultation process can also be described in terms of a curved Radon transform[15], depicted by the upper plot in Figure 1, convoluted with a step transform shown in the middle plot. For the purpose of image reconstruction, the step transform is undesirable since it is unclosed in the time domain. A high-pass Butterworth filter of the second order is then applied to the step transform to form a double-sided function with baseline at zero level, as shown in the lower plot of the figure. This filter at the same time also removes the slowly varying components (minutes to hours) of the background. The background removal effect of this filtering technique is compared with a semi-empirical background model [16] and no significant difference is found[17].

This formulation provides an important improvement over the imaging scheme implemented previously by us[1, 2]. In the previous implementation, the differentiated detector counting rates are modelled by the Radon transform. The drawback of this method is that it is highly sensitive to unwanted, rapidly varying components (∼seconds) or spikes in the background. In this new formulation, the shape of the filtered step function is used so that any spikes in the raw data would be ignored. The sensitivity of this method is also substantially improved because the Butterworth filter chosen only attenuates the signals by less than 20% while greatly reducing the background level. For the differentiation method, the signal loss is greater than 50% for a similar background level reduction.

3. Image Reconstruction Algorithm

The Maximum Entropy Method is used to perform the image reconstruction. The size of the FOV can be chosen to be between 5°×5° and 40°×40°, depending on the required angular resolution, available computer memory, and other computing limitations. The forward transform from the image space to the data space is composed of two separate transforms, as shown in Figure 1. The first one is a curved Radon transform. A numerical calculation of the curved Radon transform is implemented since no analytical form is available. The second step is a high-pass filtered step transform from the Radon space to the data space, which can be performed efficiently by FFT. The data used in the reconstruction are the raw counting rate of the LADs high-pass filtered also by the same filter. The transpose of the above two transforms are used as a backward transform from the data space back to the image space. The backward transform is physically a smearing of the original image with the forward transform function, and thus defines the direction in which the iterative process converges. The Maximum Entropy Principle is used as a stopping criteria since an exact reconstruction is not achievable due to the presence of noise and imperfect sampling.

4. Monte-Carlo Simulations

A series of Monte-Carlo simulations have been made to evaluate the image reconstruction method. Simulated sources are placed in the FOV centered around the location of GRO J1655-40. The FOV has an angular size of $10° \times 10°$ and is divided into 100×100 pixels. The true orbital parameters of the CGRO are used to determine the location of the Earth's limb for every two seconds. In the simplest case, a source with all flux confined in a single pixel is tested. The noise level is 1% of the step size in the data space. Figure 2(left) shows the input image and the data used by the reconstruction process. Figure 2(right) shows the reconstructed image and the residuals after the reconstruction. The residuals are defined as the differences between the input data and the results by forward transforming the reconstructed image into the data space. It can be seen clearly that the residuals are consistent with the Gaussian noise.

Our simulations also proved that multiple or diffuse sources can also be reconstructed properly with multiple orbital data. The sensitivity of the BATSE LADs for this new imaging scheme is determined to be around 100 mCrab for an one-day integration, by both simulations and applications of the imaging method to the real data. All those figures are ommited due to the space limitation.

5. Some Sample Images

Many regions of the sky have been imaged by this technique with the LADs data. Only a few sample images are shown to illustrate this new imaging technique. Figure 3 shows the image of the sky region containing GRO J1655-40. The difference between the peak of the image and the true location of the optical counter part, indicated by the cross near the central peak, is ~0.1°. Figure 4 shows an image of the Galactic Center region. This demonstrates BATSE's capability to closely monitor the Galctic Center region continuously with a reasonable sensitivity. In this figure, the peak near GX 1+4 is slightly offset, possibly due to systematic errors caused by the inherent fluctuations in the light curve of the X-ray pulsar GX 1+4. Despite the systematic effects, we have successfully imaged several X-ray pulsars to confirm the new hard X-ray activities of these systems[18, 19, 20]. Figure 5 shows an image of the Aquila region. The X-ray burster source is detected for the first time at hard X-ray energies upto 100 keV. Further results of detailed analysis of this source with BATSE data will be presented elsewhere[21]. In Figure 6 the X-ray transient GRO J0422+32 is imaged at a flux level of 100 mCrab with an integration time of one day, which is the sensitivity level of this imaging system.

6.   Conclusions

A new imaging technique based on a new formulation of the Earth occultation process and Maximum Entropy Optimisation principle has been developed and applied to the BATSE LADs. It is demonstrated, both by Monte-Carlo simulations and real data, that this method works successfully for the hard X-ray all-sky imaging. An all-sky image is obtainable in about every two weeks by combining many smaller images of the different sky regions together[2]. A location accuracy of $0.1°-0.5°$, angular resolution of $0.5°-1°$ and a sensitivity of 100 mCrab are routinely achieved at energies between 20 keV to 50 keV in an integration period of one day.

References

1.   Zhang, S.N. *et al.*: 1993, *Nature* vol. **366.**, 245-247.
2.   Zhang, S.N. *et al.*: 1994, *IEEE Transactions on Nuclear Science* vol.41(4), 1313-1320.
3.   Harmon, B.A. *et al.*: 1993, IAU Circular, No. 5864, 5874, 5890, 5900
4.   Zhang, S.N. *et al.*: 1994, IAU Circular, No. 6046
5.   Bailyn, C. *et al.*: 1994, IAU Circular, No. 6050
6.   Wilson, C.A. *et al.*: 1994, IAU Circular, No. 6056
7.   Campbell-Wilson, D. and Hunstead, R.: 1994, IAU Circular, No. 6052
8.   Hjellming, R.M.: 1994, IAU Circular, No. 6055, 6060
9.   Inoue, H. *et al.*: 1994, IAU Circular, No. 6063
10.   Greiner, J.: 1994, IAU Circular, No. 6078
11.   Paciesas, W.S. *et al.*: 1994, IAU Circular, No. 6075
12.   Fishman, G.J. *et al.*: 1989, *The GRO Science Workshop*
13.   Fishman, G.J. *et al.*: 1993, *Astro. and Astrophys. Suppl. Ser.* vol. **97**, 17-20.
14.   Harmon, B.A. *et al.*: 1991, *The GRO Science Workshop*, pp.69-75.
15.   Deans, S.R., 'The Radon Transform and Some of Its Applications', Wiley, 1983.
16.   Rubin, B.C. *et al.*: 1992, *AIP Conference Proceedings* vol. **280**, 1127-1131.
17.   Paciesas, W.S. *et al.*: 1995, these proceedings.
18.   Stollberg, M.T. *et al.*: 1993, IAU Circular, No. 5836
19.   Wilson, R.B. *et al.*: 1994, IAU Circular, No. 5955
20.   Wilson, C.A. *et al.*: 1994, IAU Circular, No. 6075
21.   Harmon, B.A. *et al.*: in preparation

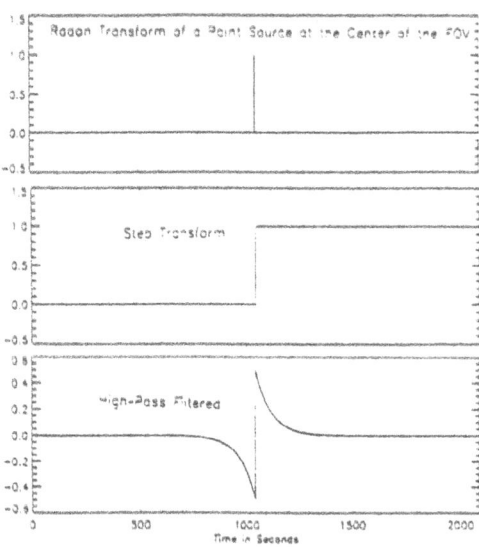

Figure 1: Formulation and Filtering of the Earth Occultation Transform, in terms Radon transform, step transform and Butterworth high-pass filter. The raw detector counting rate is also filtered by the same filter to match the filtered step function. A real occultation step after the filtering looks similar to the bottom figure except the transition width from the maximum to the minimum is finite ($\sim$seconds), and thus the reconstructed image of a discrete source in the sky will have a finite angular size ($\sim 0.5°$-$1.0°$). This is the angular resolution of this imaging system. It defines the minimum distance for separating two close sources in the sky.

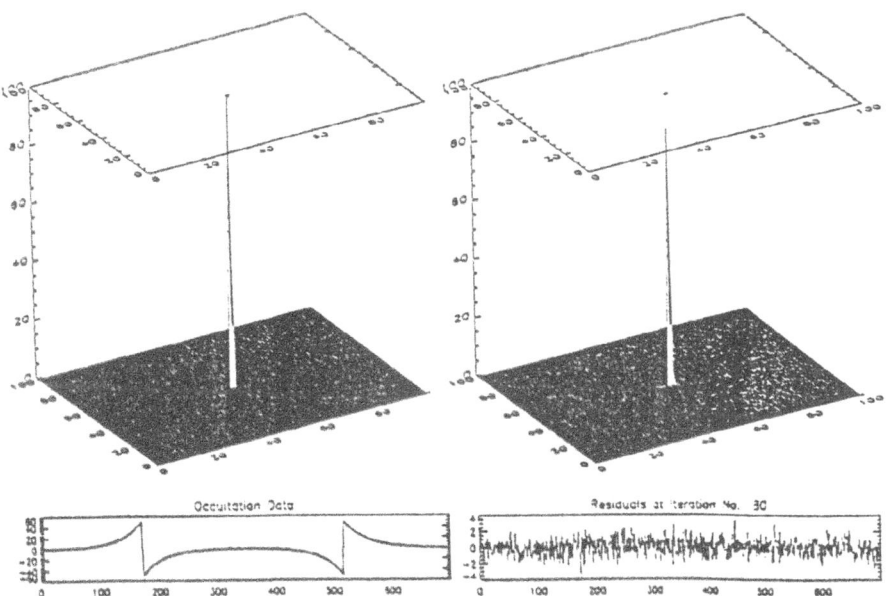

Figure 2: Simulation of a point source on TJD 9575. Only one orbital data, consisting of one rising and one setting step, are used in the simulation. Left: the simulated point source and the input data to the reconstruction code. The data are the high pass-filtered raw detector counting rates. The data points are not shown when the Earth's limb was not in the FOV. Right: the reconstructed image and the residual data. The source is at the right location with approximately the input strength. The residual data are consistent with the input Gaussian noise.

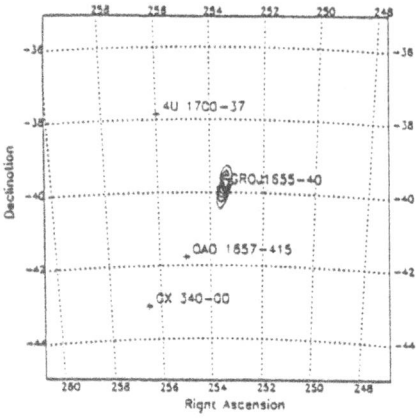

Figure 3: An image of the X-ray Nova Scorpii (GRO J1655-40) for a day of TJD 9575. The center of the peak is less than 0.1° away from the optical counter part. The hard X-ray transient 4U1700-37 was also active at about 100 mCrab level, but fitted out to improve the location of GRO J1655-40, which was about 800 mCrab at 20-50 keV.

Figure 4: An image of the Galctic Center region for a two week period between TJD 9240-9254 at energies between 50-100 keV. The identifications of the two hard x-ray sources 1E 1740-294 and GRS 1758-258 are positive. The other two peaks near Terzan-2 and GX 1+4 are consistent with these two sources with only margical significance.

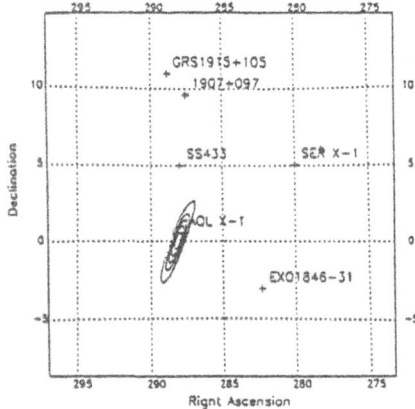

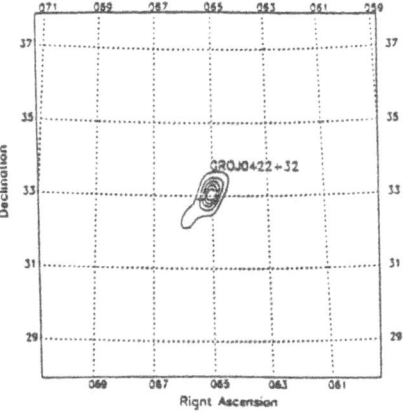

Figure 5: An image of the Aquila region for a 10 day period between TJD 8490-8499. The peak is consistent with the X-ray burster source AQL X-1. The flux is about 50 mCrab at 20-50 keV. This is believed to be the first hard X-ray detection of this source. Searchs from other periods of BATSE data also found evidence for hard X-rays from this source.

Figure 4: An image of the X-ray Nova Persei (GRO J0422+30) for a day of TJD 9028. The peak is consistent with the source. The flux is about 100 mCrab at 50-100 keV. This demonstrates the sensitivity level this imaging system has. The center of the peak is about 0.5° away from the location of the optical counter part of GRO J0422+30.

DIRECT DEMODULATION METHOD AND ITS
APPLICATION TO HARD X-RAY IMAGING

TIPEI LI

High Energy Astrophysics Lab., Inst. of High Energy Physics
Chinese Academy of Sciences, P.O.Box 918, Beijing

Abstract. High resolution reconstruction of complicated objects from incomplete and noisy data can be achieved by solving observational modulation equations or correlated modulation equations iteratively under physical constraints. Simulations and experiments show that wide field and high resolution images of space hard X-rays and soft γ-rays can be obtained by scan observation of a collimated non-position-sensitive detector.

1. Modulation Equation and Correlated Modulation Equation

An observation of an object distribution $f(x)$ by a X-ray or γ-ray telescope can be mathematically described by $\int p(\omega, x) f(x) dx = d(\omega)$, where $d(\omega)$ is the observed data, ω denotes the parameters determining the state of the observation, the modulation function $p(\omega, x)$ is the response of the instrument to incident photons from the direction x during the observation ω. Uniformly dividing object space into N bins, for M observed values $d(k), k = 1, ..., M$, the discrete observational equations constitute an algebraic equation system

$$\sum_{i=1}^{N} p(k, i) f(i) = d(k) \quad (k = 1, ..., M) \tag{1}$$

or, in matrix form

$$Pf = d \tag{2}$$

Any observation can be seen as a modulation process of an incident signal by instrument, the modulation equation is a general description for an observation with any kind of telescope. For an imaging instrument, the measured result d is an two-dimensional image of the object, $p(k, i)$ the point-spread function of the imaging system; For a coded aperture telescope consisting of position sensitive detector and coded mask, the measured $d(k)$ is an output of the kth detector element, $p(k, i)$ describes the aperture projection on the detector plane along the direction i; For a device comprising collimator and non-position-sensitive detector, e.g. a slat collimator telescope or linear scan or rotation modulation telescope with grid collimator, the data space is determined by the parameters describing the detector, orientation of the detector and collimator, $p(k, i)$ is the transmission factor of the collimator for incident photons along direction i during observation k.

Experimental Astronomy 6: 63–69, 1995.
© 1995 *Kluwer Academic Publishers.*

The measured result $d(\omega)$ is a distribution in the data space, and we can transform it into the object space by means of a correlation transformation $c(x) = \int p(\omega, x)d(\omega)d\omega$. The cross-correlation function $c(x)$, often taken as a final reconstruction, is not the object distribution. Performing the correlation transformation for both sides of Eq.(1), the correlated modulation equations relating the cross-correlation function to the object can be derived as

$$\sum_{i=1}^{N} p_1(i', i)f(i) = c(i') \quad (i' = 1, ..., N) \tag{3}$$

where $p_1(i', i) = \sum_{k=1}^{M} p(k, i')p(k, i)$, $c(i') = \sum_{k=1}^{M} p(k, i')d(k)$, or in a compact form

$$P_1 f = c \tag{4}$$

where $P_1 = P^T P$, $c = P^T d$.

2. Direct Demodulation Method

Li and Wu (1992,1993,1994) suggested to derive the object distribution from observed data d (or cross-correlation c) by solving the modulation Eq.(2) (or correlated Eq.(4)) iteratively under physical constraints. The main points of the Li–Wu algorithm are as follows: derive the background data d_b by identifying apparent discrete sources from the correlated map c and subtracting their contribution from the data; solve the equation $Pf_b = d_b$ or the corresponding correlated equation iteratively for f_b; in the iterations each approximate solution f_b is smoothed to depress sudden changes between neighbouring intensities; finally, solve Eq.(2) or (4) iteratively under the constraint condition $f(i) \geq f_b(i)$ $(i = 1, ..., N)$ to derive a reconstruction of the object f.

The convergency of the iteration process places some requirements on the matrix of coefficients of the solved equations. For the case of an imaging telescope, if the point-spread function for any object point is relatively concentrated round the corresponding image point, the iterative process to solve the modulation Eq.(4) by a normal iteration method, e.g. Jacobi or Gauss-Seidel iteration, may be convergent and satisfactory restoration with optimal resolution will be obtained provided the statistics and ratio of signal to noise are not too bad. Otherwise, if the performance of imaging system is not good enough or the measured result obtained by a non-imaging instrument is not an image of the object, reconstruction can be performed by solving the correlated Eq.(4). The coefficient matrix of a correlated modulation equation set, $P_1 = P^T P$, is symmetric and positive definite, satisfying the convergence condition of Gauss-Seidel iteration. In addition to normal

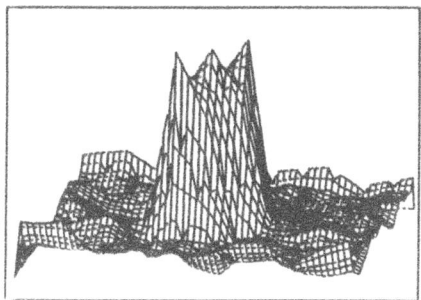 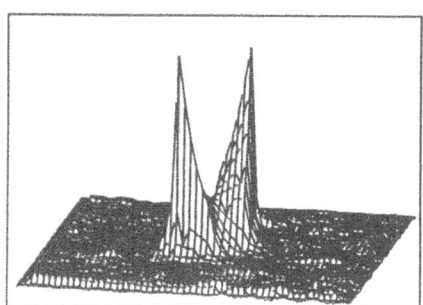

Fig. 1. Images with a coded aperture telescope. (a) Cross-correlation; (b) Direct demodulation. The program to simulate coded mask telescope was provided by Dr. F. Lei

iteration methods, the Richardson-Lucy formula (Richardson, 1972; Lucy, 1974) can also be used. It can be seen that direct reconstruction from the modulation equations without additional constraints by Richardson-Lucy iteration is somewhat equivalent to that from the correlated equations by normal iteration under the constraint $f \geq 0$. When the diffuse background is not negligible, introducing the background constraint condition $f \geq f_b$ in the iterative process is essential for obtaining a satisfactory reconstruction, no matter what kind of iteration formula is used.

3. Application to Image Restoration and Reconstruction

The direct demodulation algorithm provides a general approach to handle a large variety of image restoration or reconstruction problems. Computer simulations and analysis results for COS-B and CGRO γ-ray data show that in comparison with traditional techniques, e.g. maximum entropy method, cross-correlation deconvolution or likelihood approach, the direct demodulation method has high sensitivity, high resolution ability and capability to effectively reduce the effect of statistical fluctuations and noise in data and to simultaneously restore both the extended and discrete features in the object.

We show here the decoding ability of the direct approach for a coded aperture telescope. The telescope comprises a twin-prime mask and a position sensitive detector, the elements of both the mask and detector have a linear size of 1 cm, the separation between the mask and detector plane is 4 m. Measured data for a region containing two sources separated by 14$'$ with the telescope is produced by simulation. The image reconstructed by cross-correlation deconvolution and that by direct demodulation are shown in Fig.1(a) and (b) respectively.

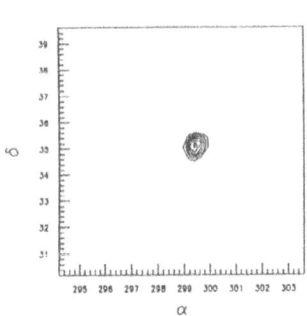 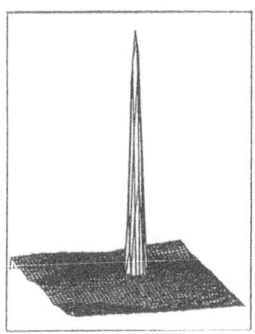

Fig. 2.   Hard X-ray image of the Cygnus region by slat collimator telescope HAPI-4

The spatial resolution of a cross-correlation map, like other traditional
restorations, is limited by the intrinsic angular resolution which is determined by the geometry of the telescope. The relation between the object
f and the cross-correlation c is presented by the correlated equation (4):
$P_1 f = c$. It is obvious that the cross-correlation function c is not really the
object distribution, but only an image of the object f produced by such an
instrument with a transmission function $P_1 = P^T P$. Taking the correlation
function c as a final reconstruction of the object f, the information contained by Eq.(4) will be completely lost. The direct demodulation technique
further restore f from c by solving Eq.(4) under physical constraints. It is
not surprising that the direct reconstruction could have a better performance
than that of a cross-correlation map.

4.   Scan Modulation Imaging

The direct demodulation method of producing images does not rely on a
position sensitive detector. From the point of view of the direct approach
restoring N objective intensities from M image intensities observed with a
position sensitive detector and that from M modulated counts measured by
scan observation with a simple collimated counter are the same mathematical problem: extract N undetermined quantities from M linear equations.

Computer simulations show that fine resolution images within a wide
field of view can be obtained with simple collimated devices through direct
demodulation (Li et al, 1993).

The feasibility of this kind of modulation imaging has been confirmed
by a balloon experiment (Lu Z.G. et al, 1994). The hard X-ray telescope
HAPI-4, constructed in collaboration of IHEP/Beijing, MEPI/Moscow and
AIT/Tuebingen, consists of multiwire proportional counter and phoswish
scintillators with a sensitive area 1600 cm^2 and slat collimator of $3° \times 3°$

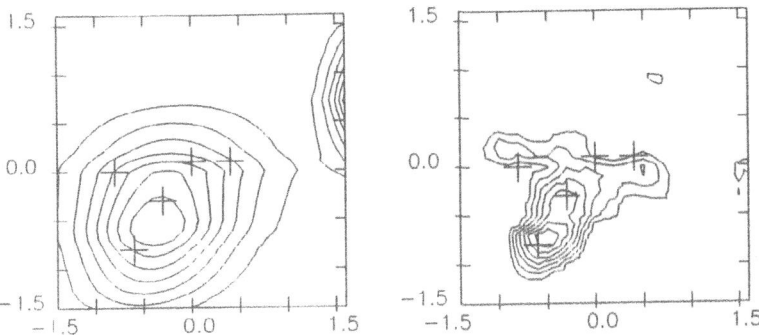

Fig. 3. Images with EXOSAT ME survey, (a) by cross-correlation and (b) by direct demodulation. Crosses denote the locations of sources detected by Skinner et al (1987). Coordinates are galactic longitude and latitude, in degree.

(FWHM). A two hour scan of the Cygnus region was made with HAPI-4 during a balloon flight launched from the Xianhe balloon site near Beijing on Sept. 25, 1993. The step size is 2.5° and 2° along right ascension and declination respectively. Fig.2 shows the direct demodulation image from the scan data, the deviation between the locations of the image and Cyg X-1 is $\leq 0.1°$ which is the attitude control accuracy of the gondola, the angular resolution (FWHM) is $\leq 0.4°$.

The performance of scan modulation imaging has been further proved by re-analysing survey data of EXOSAT ME (Lu F.J. et al, 1994) and HEAO-1 A4 (Sun et al, 1994) with the direct demodulation technique, which gave fine reconstructions, e.g. see Fig.3.

Fig.4 shows the schematic diagram of a proposed hard X-ray modulation telescope HXMT (Li T.P. and Li Q.B., 1994). The telescope consists of 6 phoswich detectors (HXD, for $10 - 600$ keV, total area 5400 cm^2) and two proportional counters (MXD, $2 - 30$ keV, 1120 cm^2). The field of view of each HXD module is defined by a slat collimator and is $5° \times 0.5°$ (FWHM) oriented with the planes of the slats inclined by 30° to the neighbouring collimator. The predicted continuum and line sensitivities of HXMT are shown in Fig.5.

A scan with HXMT over a $6° \times 6°$ region containing a 20σ point source is simulated. The step size is 0.5° over a $1.5° \times 1.5°$ region around the source and 1° for the other region. The cross section of the source image with direct demodulation is shown in Fig.6(a). It can be seen from Fig.6(a) that the point source location is much better than 1', and the angular resolution FWHM $\leq 2'$. Fig.6(b) shows a direct reconstruction map of a complicated region from a simulated 2 hours scan with HXMT/HXD. The simulated scene contains 8 sources, the strongest source intensity is 7×10^{-3} $cm^{-2}s^{-1}$,

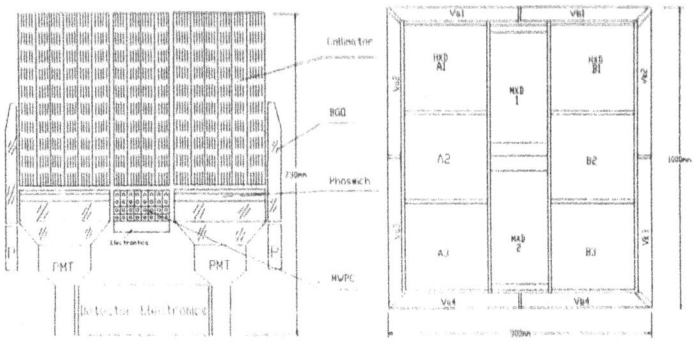

Fig. 4. Schematic view of HXMT (Hard X-ray Modulation Telescope). (a) Side view; (b) Plan view

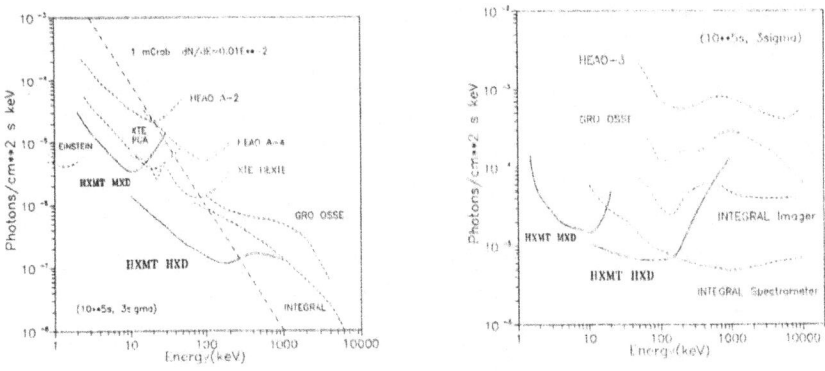

Fig. 5. The predicted continuum and line sensitivities of HXMT

the ratio of the strongest source intensity to the weakest is 100:1, the background is taken as 0.02 $cm^{-2}s^{-1}$, predicted for HXMT. More simulation studies have been done and shown that this kind of modulation imaging is able to simultaneously reconstruct both extended and discrete features in the object and is not sensitive to variation of sources and background.

The estimated mass of HXMT payload is less than 400 kg and dimensions $\sim 1 \times 1 \times 1$ m. Expanding the detection area to $\sim 10^4$ cm^2 is not difficult. With the aid of the direct demodulation technique a $\sim 10^5$ cm^2 hard X-ray imager mission might be possible. More sophisticated design of modulator and detector may achieve higher resolution imaging or precise location for transient events.

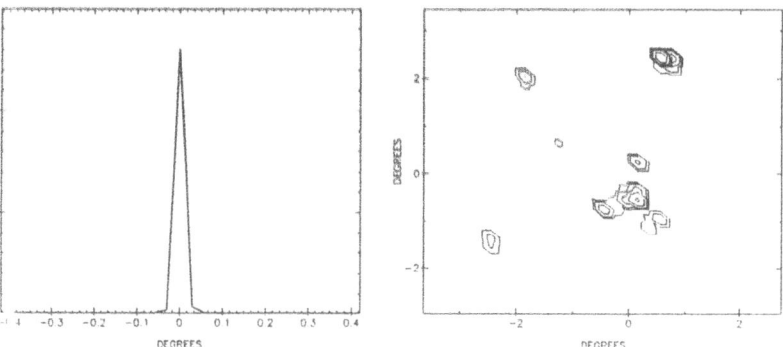

Fig. 6. Simulated images by HXMT. (a) Cross-section of image for a strong source at (0,0); (b) Map of a complicated region

Acknowledgements

This work is supported by the National Natural Science Foundation of China and UNESCO.

References

Li T.P. et al.: 1993, *Astrophys. Space Sci.* **205**, 381

Li T.P. and Li Q.B.: 1994, Space Hard X-Ray Modulation Telescope (HXMT), *IHEP, BAO*, Beijing.

Li T.P. and Wu M.: 1992, in Worrall D.M., Biemesderter C. and Barnes J. ed(s)., *Astronomical Data Analysis Software and Systems I*, 229

Li T.P. and Wu M.: 1993, *Astrophys. Space Sci.* **206**, 91

Li T.P. and Wu M.: 1994, *Astrophys. Space Sci.* **215**, 213

Lu F.J. et al.: 1994, Application of direct demodulation on Exosat galactic scan observation, to appear in *Theory and Observation of Compact Astronomical Objects*

Lu Z.G. et al.: 1994, A balloon borne hard X-ray telescope and its scan imaging to Cyg X-1, *submitted to Nucl. Inst. and Meth.*

Lucy L.B.: 1974, *Astron. J.* **79**, 745

Richardson B.M.: 1972, *J. Opt. Soc. Am.* **62**, 55

Skinner G.K. et al.: 1987, *Nature* **330**, 544

Sun X.J. et al.: 1994, A revisit of the HEAO-1 A4 all sky survey I. Images for selected regions, *(in preparation)*

THE INTEGRAL MISSION

CHRISTOPH WINKLER*

ESA/ESTEC, Space Science Department, Astrophysics Division
P.O. Box 299, 2200 AG Noordwijk, The Netherlands

November 4, 1994

Abstract. INTEGRAL, the International Gamma-Ray Astrophysics Laboratory, to be launched in 2001, is the second medium-size scientific mission (M2) of the ESA long term programme "Horizon 2000". INTEGRAL addresses the fine spectroscopy and accurate positioning of celestial gamma-ray sources in the energy range 10 keV to 10 MeV. The observational requirements will be met by a payload utilising coded mask imaging in combination with detector pixel arrays (*Imaging*) and cooled Germanium detectors (*Spectroscopy*). INTEGRAL is an ESA led mission in collaboration with Russia and USA. Most of the observing time will be made available to the general scientific community.

1. Scientific Objectives

The scientific goals of INTEGRAL - the International Gamma-Ray Astrophysics Laboratory - address the fine spectroscopy with imaging and accurate positioning of celestial sources of gamma-ray emission. The fine spectroscopy over the entire energy range will permit spectral features to be uniquely identified and line profiles to be determined for physical studies of the source region. The fine imaging capability of INTEGRAL within a large field of view will permit the accurate location and hence identification of the gamma-ray emitting objects with counterparts at other wavelengths, enable extended regions to be distinguished from point sources and provide considerable serendipitous science which is very important for an observatory class mission.

In the 15 keV - 10 MeV region, line-forming processes such as nuclear excitation, radioactivity, positron annihilation, cyclotron emission and absorption become important, and when used as astrophysical tools, are almost certain to lead to fundamental new discoveries. Unique astrophysical information is contained in the spectral shift, line width, and line profiles. Detailed studies of these processes require the resolving power ($E/\Delta E = 500$) of a Germanium spectrometer such as that employed on INTEGRAL. Lower resolution spectrometers (e.g. SIGMA, OSSE, COMPTEL) do not have sufficient energy resolution to study the parameters of these lines. The

* This paper is largely based on the INTEGRAL Phase A study report (ESA SCI(93)1), written by the INTEGRAL Phase A Science Working Team: S. Bergeson-Willis, T.J.-L. Courvoisier, A.J. Dean, Ph. Durouchoux, B. McBreen, N. Eismont, N. Gehrels, J.E. Grindlay, W.A. Mahoney, J.L. Matteson, O. Pace, T.A. Prince, V. Schönfelder, G.K. Skinner, R. Sunyaev, B.N. Swanenburg, B.J. Teegarden, P. Ubertini, G. Vedrenne, G.E. Villa, S. Volonté, and C. Winkler.

last high resolution space instrument on HEAO-3 in 1979-80, was 100 times less sensitive than INTEGRAL. Solid observational and theoretical grounds already exist for predicting detectable emission from such varied celestial objects as the Galactic Centre region, the interstellar medium, compact objects, novae and supernovae and a variety of active galactic nuclei. The scientific objectives are wide-ranging and include:

- The diffuse emission of the Galaxy will be mapped on a wide range of angular scales from arc minutes to degrees in both discrete nucleosynthesis lines, e.g. 1.809 MeV from Al^{26} and 511 keV, as well as the wideband continuum.
- INTEGRAL will provide the tools for using gamma-rays as a sensitive probe of the astrophysical processes going on within a few hundred parsec of the centre of our Galaxy.
- INTEGRAL will image compact objects in unprecedented detail at high energies and the spectroscopic capabilities of the mission will provide the first detailed physical diagnostics of these systems at gamma-ray energies.
- Measurements with INTEGRAL of the shapes of the line profiles of radioactive heavy elements, created by explosive nucleosynthesis during supernova events, will provide information about the expansion velocity and density distribution inside the envelope, whilst the relative intensities of the lines provide direct insight into the physical environment at the time of the production.
- INTEGRAL will monitor the Galactic plane regularly in order to detect high energy transient sources, determine their duty cycle and their luminosity function. Gamma-ray bursts will be studied using the very broad energy range provided by the two main instruments and the two x-ray and optical monitors.
- Because of the greatly improved sensitivity of INTEGRAL, sub-degree resolution imaging is absolutely essential to avoid source confusion from the large population of AGN and to associate gamma-ray sources unambiguously with their optical, infrared and radio counterparts.

2. Design Reference Model Payload

During the Phase A study (1992-1993) a design reference model payload has been defined: INTEGRAL consists of two main instruments (Table 1), a Germanium SPECTROMETER and a Caesium iodide IMAGER, each of which has both spectral and angular resolution but they are differently optimised in order to complement each other and to achieve overall excellent performance. The SPECTROMETER uses cooled hyperpure n-type Germanium detectors providing high spectral resolution, high line sensitivity, mapping of diffuse source emission and medium (degree) imaging capabili-

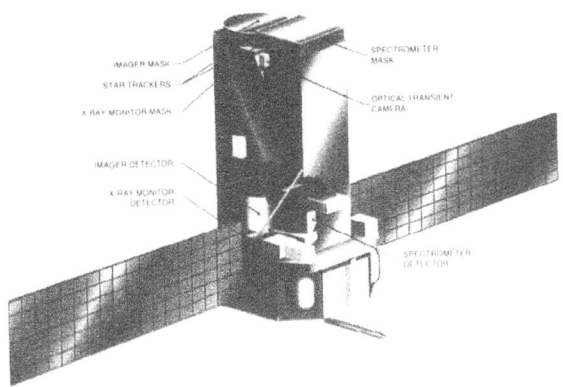

Fig. 1. The INTEGRAL spacecraft in flight configuration (multi-layer insulation not shown) and cut-away view of the payload module as studied in Phase A. Overall size of the spacecraft (bus and payload module) is 4 m x 3 m x 6 m (width x depth x height). With the solar arrays being deployed, the spacecraft spans about 16 m.

ty. The IMAGER is a high angular resolution gamma-ray telescope using a 3-dimensional detector array of CsI pixels, providing accurate source positioning, high continuum and good broadline sensitivity and medium (10's of keV) spectral capability.

In addition, two monitor instruments will provide complementary observations in other wavebands: the X-Ray Monitor (XRM) will extend the continuous spectral coverage of the payload down to 4 keV − 100 keV and will provide the best position information (3′ angular resolution), and an Optical Transient Camera (OTC) (550 nm − 850 nm) will search for, locate (20″) and study the optical counterparts (12^m for 1 s) of gamma-ray bursts.

Continuum and narrow line sensitivities of the high energy instruments are shown in Figure 2.

The SPECTROMETER, IMAGER and XRM share a common principle of operation – they are all coded mask telescopes. The coded mask technique is the key which allows imaging, which is all-important in separating and locating sources. It also provides near perfect background subtraction because for any particular source direction the detector pixels can be considered to be split into two intermingled subsets, those capable of viewing the source and those for which the flux is blocked by opaque mask elements. Effectively the latter subset provide an exactly contemporaneous background measurement for the former, made under identical conditions.

The layout of the payload has been conceived in terms of a separate science payload module containing the instruments. This will be integrated and tested independently and later incorporated onto the spacecraft as a single unit. The interface to the satellite bus has been designed to be as

TABLE I

Key performance parameters of SPECTROMETER and IMAGER as studied during Phase A

Parameter	*SPECTROMETER*	*IMAGER*
Energy range:	15 keV → 10 MeV	70 keV → 10 MeV
Detector area:	327 cm^2	2500 cm^2
Spectral resolution		
(E/ΔE @ 1 MeV):	~500	~25
Field of view:	5.6° fully coded	3.2° fully coded
	13° partially coded	22° partially coded
	10° FWHM	6° FWHM (< 300 keV)
Angular resolution:	1.4° FWHM	17′ FWHM
Point source location		
(20σ source):	10′	1′
Continuum sensitivity		
(3σ in 10^6 s @ 1 MeV,		
δE = 1 MeV):	6×10^{-8} ph cm^{-2} s^{-1} keV^{-1}	3×10^{-8} ph cm^{-2} s^{-1} keV^{-1}
Line sensitivity	Narrow (2 keV) line	Broad (20 keV) line
(3σ in 10^6 s @ 1 MeV):	1.5×10^{-6} ph cm^{-2} s^{-1}	1.2×10^{-5} ph cm^{-2} s^{-1}
Polarimetry sensitivity (3σ):	–	10 mCrab, ϕ ~ degrees
Timing accuracy (3σ):	0.1 ms	0.1 ms

simple as possible to reduce complexity, timescales and cost.

3. Mission Scenario

The spacecraft will consist of a service module (bus) containing all space-craft subsystems and a payload module containing the scientific instruments (Figure 1). The bus will be identical for the two scientific missions INTE-GRAL and XMM. The Phase A study has shown that a common service module is feasible for both missions with only very minor adaptations for the INTEGRAL bus due to the different payload.

INTEGRAL (with a total launch mass of 3.6 tons) will be launched early 2001 into a geosynchronous High Eccentric Orbit (HEO) with high perigee in order to provide long uninterrupted observation periods at nearly constant background and away from trapped radiation. As the baseline, a Russian PROTON will launch the spacecraft into a 72 hour orbit with 51.6° inclina-tion, height of perigee of 48000 km, and 115000 km apogee. As backup, an ARIANE 5 can launch the spacecraft into a 48 hour orbit with 60° inclina-tion with almost the same apogee (following a recent change in the XMM

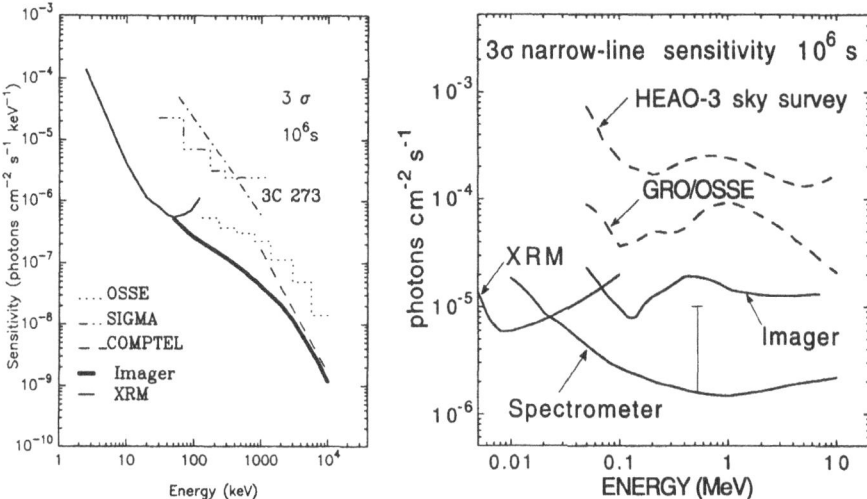

Fig. 2. INTEGRAL continuum (left) and narrow-line (right) sensitivity of the instrument configuration as studied during Phase A. The significance is 3 σ in 10^6 s of observation. Comparison with other missions is shown.

baseline.) Due to background radiation effects in the high energy detectors, scientific observations will be carried out while the satellite is above an altitude of 40000 km. This means, that 100% of the time spent on the PROTON provided orbit can be used for (real-time, 67 kbps) scientific observations, while the spacecraft will stay apx. 83% of its time above 40000 km on the 48 hour (ARIANE 5) orbit. ESA will be responsible for the overall spacecraft and mission design, procurement of bus and payload module, instrument integration into the payload module, integration of the spacecraft, system testing, spacecraft operations and acquisition and distribution of telemetry data. The PROTON launcher will be provided by Russia "free-of-charge" in exchange for observing time, or - as backup - an ARIANE 5 launcher will be provided by ESA. NASA will be responsible for one or possibly two ground stations in order to achieve full real time coverage of science data and will contribute to the scientific instruments and the data centre. Scientific instruments will be provided by the European science community through their national funding agencies and by the US science community through NASA. The INTEGRAL science operations centre (ISOC) will be provided by ESA, the INTEGRAL science data centre (ISDC) will be provided by the European science community through national funding.

4. Science Operations and Observing Programme

INTEGRAL will be an observatory type mission. Most of the observing time, the "general programme", will be awarded to the scientific community at large. Typical observations last from few minutes up to two weeks. Proposals will be selected on their scientific merit by a single review committee. A small fraction (i.e. 30% to 35%) of the total mission duration of nominal 2 years will be reserved for institutes which have developed and delivered instruments and the data centre (guaranteed PI time) and - to a smaller extent - for Mission Scientists. This fraction, the "core programme" will be devoted to (i) a survey of the Galactic plane to map its gamma-ray emission in order to detect as of yet unknown persistent sources (e.g. recent Galactic supernovae) and to facilitate the study of transient sources, and (ii) pointed observations yet to be determined by the PI's and Mission Scientists. The INTEGRAL science data centre (ISDC) will monitor the instrument performance, perform data pre-processing and spot potential targets of opportunity. INTEGRAL will be able to respond to such targets within 24 hours. All scientific data will be assembled by the ISDC into an archive and will be made available (via ISDC or ESA) to the scientific community one year after they have been collected. This guarantees the use of the scientific data for different investigations beyond the aim of a single proposal.

5. Programmatic Aspects

After selection of INTEGRAL by the ESA Science Programme Committee on 03 June 1993 as the next ESA medium-size scientific mission (M2), an Announcement of Opportunity has been issued by ESA on 01 July 1994 asking for proposals for scientific instruments and the ISDC to be provided by the science community via Principal Investigators (PI's), and for proposals for Mission Scientists. Proposals are due by 05 December 1994. After proposal evaluation and selection of instruments, ISDC and Mission Scientists (June 1995), Phase B will start 1995/1996 with a scheduled launch date of April 2001.

References

Matteson, J.L., Dean, A.J. and Winkler, C.: 1994, 'INTEGRAL - the next major space mission in gamma-ray astrophysics' in C.Fichtel, N.Gehrels, J.Norris, ed(s)., *2nd Compton Symposium*, AIP **304**: New York, 45

Pace, O., Pawlak, D. and Winkler, C.: 1994, 'INTEGRAL: Mission and Satellite', *ApJ Suppl. Ser.* **92**, 339

Winkler, C. (editor): 1993, *INTEGRAL: Report on the Phase A Study*, ESA SCI(93)1

Winkler, C.: 1994, 'INTEGRAL: Overview and Mission Concept', *ApJ Suppl. Ser.* **92**, 327

FUTURE GOALS FOR IMAGING

A. J. DEAN

Astronomy Group, Physics Department, University of Southampton,
Southampton, SO17 1BJ ENGLAND

Abstract.
The demands imposed on the imaging system of an astronomical gamma-ray telescope are numerous; it must identify and resolve individual point sources, often in crowded regions of the sky; extended emission structures must be measured on angular dimensions which can extend up to the size scale of the Galactic plane; it must achieve these goals with high sensitivity for both the wide band continuum radiation as well a for discrete spectral line emissions, and ideally have as large a field of view as possible to enhance the probability of registering the unpredictable transient events which pervade the high energy sky. True imaging systems are currently under development for operation for energies up to about 100 keV, however the most practical tool for higher energies, for the time being, remains the coded mask. Some options are briefly reviewed.

Key words: Coded Mask Telescopes, High Energy Imaging

1. Introduction - Outline Science Requirements for Imaging

During the last few years the first astronomical images of the gamma-ray sky in the energy domain (~ 0.1 to ~ 10 MeV) immediately above the classic X-ray band have been made. They have resulted from a number of balloon [1] and satellite [2,3] instruments. From an astrophysical point of view a wide variety of celestial objects have been studied, ranging from point source emitters such as compact objects which incorporate neutron stars and possibly black holes into close binary systems through line and continuum emissions which trace extended galactic features associated with gas and dust clouds and so onto the large scale structure of the Galaxy itself. Extragalactic objects are also powerful gamma-ray emitters. The forthcoming INTEGRAL mission will build upon the current rich data sets and undoubtably raise new important scientific issues of interest. The space available in this short document precludes a detailed discussion of the underlying science. Instead a number of outline scientific goals likely to be of interest in the post INTEGRAL era are discussed in the light of possible technical advances in the field of gamma-ray imaging.

1.1 POINT SOURCE LOCATION

INTEGRAL has been designed with an angular resolution commensurate with the ability to separate and hence identify at other wavelengths any sources it detects. Future increases in instrument sensitivity will impose more stringent demands on the angular resolution.

The Galactic Centre currently poses the most severe problems from a

Experimental Astronomy **6**: 77–84, 1995.
© 1995 *Kluwer Academic Publishers.*

point source confusion point of view, and, with current instrument sensitivities and associated $\sim$ 0.5 degree [4] separations, the rationale is for a system point spread function (SPSF) of typically 20 arc minutes FWHM. What are the consequences of improved sensitivity? How many more sources will there be? If the new level of gamma-ray emitting objects mirrors that of the infrared source levels, which have typical source separations of 5 – 10 arc seconds in the Galactic Centre region, then a SPSF of typically 10 arc seconds will be required for future gamma-ray telescopes.

In the context of nearby galaxies such as M31, M33 etc. INTEGRAL will be capable of detecting the integrated emission from all the compact object gamma-ray emitting sources within the overall galaxy, but is unlikely to have sufficient sensitivity to monitor these objects individually. With the likely increase in sensitivity of future gamma-ray telescopes there will quite naturally follow the capability and interest to study them on an individual basis. The increase in distance is from $\sim$ 10 kpc at the Galactic Centre to $\sim$ 600 kpc for the local group of galaxies, thus necessitating an improvement of a factor of typically 60 in the angular resolution if individual sources are to be isolated. i.e. to about 10–20 arc seconds if the same numbers are to be studied as are currently visible within our own Galaxy. In the general extragalactic context, it should be noted that, for an instrument which has been designed with an SPSF/sensitivity ratio suitable for the study of individual compact objects in the local group, then there will be no serious problems in the identification of individual AGN.

1.2 STRUCTURED EMISSIONS

Following the COS-B and SAS II observations it became clear that intense gamma-ray fluxes from our Galaxy are derived from both point sources and extended structures associated with gas, dust and energetic particle distributions throughout the structure of the Galaxy. The recent skymaps from the COMPTEL telescope on-board CGRO [5] confirm this general fact and demonstrate that these structured emissions extend down to at least 1 MeV in photon energy. Inspection of these COMPTEL galactic plane images show that the angular size scale of the gamma-ray emissions range from source regions of at least 10 degrees in extent to 'hot spots' which are typically one degree or less. More detailed measurements of these large scale structures by future telescopes clearly demands the capability of the instrumentation to map extended emissions which extend up to typically 10 degrees or even more.

Identification of some of the 'point' gamma-ray sources with objects at other wavelengths has shown that a number of them have measurable and interesting structures associated with them. It is highly probable that the gamma emission will also exhibit structured and potentially powerfully diagnostic high energy emission features. An interesting example of this type is the 1E1740 source identified by SIGMA to be a powerful, broadened and red-shifted electron-positron annihilation photon emitter situated near to the centre of our galaxy. Follow-on observations in the radio band [6] have shown this object to be highly structured, having two 180 degree opposed jets which focus on the central gamma-ray object. Do the jets periodically transport positrons to sites within the nearby molecular clouds where they annihilate and generate narrow 511 keV gamma-ray lines? The radio jets extend over an angular scale of typically one arc minute, so that the study of these highly interesting features in gamma-rays on angular scales of some tens of arc seconds may reveal all.

Jet structures are also known to exist in AGN, many of which have been identified as powerful gamma-ray sources. Most of the radio structures associated with superluminal jets are on a size scale not practicable for the next generation of gamma-ray telescope design. However the nearest/brightest examples of a quasar (3C273) and a Radio galaxy (CEN-A) are both powerful gamma-ray emitters and possess jet structures which extend up to typically one arc minute, almost exactly the same as the angular size scale of the 1E1740 jets. It is highly likely, because of the power law emission gamma-ray spectra observed from these objects, that the high energy photon emission is associated with the jets. Thus the next step should be to identify the precise location of the gamma-ray emission in the context of the jet structure. To do this, as for the galactic jets, a resolving power on the several arc second size scale is required.

1.3 SPECTROSCOPIC IMAGES

A considerable number of sources have been discovered from which the emission spectrum is dominated by discrete gamma-ray line features. As is discussed elsewhere at this meeting [7] the detailed spectroscopy of the associated line shapes, intensities etc. yields vital diagnostic information on the physical conditions at the source region. Here we briefly review the likely spatial distributions on the sky and evaluate the imaging requirements which must be associated with the fine spectroscopy measurements, if the full astrophysical information is to be extracted.

Supernovae remnants are proven emitters of discrete gamma-ray lines. [8,9] (SN1987, CAS-A) The study of real-time supernovae explosions such as SN1987A relies on the accurate point source location (PSL) characteristics of the telescope. The fact that there is a discrete line feature in the emission spectrum helps to clarify identification with the parent object, and means that the requirements are less stringent than for normal point sources. Historical supernovae events, viewed through the lines from long-lived isotopes such as ^{44}Ti, rely in the first instance upon the PSL capability to identify the emission with a specific remnant (if visible). However a finer angular resolving power will be required if the explosively created radioactive isotopes are to be unambiguously associated with specific structural features within the expanding envelope. If CAS-A can be taken as a representative example, then a PSL of ~ 1 arc minute is sufficient to identify the parent remnant, but an angular resolving power on the scale of several arc seconds is required to study internal features.

The COMPTEL observations have shown that molecular clouds are strong emitters of gamma-ray lines characteristic of the nuclei which constitute the gas. [10] Presumably the emission is caused by cosmic and sub-cosmic particle collisions with gas molecules. Clearly an accurate map of the spatial distribution of the line emissions needs to be compared with the molecular cloud structure and any 'hot-spots' need to be identified and evaluated if a deeper astrophysical significance is to be obtained. Such an exercise requires high quality mapping on the scale of several degrees and a spatial definition of enhanced regions to the arc minute level. A similar requirement is posed by the spatial distribution of very long-lived nucleosynthetic gamma-ray emitting isotopes such as ^{26}Al. The COMPTEL images [11] show a clumpy extension along the galactic plane with enhanced regions such as that from Vela and other 'hot-spots'.

2.   Technical Implications

Section 1. above shows that the next generation of gamma-ray astronomical measurements will require an advanced imaging capability on all angular scales possible, ranging from typically 10 arc seconds to 10 degrees, Furthermore this imaging performance needs to be combined with a fine spectroscopic capability. There are many techniques available for gamma-ray imaging e.g. Compton scattering, useful for large scale mapping, Bragg diffraction, useful only over limited energy ranges, and coded aperture systems. There is not space here to compare and contrast the various techniques. Perhaps future gamma-ray astronomy missions must become more specialized in their objectives and the above techniques may all be used on different missions. However this represents a more expensive solution to the problem and since gamma-ray missions will be few and far between then instruments will need to be as wide-ranging in their capabilities as is possible. To simplify matters the rest of this article explores the use of the coded aperture mask system in light of the requirements outlined in Section 1.

The performance of the coded aperture mask is not limited by diffraction effects so that, in principal, there is not an intrinsic limitation to the angular resolution which may be achieved. From a technical point of view the limit is set by the distance between the mask and detector and hence the size of the spacecraft. The size of the pixels within the detection plane also provide another technical limitation to fine angular resolution. However, both mask distance and pixel size do have an indirect influence on the large-scale imaging capabilities of a coded mask system. For a given mask diameter the mask distance defines the field of view (FOV) of the gamma-ray telescope, thus increasing the mask-detector distance automatically reduces the angular scale of the overall image and hence limits the size of structures which can be defined. Smaller detection pixels mean smaller mask elements and finer angular resolution, however since to effectively stop MeV gamma-rays the mask elements must be typically 1 to 2 cm thick, then the field of view will be autocollimated by the mask elements themselves. Thus the deployment of a mask of large diameter and large field of view with fine sized elements is not compatible with a large field of view and hence the study of extended objects.

There are, however, intrinsic reasons why a single coded mask geometry cannot provide high sensitivity images of sources over an appreciable range of angular scales. This is due to the limited bandwidth of the coding process. See for example reference [12]. As the source dimensions become comparable to the angular size of the mask elements then there is an increasing loss in coded signal to noise ratio due the increasing spillage of source events into background pixels. Ultimately as the source size becomes significantly larger than the mask elements all the coding is lost and the instrument has no sensitivity at all. The spread of angular sizes which can be studied range from the PSL which may be typically 0.1 of the mask diameter to about two or three times the mask diameter. This corresponds to a maximum dynamic range of about thirty, not enough to satisfy the observational requirements of gamma-ray astronomy, as presented by the large range (> 1000) in likely source size scales.

3.   Technical Solutions for the Size Scale Problem

If a single coded mask pattern cannot satisfy all the demands set by the wide range of gamma-ray source angular sizes, can combinations of patterns be employed to resolve the problem?

The classic solution to this problem in the optical domain (although not in astronomy) is to employ a zoom lens. This concept, adapted to the coded mask arrangement, was originally proposed for the GRASP mission [13]. The design employed an axially moving mask system such that observations of the sky could be made within a range of mask/detector separations. For a nearby mask the field of view will be large and large scale structures can be studied, as the mask is progressively moved away the field of view decreases and progressively finer imaging is possible. Hot-spots on the large scale map can be explored to identify small scale structures and point source emissions. This technique can readily provide an imaging capability over a ×1000 angular size scale, provided mask/detector separations of up to typically 30 m are possible. i.e. ×30 intrinsic plus ×30 due to mask displacement.

The INTEGRAL mission employs a more straightforward technical option - three separate instruments are provided each with different angular resolving powers. The combined set of instruments, the XRM (PSF 3′), the IMAGER (PSF 17′), and the SPECTROMETER (PSF 1.4 degrees) provide an imaging dynamic range of approximately 1000.

It is also possible to provide a larger dynamic range of imaging capability with a single fixed mask detector arrangement. This may be achieved by employing a fixed coded mask with more than one spatial scale within the pattern, [14] In this case the coarse pattern has resolution elements which are of a significantly larger size scale than the fine pattern, thus enabling an increase in overall dynamic range of angular measurement. The coarse pattern is designed to be opaque over the entire operational range of the instrument, whereas the fine pattern only operates up to a few 10s of keV and is transparent at higher energies. The angular scales upon which the imaging takes place is thus selected by energy domain.

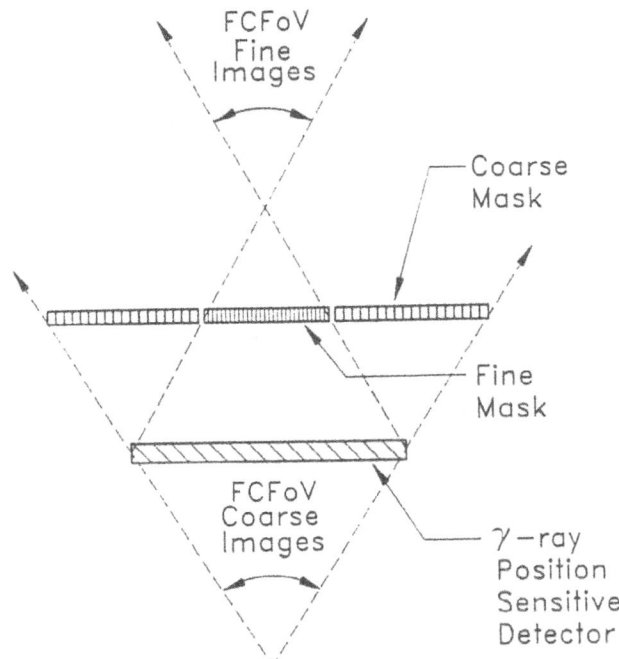

Fig. 1.   Schematic diagram of a possible combined coarse and fine concentric mask/detector arrangement, showing how coarse and fine scale images can be made with a single system over the same field of view.

For the case of antisymmetric patterned rotating masks it is possible to have more than one mask within the detector field of view and create independent simultaneous images of different regions of the sky through the separate masks. The images can be separated from within the single data set if the masks are rotated between their symmetric and antisymmetric positions in such a way that one operates at exactly twice the frequency of the other. This technique could, in principal, be extended into a somewhat ungainly system which uses a number of masks provided the individual mask/antimask rotations are time linked in a binary manner. A more elegant solution along these lines is possible by the use of two concentric antisymmetric hexagonal patterned masks as illustrated in Figure 1. The central mask provides the fine pattern and hence the fine imaging component and the external mask the coarse pattern and hence large-scale mapping. Again one of the two masks should be best operated at twice the frequency of the other. Adjustment of the relative diameter of the detection plane with respect to the two masks provides a means of adjusting the relative fields of view of the fine and coarse images. For example if the central mask is the same size as the detection plane then the fine imaging will be concentrated on a small field of view at the centre of the much larger (dependent on the size of the outer mask) field of view over which the coarse images are made. It is possible to adjust the fine and coarse fully-coded fields of view to be equal by over sizing the detection plane with respect to the central mask and undersizing it with respect to the outer mask. In the case of this geometry the full area of the detection plane is not exploited directly through the fine mask and hence the full sensitivity is not deployed in the fine image. The full area of the detector is, of course, exposed to the source, the remaining fraction through the coarse mask. The coarse images will in general suffer some degree of incomplete coding due to the fine mask 'hole' in the centre of the coarse pattern. However with careful design and exploitation of the fine resolution coded data this will not represent a serious degradation in performance.

4. Conclusions

It is clear that future gamma-ray astronomical missions will need to image celestial objects on a finer imaging scale than is currently anticipated. The next step will most likely be to the 1 – 10 arcsec size scale. It is also clear that emissions on large angular scales will also continue to be of astrophysical interest, and that reconciling the two requirements in a single mission may prove difficult. However possibilities do exist for technical improvements over and above the current single fixed distance coded mask configuration which will enable sensitive imaging to be made over a range of angular size scales of typically 1000.

A. J. DEAN

References

(1) Althouse, W.E. et al. (1985) Proc. 19th ICRC, V3. 299

(2) Roques, J.P. et al. (1989) Adv. Space Res. V10, 223

(3) Schoenfelder, V et al, (1993) Astrophys J. Suppl., V86, 657

(4) Goldwurm, A. et al. (1994) Nature, V371, 589

(5) MPE Annual Report (1993) P.25

(6) Mirabel, I.F. et al. (1992) Nature, V358, 215

(7) Von Ballmoos, P.. This Conference

(8) Matz, S.M. et al. (1988) Nature, V331, 416

(9) Iudin, A.F. et al. (1994) Preprint.

(10) Bloemen, H. et al. (1994) Astron. & Astrophys. V281, L5

(11) Diehl, R. et al. (1993) Astron. & Astrophys. suppl. ser.,V97, 181

(12) Fenimore, E. (1980) Applied Optics, V19, 2465

(13) Bignami, G.F. (1988) ESA SCI(88)2

(14) Skinner, G.K. & Grindlay, J.E. (1993) Astron. & Astrophys. V276, 673

FUTURE GOALS FOR γ-RAY SPECTROSCOPY

P. von BALLMOOS

Centre d'Etude Spatiale des Rayonnements, Toulouse, France

Abstract. Gamma-ray lines are the fingerprints of nuclear transitions, carrying the memory of high energy processes in the universe. Setting out from what is presently known about line emission in gamma-ray astronomy, requirements for future telescopes are outlined. The inventory of observed line features shows that sources with a wide range of angular and spectral extent have to be handled: the scientific objectives for gamma-ray spectroscopy are spanning from compact objects as broad class annihilators, over long-lived galactic radioisotopes with hotspots in the degree-range to the extremely extended galactic disk and bulge emission of the narrow e^-e^+ line.
The instrumental categories which can be identified in the energy range of nuclear astrophysics have their origins in the different concepts of light itself: geometrical optics is the base of modulating aperture systems - these methods will continue to yield adequate performances in the near future. Beyond this, focusing telescopes and Compton telescopes, based on wave- and quantum- optics respectively, may be capable to further push the limits of resolution and sensitivity.

Key words: gamma-ray astronomy – instrumentation

1. Introduction

With the Compton Gamma-Ray Observatory and the GRANAT/SIGMA telescope in orbit, high energy astrophysics has for the first time ever the opportunity to study all the facets of celestial gamma-rays side by side. In the energy range of nuclear transitions, the gamma-ray sky has been mapped on various angular scales and a large number of new gamma-ray sources has been discovered (see reviews by Gehrels et al. 1994 and Paul 1995). Maybe the most important scientific potential of the present generation high energy telescopes is their extremely broad common energy coverage. Together with the operating X-ray and high energy gamma-ray telescopes, a quasi-continuous coverage has opened the possibility for multi-wavelength studies of continuum spectra spanning from the keV- to the GeV-range. Figure 1 indicates that the present area once might be called the "golden age of gamma-ray astronomy": never before has the high-energy sky been examined so thoroughly and over such a broad energy range. For many of the high energy sources, multi-wavelength studies may be the only way that leads to an understanding of their complex source mechanisms. A model case is the spectrum of the quasar 3C273 (Lichti et al. 1995) that has been observed - partly simultaneously - from radio to gamma-ray energies. Nevertheless, the gamma-ray telescopes on the Compton Gamma Ray Observatory and on GRANAT also have raised new astrophysical questions and highlighted those which remain unanswered. The future goals of gamma-ray astronomy must be defined in this context. The progress of SIGMA, BATSE, OSSE and

Experimental Astronomy 6: 85–96, 1995.

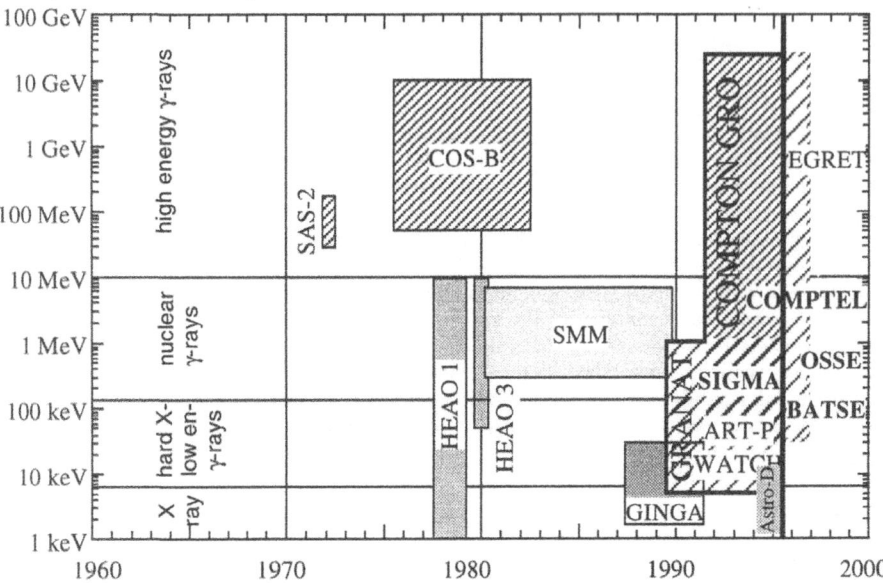

Fig. 1. The "golden age of gamma-ray astronomy" ? Never before the high-energy sky has been examined so thoroughly and over such a broad energy range.

COMPTEL is based primarily on skymaps, excellent timing analysis, and moderate to fair spectral resolution. The observations have revealed certain aspects of the morphology of celestial gamma-ray emitters, yet, the physical processes at work are often only poorly understood. Frequently, the observed spectra do not sufficiently constrain the emission mechanisms : explaining a relatively simple, featureless continuum with a complex multiparameter model can be ambiguous, moreover, different components may blend into one another, each of them can depend on various physical parameters in the emitting region.

In many ways, the present situation resembles the situation of optical astronomy in the beginning of the last century: Back then, the available observational data mainly consisted in images, starcounts, variability's, and colors. Astrophysics was born in 1859 when G. Kirchhof and R. Bunsen developed the spectral analysis and explained the Frauenhofer-lines in the spectrum of the sun. The exploration of atomic and molecular lines has since turned out to be the most powerful tool for the study of the physical conditions in celestial sources.

Up to today, only little advantage has been taken from the fundamental astrophysical information contained in gamma-ray lines. The reason for this is the modest energy resolution of the existing instruments. High resolution spectroscopy will therefore be one of the major goals of the next generation of space borne gamma-ray telescopes.

The scientific potential of nuclear gamma-ray astronomy is outlined in section 2. In sections 3 instruments for spectroscopy in the low and medium gamma-ray channel are presented: modulating aperture systems, Compton telescopes and diffraction lens telescopes. The three techniques actually reflect our current perception of light itself - they are based on the principles of geometrical optics, quantum optics and wave optics, respectively.

2. The promise of nuclear gamma-ray astronomy

Ultraviolet, visible and infrared-spectroscopy has long become an extremely valuable technique in astronomy. While optical lines reflect structural changes in the electron shell of atoms, caused by collisions with energies of the order of 10 eV ($T \sim 10^3$K), transition between discrete nuclear energy levels imply MeV energies ($T \sim 10^7$ to 10^9K), corresponding to the binding energy of nucleons. Collision energies of this order are characteristic for the temperatures inside stars, particles accelerated by electromagnetic fields in solar flares, or interactions of cosmic ray particles with the interstellar medium. Gamma-ray lines are the fingerprints of nuclear transitions, carrying the memory of high energy processes in the universe. High resolution gamma-ray spectroscopy is a unique tool to identify the presence of exited nuclei, quantitatively determine their abundance's, and provide insight into the physical conditions of the source regions (temperature, density, gravitational or cosmological energy shifts).

The scientific potential of gamma-ray spectroscopy includes better understanding of the chemical evolution of the Galaxy by mapping the sites of recent nucleosynthetic activity (novae, SN, massive stars). The observation of supernovae in the galaxies of our local cluster and closeby superclusters (Virgo cluster) will verify our models of nucleosynthesis in massive stars. Spectral features in solar flares, cyclotron lines in gamma-ray bursts and pulsars are among the further challenges of gamma-ray spectroscopy.

An inventory of observed gamma-ray lines is presented in table I, it reflects the current status of nuclear gamma-ray astronomy. Note that most of the knowledge on gamma-ray lines is based on observations made with scintillation detectors that typically feature energy resolutions of 10%. Nevertheless, these elementary spectroscopic measurements already indicate the tremendous potential of gamma-ray line emission. Performance requirements for future spectroscopy missions become conspicuous when the measured/-anticipated line fluxes are compared to the expected angular scales. Figure 2 indicates that emissions with a wide range of angular and spectral extent are expected, varying in intensity by several orders of magnitude. The scientific objectives for gamma-ray spectroscopy are spanning from compact

TABLE I

Observed celestial gamma-ray line features

Physical Process	Energy [keV]	Source	Flux ph cm^{-2}s^{-1}	Instrument, Detector
Nuclear deexcitation				
^{56}Fe(p,p',γ)	847	Solar flares	≤ 0.05	SMM, NaI
^{24}Mg (p,p',γ)	1369	Solar flares	≤ 0.08	SMM, NaI
^{20}Ne (p,p',γ)	1634	Solar flares	≤ 0.1	SMM, NaI
^{28}Si (p,p',γ)	1779	Solar flares	≤ 0.08	SMM, NaI
^{20}12C(p,p',γ)	4439	Solar flares	≤ 0.1	SMM, NaI
	4439	Orion Comp.	$\leq 5 \cdot 10^{-5}$	COMPTEL, sc)
^{16}O(p,p',γ)	6129	Solar flares	≤ 0.1	SMM, NaI
	6129	Orion Comp.	$\leq 5 \cdot 10^{-5}$	COMPTEL, sc)
Radioactive decay				
^{56}Co(EC,γ)^{56}Fe	847, 1238	SN 1987A	$\approx 10^{-3}$ b	Ge det's and
	2598			var. scint's,
	847,1238	SN 1991T	$5 \cdot 10^{-5}$ b	COMPTEL, sc)
^{57}Co(EC,γ)^{57}Fe	122, 136	SN 1987A	$\approx 10^{-4}$	OSSE, NaI-CsI
^{44}Ti(EC)^{44}Sc($\beta^+\gamma$)	1157	Cas A SNR	$7 \cdot 10^{-5}$	COMPTEL, sc)
^{26}Al($\beta^+\gamma$)^{26}Mg	1809	gal. plane	$4 \cdot 10^{-4}$	COMPTEL, sc)
	1809	Vela SNR	$1\text{-}6 \cdot 10^{-5}$	COMPTEL, sc)
e$^-$e$^+$ Annihilation				
	511	Gal. bulge	$1.7 \cdot 10^{-3}$	OSSE, NaI-CsI
	511	Gal. disk	$4.5 \cdot 10^{-4}$	OSSE, NaI-CsI
	480±120 a)	1E 1740-29 c)	$1.3 \cdot 10^{-2}$	SIGMA, NaI
	511	Solar Flares	≤ 0.1	SMM, NaI scint.
	479±18 a)	Nova Muscae	$6.3 \cdot 10^{-3}$	SIGMA, NaI
	400-500 a)	Bursts ? c)	≤ 70	various scint.
	440±10 a)	Crab PSR ? c)	$3 \cdot 10^{-4}$	FIGARO, NaI
Neutron Capture				
^{1}H(n,γ)^{2}H	2223	Solar flares	≤ 1	SMM, NaI
^{56}Fe(n,γ)^{57}Fe	5947 a)	6/10/1974 tr.	$1.5\ 10^{-2}$	SMM, NaI
Cyclotron Lines				
	20-58	Hercules X-1	$\leq 3 \cdot 10^{-3}$	various scint.

legend: sc) liquid scintillator/NaI scintillator; a) Redshifted line; b) Maximum emission, c) detection uncertain, the feature has not been confirmed yet

sources as broad class annihilators, over long-lived galactic radioisotopes with hotspots in the degree-range to the extremely extended galactic disk and bulge emission of the narrow e$^-$e$^+$ line.

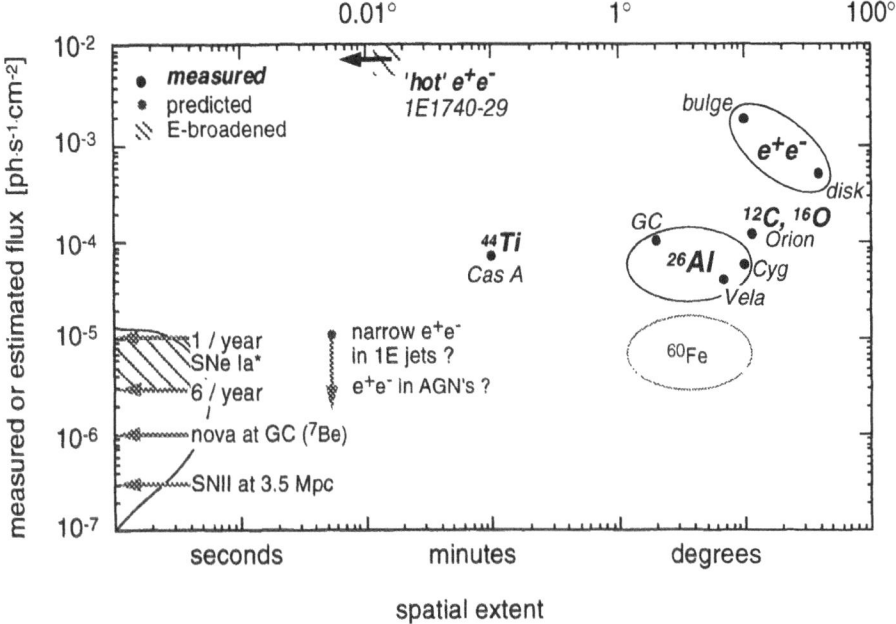

Fig. 2. Future spectroscopy missions have to face emissions with a wide range of angular extent, and with intensities different by several orders of magnitude. The anticipated flux for extragalactic SNe of type 1 has been deduced from the COMPTEL detection of SN1991T (Morris et al. 1995) and by scaling its ^{56}Co 847 keV gamma-ray flux with the optical peak magnitude of observed SNIa.

According to our present view of celestial gamma-ray sources in the energy range of nuclear transitions, narrow lines seem to be generally emitted from extended distributions while broad lines tend to be radiated by compact sources.

3. Instruments for gamma-ray spectroscopy

A natural first objective for a future gamma-ray spectroscopy mission is the mapping of the relatively intense sources (on the upper right of figure 2) which are typically emitting 10^{-4} ph cm^{-2}s^{-1} to a few 10^{-6} ph cm^{-2}s^{-1}. Candidate sources of this intensity are mostly galactic and include the sites of recent nucleosynthesis, regions of e$^-$e$^+$ annihilation and clouds where nuclear de-excitation by energetic particles takes place. Some of them might appear as extended structures: either because of their apparently diffuse origin - such as the narrow 511 keV line (Purcell et al. 1994) - or because they are relatively closeby as the nucleosynthesis sites in the local spiral arm (del Rio et al. 1994). An instrument that is adequate for this kind of objectives should provide a sensitivity of several 10^{-6} ph cm^{-2}s^{-1}, a wide field of view and an angular resolution in the degree range. Such a profile corresponds to the expected performances of the SPECTROMETER on ESA's INTEGRAL mission.

3.1. SPECTROSCOPY WITH MODULATING APERTURE SYSTEMS

Our present understanding of the low energy (<1 MeV) gamma-ray sky has been acquired mainly by modulating aperture systems. The underlying concept of this type of instrument is geometrical optics, this is, the source photons are considered as traveling on rectilinear paths only. Light is passing through open elements of an otherwise opaque aperture. The aperture systems - consisting of masks or collimators - modulate the signal which then reaches the detection plane designed to discern shadow patterns of some kind (spatial or temporal). Since the photoelectric effect is the predominant mode of gamma-ray interaction in medium- and high-Z materials, these systems are well adapted to the low energy gamma-ray channel. Pinhole cameras, rastering collimators, coded masks and modulation collimators belong to this category of instruments.

Two main classes of modulating aperture systems can be identified, according to whether the signal is encoded by spatial modulation (coded mask telescopes) or by temporal modulation (modulation collimators). These two types stand for a whole spectrum of devices mixing the basic concepts of spatial and temporal modulation. Modulating aperture systems of both classes can be used to produce images of the sky if the system allows to observe/resolve an extended region simultaneously (or quasi simultaneously). This multiplex advantage improves the survey sensitivity with respect to plain rastering collimators ('on-off' techniques).

Operational satellite instruments based on the principles of geometrical optics include SIGMA, OSSE and BATSE (Paul et al. 1991, Johnson et al. 1993, and Fishman et al. 1989). While the detection plane of all three of them is based on scintillators ($\Delta E/E \approx 10$), different aperture systems are used. Whereas OSSE and BATSE perform temporal modulation with collimators and the earth (an "anticollimator") respectively, SIGMA is a multiplexing device using spatial modulation with a coded mask.

High resolution spectroscopy with a coded mask instrument will be available by the beginning of the next decade with ESA's INTEGRAL mission. The INTEGRAL SPECTROMETER described below corresponds to the ESA proposal by a consortium between the CESR (Toulouse), MPE (Garching), CEA (Saclay), IFCTR (Milan), IEEC (Barcelona), Univ. of Louvain, Univ. of Valencia, Univ.of Birmingham and US Institutes. Figure 3a) gives a cutaway view of the SPECTROMETER. The instrument features a compact array of high purity Germanium detectors entirely shielded by a BGO/CSI anticoincidence system.

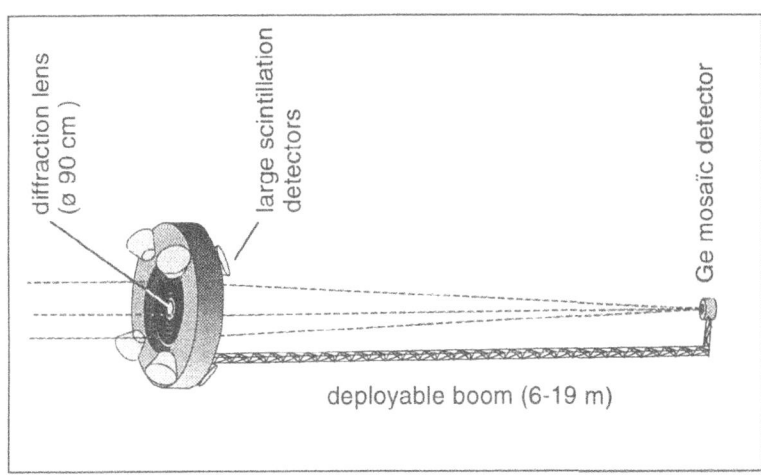

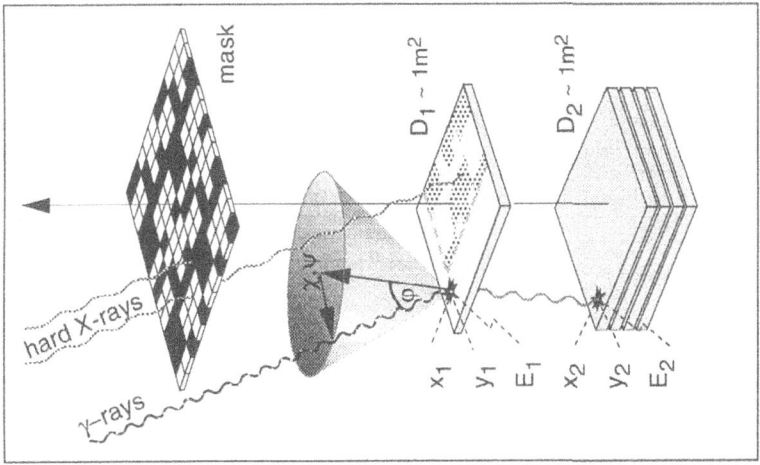

Fig. 3 a) Cutaway view of the proposed
INTEGRAL SPECTROMETER.

Fig. 3 a) Schematic view of a solid state Compton telescope (ATHENA type) with a coded aperture mask added for hard X-ray imaging.

Fig. 3 c) "artists view" of a space borne crystal diffraction telescope. As a counterpart to the extremely directional lens telescope, a full sky monitor (large area scintillators) could complement the payload.

The Ge detector assembly consists of 19 high purity n-type germanium detectors cooled to 85K via pulsed tubes supplied by two cryogenic compressors. With a total geometric detection area of ≈ 500 cm^2 the detection plane is the largest one that could be fitted into the overall constraints of the INTEGRAL mission. The aperture subsystem features a tungsten coded mask (HURA with 127 elements) 171 cm above the detector. The coded mask together with the detector plane defines a fully coded field of view of $16° \times 16°$.

The narrow line sensitivity shown in Fig. 4 has been obtained by completely modeling the SPECTROMETER for the radiation environment conditions expected outside the magnetosphere. The sensitivity is based on models for the instrument efficiency (GEANT) and background in ^{70}Ge detectors (Naya et al. 1995). The anticipated performance for narrow line spectroscopy is characterized by an energy resolution in the parts per thousand range (typically 2 keV at 1 MeV), an angular resolution of order 2°, and a sensitivity 2-$5 \cdot 10^{-6}$ ph cm^{-2}s^{-1} in the energy range relevant for nuclear astrophysics. The 511 keV line sensitivity is of the order of $2 \cdot 10^{-5}$ ph cm^{-2}s^{-1}.

3.2. SPECTROSCOPY WITH COMPTON TELESCOPES

While the total cross section has its minimum at a few MeV, the energy range between several hundred keV and about 30 MeV is dominated by the Compton effect. At these energies modulating aperture systems run into several problems: the efficiency of the modulation decreases and various background problems related to heavy shielding become increasingly prohibiting. Hence, making use of the quantum nature of the photon interactions is an inevitable choice for instruments at medium and high energy gamma-rays. The idea to make use of the Compton effect instead of fighting it with thicker shielding and modulators has stimulated several groups and resulted in a distinct class of imaging instruments.

In a Compton telescope the incident gamma-ray is identified by successive interactions in the two detector layers D_1 and D_2. Compton scattering in D_1 is favored when low Z material is chosen. Total absorption of the scattered photon in D_2 can be expected when high Z materials are used. Since the $D_1 \wedge D_2$ coincidence condition discriminates against most of the internal nβ background, a Comptom telescope has an extremely low background. At the same time, this coincidence condition causes the relatively low detection efficiency. In order to further reduce the background due to upward scattered events, the time-of-flight between the two detectors is usually measured in Compton telescopes. Neutron interactions can be identified by measuring the pulse shape of the scintillation pulse.

From the interaction location in D_1 and D_2, the direction χ, ψ of the scattered gamma-ray is obtained; the energy deposits in D_1 and D_2 are E_1 and E_2 respectively. The scatter direction χ, ψ, together with the amounts of energy deposited in the two interactions can be used to reconstruct the arrival direction of the gamma-ray. The Compton equation allows to express the scatter angle φ as a function of the energy-deposits E_1 and E_2:

$$cos\overline{\varphi} = 1 - \frac{m_e c^2}{E_2} + \frac{m_e c^2}{E_1 + E_2} \quad \text{with } m_e c^2 = 511\text{keV}$$

If E_1 and E_2 are measured without systematic errors ($E_{tot} = E_1 + E_2 = E_\gamma$), the derived scatter angle $\overline{\varphi}$ equals the true Compton scatter angle φ. The arrival direction of the incident gamma-ray can then be confined to lie on a cone-mantle with axis χ, ψ and opening angle $\overline{\varphi}$ (see fig 3b).

With COMPTEL on the Gamma-ray Observatory, the concept has definitely proven its unique potential for MeV gamma-ray astronomy (Schönfelder et al. 1993).

The ATHENA concept (Kröger et al. 1995, Johnson et al. 1995, Kurfess et al. 1994) is based on D_1 and D_2 layers using Germanium planar strip detectors providing 2-3 keV spectral resolution and spatial resolution of ~2 mm. Such detectors, typically 5 cm × 5 cm × 1 cm, are available today and might be integrated into large panels in the future. The ATHENA concept foresees a 1 m² D_1 layer consisting of one panel, and a 1 m² D_2 layer of four panels, each panel containing 400 Ge strip detectors. Figure 3b) shows a schematic diagram of a solid state Compton telescope combined with a coded mask for low energy gamma-rays.

In the Compton mode (300 keV-10 MeV) such an instrument can achieve angular resolutions of 0.2°- 0.3° within a field of view of typically one steradian, and a narrow line sensitivity of several times 10^{-7} ph cm^{-2}s^{-1} above 1 MeV (see fig. 4).

3.3. SPECTROSCOPY WITH DIFFRACTION LENS TELESCOPES

Pushing the performances even further leads to several apparently unsolvable problems. Up to now, better instruments generally were bigger instruments. Yet, physical limits for the size of spaceborne instruments are being attained today. Achieving sensitivities better than 10^{-6} ph cm^{-2}s^{-1} and resolutions better than fractions of a degree seems to be principally impossible with the presently practiced instrumental concepts: even larger collection areas are synonymous with larger detectors and thus higher background noise. The ensuing mass/sensitivity dilemma can ultimately only be overcome by a radically different approach for detecting gamma-radiation.

Being used to accept that it is 'impossible to reflect or refract gamma rays', high energy astronomy has never considered relying on the wave-quality of gamma-ray photons. The focussing diffraction lens telescope (Smither et al. 1995, von Ballmoos et al. 1995) makes for the first time use of the periodic nature of light and its interference with the periodic structure of a crystal. Featuring a Laue diffraction lens, it is designed to collect gamma-

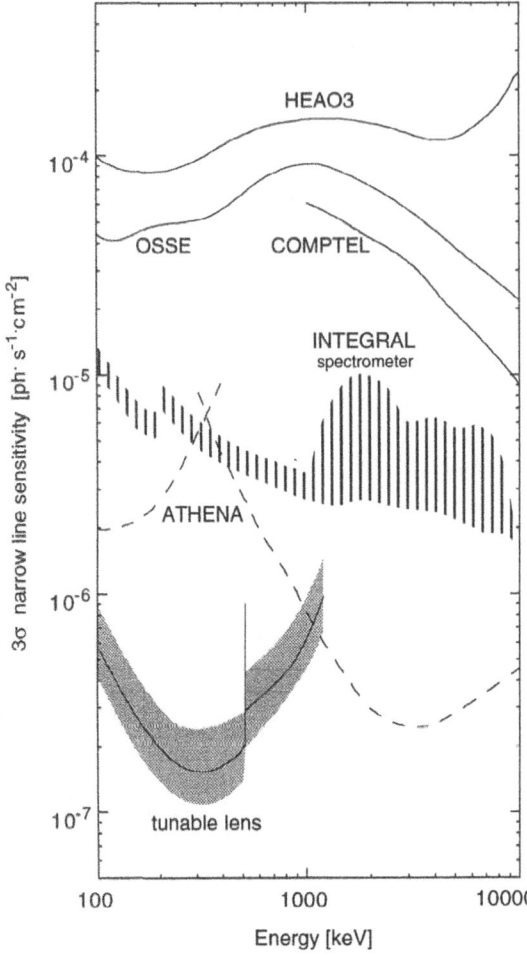

Fig. 4. Comparison of 3 σ narrow line sensitivities: HEAO3 (achieved sensitivity). For all other instruments $T_{obs}=10^6$ sec. The hatched area outlines the sensitivity of the INTEGRAL SPECTROMETER. Lower and upper limits are for an optimistic(o)/pessimistic(p) model: o) including the multiplexing advantage (i.e. sensitivity for any galactcic source during a 10^7 sec survey of the inner galaxy); p) including a β^+ background component. The shaded area indicates different background estimates for the diffraction lens telescope.

ray line photons on a large effective area and focus them onto a Germanium detector matrix with a small equivalent volume for background noise.

The instrument has first been proposed as a balloon-borne telescope with the lens tuned to diffract 511 keV photons only (von Ballmoos and Smither 1994). Such a configuration makes possible the study of galactic 'microquasars' and other broad class annihilators in the light of e^-e^+ annihilation during a balloon flight. The performances of this Ge-lens/Ge-matrix system have been verified in summer 1994 during laboratory measurements with a ground based prototype (Naya et al. 1995).

Ultimately however, the concept should be put to use in space where longer exposures and steady pointing would result in outstanding sensitivities. A 'tunable crystal diffraction lens' (von Ballmoos et al. 1995) can observe any identified source at any selected line-energy in the cardinal range of nuclear transitions (200 keV - 1300 keV). The 'sequential' operation mode resulting from such a concept makes the sites of explosive nucleosynthesis natural targets for a tunable crystal diffraction telescope.

While the evidence for point like sources of narrow gamma-ray line emission has been mostly implicit at this point – besides of the supernovae 1987A (Matz et al. 1988) and 1991T (Morris et al. 1995) – various objects like galactic novae and extragalactic supernovae are predicted in the area at the lower left of figure 2. These sources should have small angular diameters but very low fluxes - mostly because such objects are relatively rare and therefore are more likely to occur at large distances. The instrumental requirements for exploring this kind of sources match with the anticipated preformance of a crystal diffraction telescope:

The imaging capabilities of a crystal diffraction telescope are defined by its beamwidth which is identical to the field of view of the lens: for compact sources discrete pointings of the object will be an appropriate observation mode, while extended structures as for example the jets of galactic microquasars will be scanned with the narrow beam. The field of view/angular resolution will typically be $\sim 15''$ FWHM. Below 1 MeV, the narrow line sensitivity is expected to be a few 10^{-7} ph cm^{-2}s^{-1} (see figure 4).

4. Conclusion

With the spectrometer on ESA's INTEGRAL mission, a coded mask instrument for gamma-ray spectroscopy will be available by the beginning of the next decade. The foremost objectives of the INTEGRAL spectrometer will be the mapping of gamma-ray line sources emitting 10^{-4} ph cm^{-2}s^{-1} to a few times 10^{-6} ph cm^{-2}s^{-1}. Many of these potential sources will be galactic. Some of them might appear as extended structures - either because of their truly diffuse origin or because they are relatively closeby as the nucleosynthesis sites in the local spiral arm. An energy resolution of $E/\Delta E \approx 500$, a wide field of view and a mid-scale angular resolution make the INTEGRAL spectrometer adequate for such objectives.

On a more distant horizon, experimental gamma-ray astronomy has to find ways to further push the limits of resolution and sensitivity: At energies above one MeV, Compton telescopes can provide angular resolutions of fractions of degrees and sensitivities of a several 10^{-7} ph cm^{-2}s^{-1}. At energies below the MeV, tunable crystal diffraction telescopes can achieve sensitivities of a few times 10^{-7} ph cm^{-2}s^{-1} and angular resolutions of the order of 15''.

References

del Rio E. et al., 1994, AIP Conf. Proc. 304, p. 171, (ed. Fichtel C. et al., New York)

Fishman G. et al., 1989 in Johnson, W.N. (Ed), Proc GRO Science Workshop, p.239

Gehrels N. et al., 1994, ApJ Sup. S., 92, 351

INTEGRAL ESA SCI(93)1, ESA phase A study report, April 1993

Johnson, W.N. et al., 1993, ApJ Sup. S., 86, 693

Johnson, W.N. et al., 1995, proc. of workshop 'Imaging in High En. Astron.', Capri 1994

Kurfess J.D. et al., NASA proposal for new mission concepts in Astrophysics, NRA 94-OSS-15, 1994

Kröger R.A. et al., 1995, proc. of workshop 'Imaging in High En. Astron.', Capri 1994

Lichti G. et al., Astron Astrophys 1995, in press

Matz S.M. et al., 1988, Nature 331, 416

Morris D. et al. 1995, proc. 17th Texas Symposium on Relativistic Astrophysics

Murphy R.J. et al. 1990, ApJ ,358, 290

Naya J., et al., 1995a, submitted to NIM

Naya J. et al. , 1995b, proc. of workshop 'Imaging in High En. Astron.', Capri 1994

Paul J. et al., 1991, Adv.Space Res., 11 (6), 289

Paul J. et al., 1995, in Signore M. et al (Ed), The Gamma Ray Sky with Compton GRO and SIGMA, p. 15

Purcell W.R et al., 1994, AIP Conf. Proc. 304, p. 403, (ed. Fichtel C. et al., New York)

Schönfelder, V., Hirner, A. and Schneider, K.: 1973, Nucl.Instrum. Meth., 107, 385.

Smither R.K. et al., 1995, proc. of workshop 'Imaging in High En. Astron.', Capri 1994

von Ballmoos P. and Smither R.K. 1994, ApJ Sup. S., 92, 663

von Ballmoos P. et al, 1995, proc. of workshop 'Imaging in High En. Astron.', Capri 1994

MAXIMUM ENTROPY IMAGING OF COMPTEL DATA

A. W. STRONG

Max-Planck Institut für extraterrestrische Physik
Postfach 1603, 85740 Garching, Germany

Abstract. The Maximum Entropy method is a practical technique for generating intensity skymaps from Compton-telescope data. The application of the method to COMPTEL data for point sources and large-scale emission is described. New developments in the method are illustrated.

Key words: gamma-ray astronomy – image processing – maximum entropy

1. Introduction

Compton telescope data provide an interesting challenge in astronomical data analysis. Unlike a spark-chamber telescope it is not possible to generate an image by simply plotting the arrival directions of the photons: 'indirect imaging' methods are necessary if we wish to make intensity maps of the sky. This implies methods in which the 'data-space' and 'image-space' are logically separated from the beginning, the two spaces being related by the instrumental response function.

One method in which this division is explicit is maximum entropy (Max-Ent) which has proved to be very effective in producing images from the COMPTEL instrument on the Compton Gamma Ray Observatory CGRO. In this paper I briefly describe MaxEnt and illustrate applications to COMP-TEL data for imaging of continuum emission from point sources and for imaging the full sky. Applications to line emission are covered by R. Diehl in an accompanying paper. I also mention some new developments in Bayesian methods.

2. Why MaxEnt ?

The ultimate aim (not yet achieved) is an image-reconstruction algorithm which is not *ad hoc*, i.e. one which is based on basic principles and is in some sense optimal. One form of the principle of maximum entropy may be stated as follows: given constraints such as $\sum c_i p_i = constant$ on a set of proportions p_i then the only *self-consistent* assignment of p_i in the absence of other information is that which maximises $S = -\sum p_i log p_i$ under the constraints. For image-processing this has been applied to a constraint of form '$\chi^2 =$ Number of data points' or its likelihood equivalent for Poisson statistics. (For an extensive introduction see Gull and Skilling 1984). This method

Experimental Astronomy **6**: 97–102, 1995.

was implemented in MEMSYS2 software package (Gull and Skilling) in the early 1980's. More recently a completely *Bayesian approach* has been developed (see Sec 7).

Why use MaxEnt for COMPTEL ? The main reason is that for a COMPTEL-like response with its unusual point-spread-function and 3-D data space, we know of no other way to make true intensity images. Note that other methods such as maximum-likelihood are extremely important in searches for point-sources, but they do not yield intensity distributions, rather they give maps of significance or point-source flux under the assumption of a single or at most a few sources. Especially for *diffuse, extended* emission the MaxEnt method is valuable in providing quantitative intensity distributions.

It should be emphasized that this is a subject in rapid development; the proceedings of the annual workshops under the title of 'Maximum Entropy and Bayesian Methods' (see references) are an invaluable source of information on the current state of the art, and the reader is recommended to consult these.

3.   Maximum Entropy Method

The entropy of an image is defined as $S = -\sum f_i log f_i$ where f_i is the intensity in pixel i relative to a default or model value. The statistic used is the Poisson log-likelihood L=log Prob(data|model) = $\sum ln\ e^{-x}x^n/n!$ The MaxEnt algorithm *maximises S over all images which are consistent with the data*; since S is a measure of structure in the image, the method is often characterised as finding the 'flattest' image consistent with the data. The method is basically as follows: define $C = S - \lambda L$; then, starting with a flat map $f_i=$ const corresponding to $\lambda = 0$ progressively increase λ, at each step maximising C. λ is a Lagrangian multiplier so we obtain images with maximum S for a given L. The process stops when L indicates that the model is first 'consistent with the data ' at some 'confidence level'; in practice more subjective criteria, often just the appearance of the image, are applied; when the whole series of iterations is displayed it is usually evident at what point real features give way to artefacts due to over-fitting the data.

4.   The COMPTEL Data

4.1   COMPTEL RESPONSE REPRESENTATION

The measured quanitites are the scattered photon direction (χ, ψ) and the energy loss in the upper and lower detectors from which is computed the Compton scatter angle $(\bar{\varphi})$. We use the 3-D dataspace $(\chi, \psi, \bar{\varphi})$ to bin the data. The response to a sky intensity distribution $I(\chi', \psi')$ can be represented by the formula:

$$n(\chi, \psi, \bar{\varphi}) = g(\chi, \psi) \int \int I(\chi', \psi') X(\chi', \psi') f(\chi, \psi, \chi', \psi', \bar{\varphi}) d\Omega \qquad (1)$$

where $n(\chi, \psi, \bar{\varphi})$ are the expected counts per steradian, $g(\chi, \psi)$ is a geo-metrical factor related to the upper and lower detector layers, $X(\chi', \psi')$ is an exposure factor, $f(\chi, \psi, \chi', \psi', \bar{\varphi})$ is a point-spread-function and Ω is sol-id angle. A complete description of the instrument and its reponse can be found in Schönfelder et al. (1993).

A good computation of the response is now available based on extensive Monte-Carlo simulations of the COMPTEL instrument for various assumed input spectra. The convolution in equation (1) is time-consuming; it can be done by FFT methods but only when the field studied is not too large (see discussion on large-scale imaging below).

4.2 BACKGROUND ESTIMATION

COMPTEL data are background-dominated, this component contributing at least 95% of the counts. For the case of *continuum emission*, which is the subject of this paper, the problem of estimating the background is difficult and the solutions not entirely satisfactory; in contrast for *line* emission more satisfactory methods are available, as described by R. Diehl in these proceed-ings. Two different methods for estimating the continuum background have been applied. Method 1 uses a smoothed version of the data themselves, so that source-like signals are removed from the background dataset. This has the advantage that it is not affected by time-dependence of background and is easy to apply, but the disadvantage that part of the signal is discarded and extended emission is suppressed. Method 2 uses independent data in the form of an average over many 'empty field' observations (e.g. high Galac-tic latitudes); this has the advantage that no signal is lost and is therefore preferred for extended emission and for model-fitting. Its disadvantages are that it is affected by time-dependence of background, and that it is quite complicated to apply.

5. Point sources

Fig 1 shows an image of the Crab in the 1-3 MeV range, illustrating the very good image attainable for a strong source. Fig 2 shows the same region in 10-30 MeV showing in addition to the Crab the QSO PKS05128+134 which was active in early 1991 (see Collmar et al. 1994). At this time the QSO was on average only about 20% of the Crab flux, showing that MaxEnt is capable of resolving very unequal sources at a few degrees separation. Both these images were obtained with background method 1.

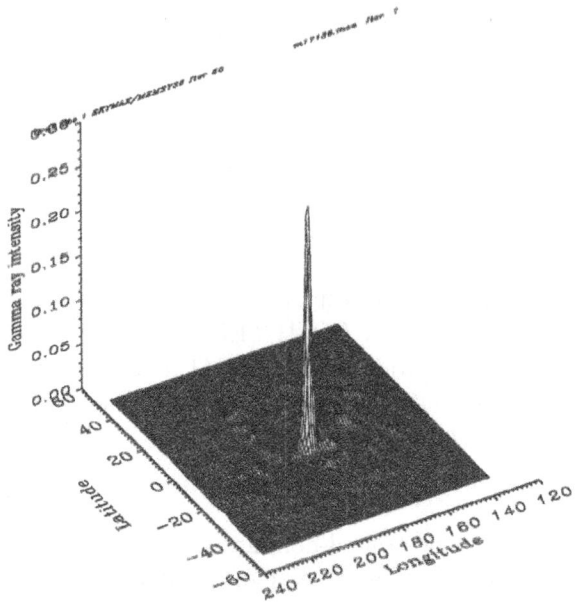

Fig. 1.   Crab image in energy range 1-3 MeV

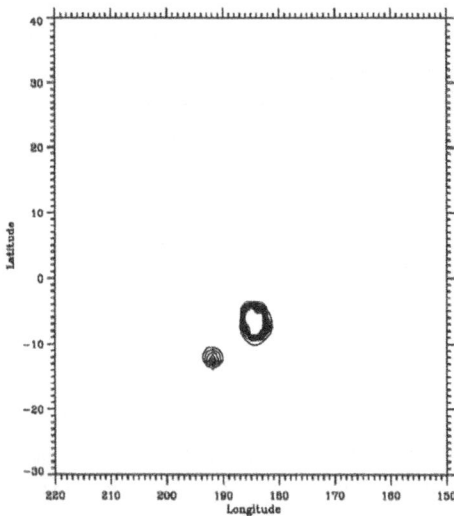

Fig. 2.   Crab and PKS05128+134 in energy range 10-30 MeV

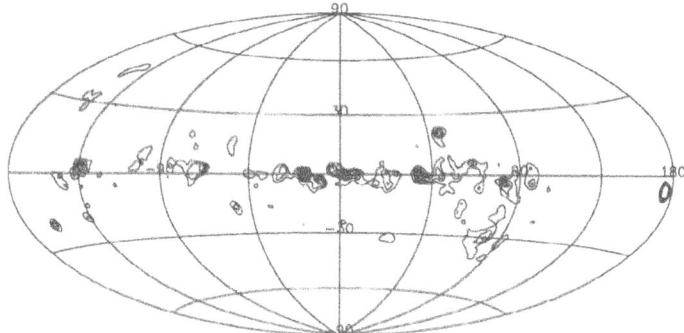

Fig. 3.   Full-sky image in 3-10 MeV energy range, in Galactic coordinates centred on l=0.

6.   Maps of the whole sky

Maps of the whole Galactic plane are useful especially in connection with diffuse Galactic emission; such a map, generated using background method 2, can be found in Strong et al. (1994). More recently it has been possible for the first time to produce maps of the entire sky. The advantages of this method are that it uses *all* available data, there are no 'edge-effects' and the result can be transformed to other projections from the original Mercator. A disadvantage is that it is very computer-time consuming; also any time-variations are smeared out so that transient sources may not be detectable. In a 'Cartesian' system one can do the convolution by FFT since the point-spread-function is shift-invariant. This works only for maps which are restricted in size at least in latitude. For *full-sky imaging* we have to use spherical coordinates; the convolution cannot then be done by FFT but has to be done explicitly. This is however much more time-consuming. For a full-sky image with $1°$ binning and 20 $\bar{\varphi}$ values we require $360 \times 180 \times 60^2 \times 20 \approx 4 \ 10^9$ operations per convolution. The method requires typically 36 such convolutions to reach a reasonable image. For the 3-10 MeV full-sky image shown in Fig 3, a large fraction of the sky survey data were combined into one dataset, and the background was estimated by method 1. The most striking feature is the band of emission following the Galactic plane; some point sources are visible including the Crab, Cen A and 2CG135+1.

7.   Quantified Maximum Entropy

Among the defects in MaxEnt as described above is the fact that the stopping criterion $\chi^2 = N$ (see Section 2) lacks justification and there is no way to judge significance of features (there are no 'error bars'). A new development in the framework of Bayesian statistics (Skilling 1989, 1991;

Gull 1989) addresses these problems: 'classical' or 'quantified' MaxEnt (in contrast to the previously described method, now dubbed 'historical'). It has been implemented in the MEMSYS5 package. Imaging is treated as a Bayesian parameter estimation problem for the pixel intensities. The prior probability has the form $P(\mathbf{f}) \propto e^{\alpha S(\mathbf{f})}$, the posterior probability for data D is given by $P(\mathbf{f}) \propto e^{\alpha S(\mathbf{f})+L(D|\mathbf{f})}$, with the notation of Section 3. α is a parameter controlling the relative effect of entropy and data; for $\alpha \to \infty$ entropy dominates giving a flat image, while for $\alpha \to 0$ the data dominates giving a maximum likelihood image. MEMSYS5 determines the full posterior probability $P(\mathbf{f}|D)$ from which error bars on intensities can be derived. The only unknown is α; in general this can be treated as a parameter with prior $P(\alpha)$. In practice we can choose one best α determined by maximising the 'evidence' $P(D|\alpha)$.

In view of these advances, some efforts have been made to evaluate this method for COMPTEL data using MEMSYS5. The same data as used for the 1-3 MeV Crab image in Fig 1 were used to make an intensity scan with error estimates; the intensities are for 10° bins centred on the Crab:

b / l	195°	185°	175°	
+ 5°	104±83	99±80	229±106	Average intensity
- 5°	144±86	3148±117	261±109	$10^{-5} cm^{-2} sr^{-1} s^{-1}$
-15°	11±40	210±93	221±96	

The Crab flux and error are in good agreement with those obtained by direct fitting using maximum likelihood (Strong et al. 1993). This only illustrates the technique: it will be most useful for deriving errors on extended emission, for example in longitude or latitude profiles.

References

Collmar W. et al.: 1994, 'COMPTEL Observations of the Quasar PKS 0528+134' in C.E. Fichtel, N. Gehrels, J.P. Norris, ed(s)., *Proc. 2nd Compton Symposium, AIP Conf. Proc. 304*, AIP Press: New York, 659

Gull, S.F.: 1989, 'Developments in maximum entropy data analysis' in J. Skilling, ed(s)., *Maximum Entropy and Bayesian Methods*, Kluwer:Dordrecht ISBN 0-7923-0224-9, 53

Gull, S.F., Skilling J: 1984, ' Maximum entropy method in image processing', *IEE Proc.* **131(F)**, 646

Schönfelder V. et al.: 1993, *ApJ Supp* **86**, 657

Skilling, J.: 1989, 'Classic maximum entropy ' in J. Skilling, ed(s)., *Maximum Entropy and Bayesian Methods*, Kluwer:Dordrecht ISBN 0-7923-0224-9, 45

Skilling, J.: 1991, 'On parameter estimation and quantified MaxEnt ' in W.T. Grandy, Jr. and L.H. Schick , ed(s)., *Maximum Entropy and Bayesian Methods*, Kluwer:Dordrecht ISBN 0-7923-1140-X, 267

Strong, A. W.: 1993, 'The Crab and Galactic anticentre region observed by COMPTEL', *A& A Supp.* **97**, 133

Strong, A. W.: 1994, 'Diffuse Galactic continuum emission measured by COMPTEL and the cosmic-ray electron spectrum', *A& A* **292**, 82

IMAGING DIFFUSE EMISSION WITH COMPTEL

Roland Diehl

Max-Planck-Institut für extraterrestrische Physik,
Postfach 1603, D-85740 Garching, Germany

Abstract. The imaging telescope COMPTEL aboard the NASA Compton Observatory satellite has been demonstrated to be capable of imaging diffuse emission along the Galactic plane. Here we describe details of the imaging data spaces and methods to model the background. Different methods of imaging analysis are compared. Verification of consistency among these methods is a key factor in overall assessment of diffuse emission results. Source simulations and statistical analysis through the bootstrap method are applied to verify the significances of image structures.

Key words: gamma-ray astronomy - diffuse emission - imaging - maximum entropy method-maximum likelihood method - bootstrap method - background

1 Introduction

The imaging problem for a Compton telescope is typical for any general event type recording telescope (Prince, this conference):

Measured Data = Instrument Response × Sky Intensity Map + Background

Issues for the case of a telescope with a complex response translating two-dimensional sky information into a multi-dimensional data space of the measurement are in general: Various *background* uncertainties from spatial or temporal variations or normalization problems, *blurring* of the instrument response itself, the generally sparse data, and how to include *prior knowledge* and known constraints into the analysis algorithms.

In the case of COMPTEL imaging analysis, the determination of a proper background description in the multi-dimensional data space is essential for the imaging results due to the low signal relative to background. In addition the prior knowledge about instrument and expected sky distributions are manifold and only a few of these have been attempted to be translated into constraints for the imaging analysis algorithms.

A full desciption of the instrument can be found in (Schönfelder et al. 1993). The imaging response had been described in detail by Diehl et al. (1992), the primary analysis methods by deBoer et al. (1992), and Strong et al. (1992); in this conference, an updated description of Maximum Entropy Imaging with COMPTEL is given in an accompanying paper by A. Strong. This paper addresses details that are relevant in imaging of diffuse emission over extended regions of the sky.

2 Imaging Usage of Raw Measurement Parameters

The primary measurements per recorded photon event in the telescope are the energy deposits and the interaction locations in upper and lower detector: E_1, E_2, $(x/y)_1$, $(x/y)_2$. With these 6 raw measurements, we define the summed energy deposits as a useful approximation for the energy of the primary photon, and translate the energy deposit measurements via the Compton formula into an estimate for the scatter angle φ of the

Experimental Astronomy **6**: 103–108, 1995.

primary in the upper detector plane; the direction of the scattered photon (χ, ψ) completes the imaging event parameters, and defines an event point in 4-dimensional data space $(E_{total}, \varphi, \chi, \psi)$. The imaging options that emerge are determined by the different degree of including the instrument response in the analysis. Note that full response treatment requires at least 7-dimensional response handling (from four dimensions in data space $N(E_{total}, \varphi, \chi, \psi)$, and sky parameters $I(E_\gamma, \alpha, \delta)$). In order to retain visibility of stages and signal/background interferences of data analysis, different compromises have been made for different methods.

A simple straightforward analysis can be done with the method of *'software collimation'*: Here the imaging parameters per event are used to select data that most likely originate from a specific region in the sky. If we interpret the measured φ as being the true Compton scatter angle, the scatter direction (χ, ψ) defines the centre of an 'event circle' of diameter φ, which includes all photon incidence directions that are compatible with such an idealized event. In turn, the compatibility of 'event circles' with any point in the sky can be employed as data selection criterion. In 'software collimation' we accept events that are compatible with a specified direction on the sky within a range of 2 or 3°, in order to account for the widening of the φ distribution from the detector energy resolution. This results in an effective acceptance 'beam' about 12° wide (FWHM; determined from simulations at 1.8 MeV) around the pivot direction of the sky, that broad due to incompletely absorbed events and the azimuthal ambiguity of the event circles. The 1 sr field of view of the telescope can thus be subdivided into largely independent fields of size ~0.2 sr. Alternatively we can create signal profiles, shifting the selection direction across the field of view in steps of a few degrees. Here we have to bear in mind the dependence of neighbouring profile points due to their overlapping acceptance cones. (For examples of this method see the spectra in Diehl et al 1993b Figure 1, and the 1.8 MeV signal profile along the Galactic plane in Diehl et al 1993a Figure 5).

Event circles may also be projected onto the sky directly. Then the circle intersections have a tendency to cluster in the direction of strong sources (see von Ballmoos et al. 1989 Figure 5a). This had been the imaging method applied in the first generation of Compton telescopes (Graser and Schönfelder 1982). The azimuthal ambiguity of each measured event circle, and the blurring from the imperfect instrument response, however, result in complex patches of event circle intersections. Their interpretation is complicated by the bias in possible event circle center directions due to the finite extent of the lower detector plane of the telescope, and by variations of signal and background contributions versus event circle diameter φ. This can be overcome only if selection criteria reduce systematic effects for one such map, and sets of maps with different criteria are then made to extract the celestial signature. Mathematically refined algorithms of this method, combined with Monte Carlo simulations for reference, have successfully been applied to first-generation Compton telescope data for scientific analysis (von Ballmoos et al. 1989).

Imaging algorithms utilizing the full imaging response in specified energy bands form the basis of COMPTEL data analysis. Several methods can be applied, depending on the objective: *Hypothesis testing* with the 'maximum likelihood' method folds a sky model through the instrumental response into a model data distribution, which is compared to the actual measurement and yields a likelihood value for the sky model parameters chosen; variation of the model parameters yields likelihood distributions per model parameter. One implementation of this method is the point source scan, providing a sky map of likelihood values per assumed source position, maximizing the likelihood for each position by variation of the point source intensity. Another implementation of this method is the generic sky model test, where complete 2-D skymaps (e.g of diffuse emission, such as obtained from CO or H1 observations) are tested against the data, and intensity parameters of these skymaps are varied in the maximisation process. Yet another implementation of imaging analysis in 3-D data space is the *'maximum entropy'* algorithm (see Strong, this conference), where a skymap is obtained by iteration of trial maps until satisfactory agreement with the measured data is achieved. The solution to this deconvolution problem in general is ill-determined (large families of solutions exist), so the image entropy is introduced as additional criterion in the maximisation process in order to select among the manifold of acceptable images. This method provides a deconvolution of the data into a sky map when no specific source hypothesis is available.

In the following sections, we address specific aspects of the 3D dataspace imaging of diffuse emission, using the 1.809 MeV gamma-ray line from ^{26}Al as example.

3   Diffuse Source Imaging and Background

The signal-to-noise ratio in COMPTEL data is below 5% even for the strongest sources, such as the Crab, or diffuse Galactic emission. Therefore characteristics of the background model have a substantial impact on the result. In the case of a point-like source, the sky region outside the source of interest provides an area where background model imperfections can be inspected for reference; or vice versa, the absence of spurious features in the sky region away from a point-like source serves as a proof of background model adequacy. The background underlying the source signal can readily be obtained by interpolation in data space, excluding the well-determined cone regime of the point source response (Diehl et al. 1992). The accuracy of the background model can be high if assumptions about smoothness of background variations are employed to include regimes of the order of ~10° in the background model; the statistical uncertainty in this case is determined by the regime where the model was interpolated, i.e. a ~2° wide cone regime of the data space. In contrast, for diffuse emission, no independent 'off-source' region can generally be found. Determination of a background model is challenging, as even small large-scale imperfections will impact on the final result. Note that the diffuse emission result derives from many more data space cells, namely the composite volume encompassed by the response cones of the entire emission region. Correspondingly, the background underlying the diffuse source is substantially higher than for the case of a point source. Two main questions arise: How can an 'independent but representative' background measurement be defined, and how can it be validated?

The COMPTEL background counts originate predominantly from activation of the instrument and spacecraft material due to cosmic ray and solar wind particle interactions. Changes in orbital and source aspect parameters from observation to observation, as well as the variations of the cosmic ray spectrum and intensity with time, imply complex variations of the background pattern in COMPTEL data, which are far from being understood. Therefore all attempts to use observations from another epoch or pointing for background estimates have to be taken with care. Exept for the Galactic continuum emission result (Strong et al. 1995), where observations from high Galactic latitudes have been transformed to background model for the Galactic plane observations, all COMPTEL analysis to date employs the source observation itself for efficient background modelling. Separation of source and background signals in data space must be achieved for these approaches, however.

For the case of diffuse gamma-ray line emission, energy domains outside the gamma-ray line of interest can be used as 'sufficiently independent'. Most of the signal of a gamma-ray line appears in the photopeak part of the instrument response, which has a typical energy resolution ~10% (FWHM): for a 1.8 MeV line, ~50% of the measured events have an energy deposit within 100 keV of the incident energy, while the remaining ~50% of the signal spread in the Compton tail, basically from the photopeak energy down to the detection threshold of ~700 keV. Therefore the signal-to-background ratio is best in a ~3σ window around the photopeak energy, at other energies the total signal is essentially free from source contributions. For the 1.809 MeV line, we choose a 'source' energy band 200 keV wide from 1.7 to 1.9 MeV, and basically extract the background from two 100 keV wide bands above and below this band, 1.6-1.7 and 1.9-2.0 MeV. Actually we found that we can make use of wider energy bands for assembly of a background model less sensitive to statistical fluctuations, provided we account for the energy dependencies: Signatures of scattered photon directions (χ, ψ) are decoupled from signatures of the Compton scatter angle (ϕ), and only the latter encodes the energy dependence of the background signatures. This is plausible as instrumental background is dominated by activation of matter of the instrument and spacecraft by very energetic particles; since that matter mostly is located at large off-axis angles, deactivation photons find their way into data space through a variety of multiple interaction signatures, unlike the Compton forward scattering that characterizes the events from sources within the field of view. The distribution of scattered photon directions in imaging data space does not significantly vary from threshold up to at least 10 MeV. Therefore we include all events from threshold to 10 MeV, exclusive the line energy band 1.7-1.9 MeV, in modelling the scatter direction background. The scatter angle distribution ϕ however varies substantially with energy, due to e.g. convolutions of cascade line signatures in both detector planes; this is accounted for by extracting it from the narrow bands close to the line energy only (for details see Knödlseder et al. 1995). The final background model thus is a combination of signatures in χ, ψ and ϕ, derived from energy bands excluding the energy band of the gamma-ray line. An additional advantage of the so-constructed background model is that it includes all source signatures that are identical in the energy bands adjacent to the gamma-ray line of interest; hence it effectively removes celestial continuum sources such as Galactic continuum emission from the imaging result.

Validation of the background model is comparatively difficult in the case of diffuse emission, as no apparent 'off-source' regime of the result can be identified *a priori*. We simulate 'observations' with Poisson samples of the background model: The imaging results are demonstrated to be free from any 'source' signatures within statistical uncertainty. Further tests in the case of the 1.809 MeV line include imaging of individual high latitude observations (no [26]Al sources found, as expected), and imaging of the Crab observations (we confirm the suppression of continuum sources). Additionally, the data spaces of observation and background model are found identical with respect to their χ/ψ, and ϕ signatures - with the expected exeption of celestial 1.809 MeV signature.

4   Imaging Consistency Checks

Images of diffuse emission are more sensitive to large-scale uncertainties in the background model, data selection biases, or systematic effects of the imaging methods.

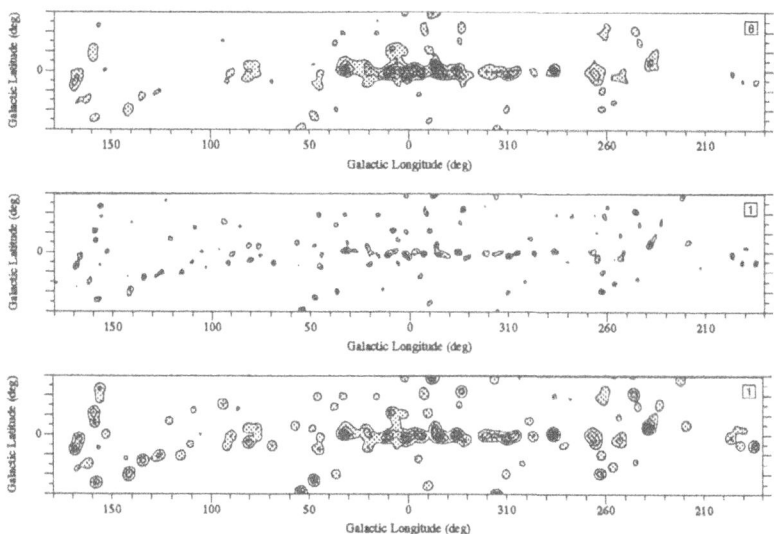

Figure 1: Maximum entropy image consistency checks: The top image shows the skymap after the 5th iteration, the centre image after 30 iterations. In the bottom image, Gaussian smoothing with the instrumental resolution has been applied to remove the super-resolution spikiness of the 30-iteration map, which then is demonstrated to show emission structures very similar to the 5th iteration map.

Therefore different analysis methods have been used and checked for consistent results. Shapes and extents of the diffuse emission features, and the derived fluxes and detection significances have been used as criteria.

First-order consistency of the Galactic diffuse 1.809 MeV emission is evident by comparison of the software collimation result, the likelihood map in three adjacent energy bands for point source hypothesis testing with background filtering method, and the maximum-entropy imaging modelling the background from adjacent energies (see Diehl et al. 1995a, Figures 7, 4 and 5, respectively). All methods show substantial 1.8 MeV emission along the plane of the Galaxy, with a narrow extent of the emission in latitude. The variation of intensity along the plane of the Galaxy is present in all methods, as well as the remarkable intensity drop beyond 35° longitude. The narrow extent in latitude is consistent with the instrinsic angular resolutions of the methods (~10° for the software collimation versus ~2° for the 3-D data space imaging).

In maximum entropy imaging, iterative map refinement starts out from a 'default' map, chosen by us as a flat image of an average intensity close to the average intensity of the final map. A choice of particularly high and low default intensity, respectively, only modifies the map in regimes where data are sparse, with a trend to retain the default intensity in those regions. Those tests, on the other hand, indicate the regions of the map that are best constrained by the measurement itself - the image structure in the central part of the Galaxy essentially is unaffected by the choice of default map level.

The different image iterations reflect the changes driven by the degrading importance of the entropy criterion in comparison to the fit quality (likelihood of image consistency with data). After about 5 iterations the image essentially remains unchanged, although the images of very late iterations are characterized by 'super-resolved' spiky features with much higher flux value per pixel: if smoothed with a Gaussian representing the instrumental resolution, the overall image appearance of a late iteration is quite similar to early iterations (see Figure 1). Late iterations (=maximum-likelihood determined) reveal proper fluxes for all source features independent of their significance, while we chose early iterations (before the super-resolution sets in) as more representative of the true (diffuse) sky intensity distribution.

The significance of diffuse emission features was investigated both by simulations and by fluctuation analysis through the bootstrap method. In simulations we superimposed sources of known intensity (convolved with instrumental response) to our background model, and performed our image analysis on those 'artificial' datasets. Maximum entropy deconvolution shows a trend for some clumpiness in the case of weak extended emission close to the sensitivity of the instrument, as well as some bias towards emission in well exposed regions of the sky.

The bootstrapping method may be applied in cases of sparse data in order to analyze the impact of statistical fluctuations of the measurement on the final result, without additional assumptions about the structure of fluctuations. The measurement itself is adopted as the only source of information, assuming however that each measured event is independent from the others. The measured event set is re-sampled, extracting 'bootstrapped events' one by one from the full measurement in a random manner until an event population identical to the actual measurement is obtained. With many bootstrap samples averaging over population fluctuations is achieved, with a distribution determined by the photon source fluctuations as convolved with possible biases of the instrumental detection process. We repeated the complete imaging process (including background modelling) on a set of 30 bootstrap samples. We analyse the distribution of flux values that results from the bootstrap samples in terms of a significance (i.e. how far above the zero level is the detected intensity, measured in units of the bootstrap-determined width of the intensity variations). Comparisons to maximum likelihood determined significances for point-like image features shows consistency of both significance determination methods. Taking this as justification to apply the bootstrapped significance determination also to the structured and more complex regions of the sky, we evaluate the significance of any diffuse emission feature in the map, without prior assumptions about the feature's shape: the flux variations within a specified region of interest in the bootrstrap analysis set the scale. Application of the method to construct a band of confidence for the 1.809 MeV emission profile along the plane of the Galaxy is shown in Diehl et al. 1995b, Figure 3.

In summary, with multi-dimensional data analysis reliable background models can be obtained from the source measurement itself in the case of diffuse gamma-ray line emission. Simulations and bootstrapping are employed to ensure proper imaging results.

5   References

von Ballmoos P., Schönfelder V., and Diehl R., 1989, A&A 221, 396-406
de Boer H. et al. 1992, in: Data Analysis in Astr. IV, ed. V. diGesu et al., Plenum N.Y., 241
Diehl R. et al. 1992, in: Data Analysis in Astr. IV, ed. V. diGesu et al., Plenum N.Y., 201
Diehl R. et al. 1993, Adv.Sp.Res. 13, 12, 723-726
Diehl R. et al. 1995a, A&A, in press
Diehl R. et al. 1995b, Adv. in Sp.Res., 15, 5, 123-126
Graser U. and Schönfelder V., 1992, ApJ 263, 677-689
Knödlseder J. et al. 1995, in preparation (see also Knödlseder J., Diploma Thesis, MPE 1994)
Schönfelder V. et al., 1993, Ap.J. Suppl. 86:657-692
Strong A.W. et al., 1992, in: Data Analysis in Astr. IV, ed. V. diGesu et al., Plenum N.Y., 251

FOCUSSING IN THE HARD X-RAY BAND

P. GORENSTEIN

Harvard-Smithsonian Center for Astrophysics, Cambridge Massachusetts, USA

and

K. JOENSEN

Harvard-Smithsonian Center for Astrophysics, Cambridge Massachusetts, USA

28 September 1994

Abstract. Several novel techniques can raise the upper limit of grazing incidence focussing to 100 keV, making it possible to extend the domain of focussing to studies of non-thermal processes. With an angular resolution of an arcminute or better, focussing reduces background to low levels. Consequently, a focussing instrument can be significantly more sensitive than a scanning collimator or coded masked system of similar aperture. In addition, the small area of the image facilitates the use of high spectral resolution detectors. Promising techniques for focussing hard x-rays include: reflecting x-rays more than twice on their way to the focus; non-traditional optical configurations based upon micro-channel plate arrays and capillaries, and two types of two-element, multilayer coatings, one with a uniform period and the other with a depth varying period. Long focal length and modular multiple focus configurations will augment the effective area in all cases. Telescopes which focus to 100 keV are strong candidates for the next generation of large x-ray observatories beyond AXAF and XMM.

Key words: Focussing, Hard X-rays, multilayer

1. Introduction

The image on the publicity poster of this workshop and most of the papers presented here imply that the subject of the title, "Imaging in High Energy Astronomy", can only be accomplished with a coded aperture. Indeed that is the technique used by all of the instruments aboard INTEGRAL, the primary mission for most of this workshop's participants. However, as we look beyond INTEGRAL, the limitations of a coded aperture system become apparent. Attaining an order of magnitude more sensitivity than INTEGRAL, in the 10 to 100 keV band, with a coded aperture method requires detectors that are two orders of magnitude larger. Clearly this will not be possible for a long time.

This paper discusses several focussing techniques based upon grazing incidence reflection that are applicable to 100 keV. We do not consider telescopes based on Bragg or Laue crystal diffraction. They are discussed at this workshop in papers by F. Frontera, N. Lund, R. Smither, and P. Van Ballmoos.

Experimental Astronomy **6**: 109–117, 1995.
© 1995 *Kluwer Academic Publishers.*

2. Advantages of Focussing

Grazing incidence reflection offers the following advantages:

- − 10 times, or more, greater sensitivity in the 10 to 100 keV band.
- − high angular resolution for imaging of extended sources and precise positioning of unidentified objects,
- − three decades, 0.1 to 100 kev, of spectrophotometry,
- − better spectral resolution with smaller detectors,
- − lower mass detectors, cooling systems, and shielding,
- − substantially lower data storage, and transmission requirements.

Since focussing has not been employed previously above 10 keV an explanation of its sensitivity advantage over that of a coded aperture is needed. We compare the sensitivity in background limited observations, of the INTEGRAL Imager at 50 keV with that of a hypothetical focussing instrument that fills the same geometric aperture. According to the INTEGRAL Phase A Study Report the gross aperture is 2500 cm^2. Hence, the effective area of the INTEGRAL Imager is half, or 1250 cm^2. We select an example from the list of techniques discussed in section 3, the graded d-spacing or EBB multilayers. From laboratory measurements upon sample reflectors, the effective area is estimated to be about 200 cm^2 for a four telescope modular system that fits within an aperture of 2500 cm^2 and has a focal length of 10m. The 10m focal length can be compacted for launch to the 3.7m length of INTEGRAL. If the angular resolution of the system is one arcminute in two dimensions, the physical area of the image is about 0.4 cm^2 summed over the four telescopes.

Table 2-1

Instrument	Effective Area, (cm^2)	Background Area, (cm^2)	Relative Sensitivity, $A\sqrt{B}$
INTEGRAL Imager	1250(50 keV)	2500**	25
Focussing Telescope*	200(50 keV)*	0.4	316

* Four modules, 10m focal length, EBB Multilayers.
** From the Report on the Phase A Study for INTEGRAL.

We assume that the four detectors of the focussing system are collimated and shielded to the same field of view as the INTEGRAL Imager. The uncertainty in the background level of the INTEGRAL Imager is proportional to the square root of the total detector area. For the focussing instrument it is proportional to the square root of the image area. The position sensitive detectors measure the mean background level with good statistics from the larger area of the detector outside of the image. Table 2-1 compares the areas and sensitivity of the focussing with that of the non-focussing system at 50 keV. According to these figures the focussing instrument has an order of magnitude more sensitivity. However, it has a small field of view and cannot be used to survey. Like most telescopes, it is used for pointed studies.

The principal limitation of grazing incidence focussing is that it is applicable only to 100 keV. However, the photon number-density spectrum of virtually all objects fall with increasing energy. Therefore it is often more effective to observe below 100 keV where photons are more abundant than above 100 keV. This is particularly true of AGNs whose spectra appear to be exponentially cut off at a characteristic energy of 100 to 400 keV. In fact, with good photon statistics, good spectral resolution, and low background the precise value of an exponential cutoff around 200 keV, can be determined from the analysis of 50 to 100 keV data. The capability is illustrated by ROSAT, measuring a kT value of 7.5 keV correctly with a telescope that cuts off at 2 keV (Briel, and Henry, 1994).

The 100 keV cut off is even less limiting when the objects have a significant redshift like many QSOs. For a QSO at $z = 2$ a photon emitted at 200 keV photon is detected at 66 keV which is well within the bandpass. In light of the intrinsic cut off of QSOs, many redshifted objects will have very few photons at higher energy. Hence, the ability of a grazing incidence telescope to detect redshifted objects falls much less rapidly with distance than it does for a non-focussing instrument.

3.   Techniques for Grazing Incidence Reflection of Hard X-rays

3.1   Introduction

A variety of techniques have been proposed for reflecting x-rays above 10 keV. All of them involve optical designs where the angle of incidence is small. The methods are listed in Table 3-1. Some optical designs allow several of these techniques to be employed simultaneously within the same instrument with a combined effect. In fact, the first two entries, long focal length and modular multiple telescope systems, can be combined with any of the others. The last two concepts involve the use of multilayer coatings consisting of two alternating materials with different indices of refraction. One type is very similar to the multilayers of uniform periodicity that reflect soft x-rays in a narrow range of wavelength at normal incidence. A more recent type of multilayer has a periodicity that varies with depth.

Table 3-1
Techniques for Grazing Incidence Reflection of Hard
X-Rays

- Long Focal Lengths, 10m or more
- Modular Multiple Telescope Systems
- Multiple Reflections, 4 or more
- Kirkpatrick/Baez Multi-Focus Systems
- Channel Plate Optics (Lees et al. this workshop)
- Microcapillaries
- Periodic Multilayer Coatings
- Graded d-Spacing (or EBB) Multilayer Coatings

Although no instruments of significant size have been built to date employing any of these techniques their performance can be estimated from laboratory measurements upon small reflectors or from simulations and analyses. Several techniques are discussed in more detail below. Channel plate optics are described at this workshop in the paper of Lees et al.

3.2   Multiple Reflection Optics

The simplest method of extending the bandwidth above 10 keV is to decrease the graze angle by increasing the number of reflections. This method succeeds because, at an energy which is far from an absorption edge, the reflectivity is high at angles that are well below the critical angle. With respect to the critical angle, θ_c, the reflectivity R behaves as follows: $R^4(\theta_c/2) >> R^2(\theta_c)$. This allows a larger diameter aperture to be effective with a fixed focal length. In principle a four reflection telescope could be made by replication methods similar to those used to make two reflection telescopes. For example, XMM intends to fabricate the three telescope modules by electroforming. In this process both the front and rear sections of each telescope shell are replicated as an integrated unit. It should not be much more difficult to replicate four sections as an integrated unit in the same way. However, the four-reflection mirror has several drawbacks. It is more massive than the two reflection mirror, particularly if the shells are nested with high density. Also, unless the surfaces are very smooth there is more loss due to scattering.

The effective area as a function of energy of a four reflection telescope system with gold surfaces is shown in comparison to a two reflection gold telescope of the same aperture and focal length in Figure 1. Theoretical values of gold reflectivity are assumed. The effective area of the conventional two reflection mirror is larger below 35 keV, but is falling rapidly above 25 keV. The four reflection mirror, which is 30% heavier in this particular design, has effective area to 80 keV.

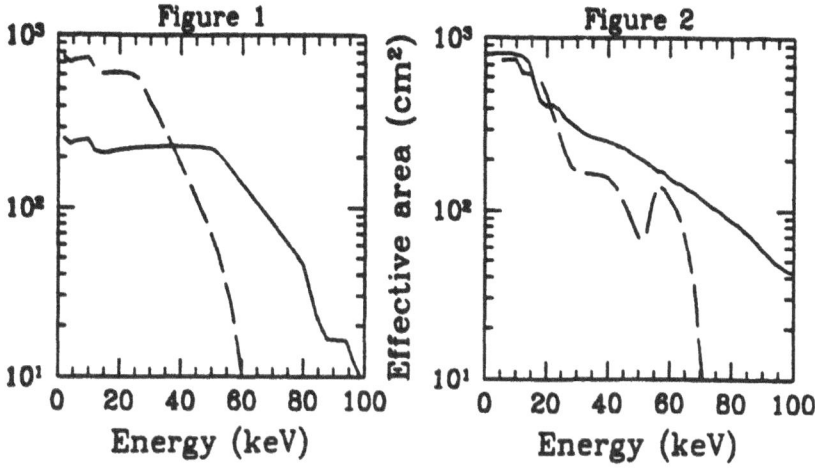

Figure 1. Effective area of two grazing incidence mirror systems. Both contain four modules, each with a 24cm diameter, 10m focal length and 80 cm long gold reflectors. The shell thickness is 0.7mm and the minimum spacing between reflecting surfaces is 1.2mm. One type (dashed line) is a conventional two reflection conical (or Wolter Type I) system of 45 shells. The other is a four reflection conical mirror that contains 58 shells.

Figure 2. Effective area of two mirror systems with the same mechanical configuration as the two reflection mirror system of Figure 1 but with different coatings. One system (dashed line) contains gold on the innermost 20 shells and two groups of uniform period Ni/C multilayers with 20 periods (40 or 50 Å) on the outer 25 shells. Random errors of 4% FWHM in layer thickness and 0.5 arcmin in angle are included. The other coating is a graded d-spacing or "EBB" Ni/C multilayer with 500 layers; the period ranges from 26 to 160 Å.

3.3 The Multifocus Kirkpatrick-Baez System

The multifocus K-B system was described by Elvis, Fabricant, and Gorenstein, 1988. All the reflectors are identical flats which extend through an entire row or column of telescope cells. They can be stacked to very small graze angles. A system with a focal length of 3.3m, an aperture of 2500 cm^2, and an 8 × 8 matrix of telescope cells would have an effective area of about

175 cm^2 at 50 keV with EBB multilayers of the type described below. The detector could be a matrix of small format devices, i.e. 1cm × 1cm, with one detector at each of the 64 foci, perhaps cadmium telluride pixel detectors like those proposed for the front face of the INTERGRAL Imager.

We have investigated a number of candidate reflectors but have not been able to find reflectors with the right combination of thickness, surface quality for good multilayer deposition, and flatness. One candidate was a thin glass material from Deutsche Specialglass, obtainable in thicknesses of 0.2-0.4 mm, that W/C multilayers of 4.1 Å interface roughness could be fabricated on. Unfortunately the best figure, after screening, was 3.5 arcmin. An additional problem with multilayers on thin flat reflectors is reflector deformation caused by multilayer stresses. Developing stress-free multilayers, novel figuring schemes, and simultaneous back-side coating are some of the techniques that could be employed to solve this problem.

3.4   Capillary Concentrators

A capillary concentrator is an extreme multiple reflection telescope. A glass made entirely of light elements has no absorption edges above 2 keV. If its surface is very smooth the dependence of the reflectivity upon angle will be nearly a step function that falls rapidly at the critical angle. In theory an x-ray can be transmitted with finite efficiency with as many as 100 small angle reflections through a smooth capillary. Small capillary concentrators and beam deflectors have received a great deal of attention in x-ray synchrotron facilities following the success of Kumakhov and co-workers in Russia in fabricating "polycapillaries", millimeter diameter bundles of capillaries with a diameter of a few microns. Small capillary concentrators are sold in Russia by Kumakhov's group and in the USA by X-Ray/Optical Systems of Albany, NY and Collimated Holes, Inc. of Campbell CA.

A capillary telescope is a concentrator rather than a true imager. The theoretical performance of a large area device consisting of many polycapillary bundles was described by Gorenstein, 1992. According to those projections, a four module capillary concentrator with a 10m focal length and 2500 cm^2 aperture would have more effective area than any of the other telescope types discussed in this paper. However, it is also the most speculative of all the telescopes discussed here as there are no hard x-ray measurements to support the simulations.

3.5   Multilayer Reflectors

3.5.1   Uniform Period Multilayers

Multilayer reflectors are periodic structures of alternating layers of high and low index of refraction materials, typically a metal such as nickel and a lighter element such as carbon. The multilayer reflects x-rays when constructive interference takes place among the radiation reflected from the boundaries between the high-low index materials. The final result is similar to that of Bragg-reflection in crystals. Multilayer x-ray reflectors with a constant period have been employed in several applications. The most familiar ones are the rocket borne normal incidence soft x-ray telescope known as NIXT, and EUV telescopes that have obtained high resolution images of the Sun in a narrow band of wavelength (e.g. Herant, Pardo, Spiller and Golub, 1991).

Multilayer reflectors can be used in a grazing incidence telescope as well. The resonant wavelength is much shorter because it varies with the sine of the graze angle. With a multilayer period of 100 Angstroms and a graze angle of 3 arcmin (10^{-3}rad.) the resonant wavelength is 0.2 Angstroms or 62 keV. Multilayer reflectors with uniform periods are a good match to the geometry of a conical or Wolter Type 1 telescope like XMM or Spectrum-X-Gamma which contains many concentric shells covering a range of graze angles. Each reflects a narrow range of energy. This system should also work well with multiple reflection conical telescopes since the graze angle is the same for all reflections. When an x-ray survives the first reflection it has a high probability of surviving subsequent ones because its energy has been selected as being in the correct range for that angle. Figure 2 shows the theoretical effective area of a two-reflection conical telescope system where the surfaces consist of gold on the twenty innermost mirror shells and twenty periods of nickel and carbon on the outer 25 shells. The telescope parameters are: four modules of 24cm diameter and 10m focal length. Attaining these effective area values requires that the layers be relatively uniform in thickness.

3.5.2   Extremely Broad Band (EBB) Multilayers
Better performance, may be obtained from a newer type of multilayer where the period varies with depth. Over the past three years we have designed, fabricated and tested such coatings, (Joensen et al., 1993, 1994a, 1994b, 1994c).

Different wavelengths are reflected at different depths. An x-ray penetrates the stack of layers until it arrives at the depth where the period resonates with its wavelength and angle. In this manner, a very broad band of reflection can be obtained, therefore, we call the reflectors "extremely broad band" or EBB multilayers. This system is not as critically dependent as the uniform period multilayer upon adhering to the optimum prescription of thickness versus depth. They were originally conceived for the reflection of slow neutrons (Mezei, 1976, and Mezei and Dalgleish, 1977). Christensen

et al 1992 suggested that they could be used for harder x-rays. He showed
how they could increase the 30 to 40 keV collecting area of a telescope in
the Spectrum-X-Gamma geometry.

The primary limiting factor in an EBB multilayer is the absorption experienced by the x-rays as they pass through the overlying layers, before and
after reflection. The absorption is small for hard x-rays. Hence, reflection
becomes more efficient with energy up to a point. The high energy limitations of this approach are caused by the minimum attainable multylayer
periodicity (15-20Å). The reflectivity is also limited by interface imperfections, either roughness or interdiffusion of the two materials, whose effect can
be modeled through a roughness parameter, i.e. the modified Debye-Waller
factor (Sinha et al. 1988). From measured reflectivities of EBB multilayers,
we obtain a best fit roughness parameter of, typically, 5Å which is used in
calculations of telescope performance.

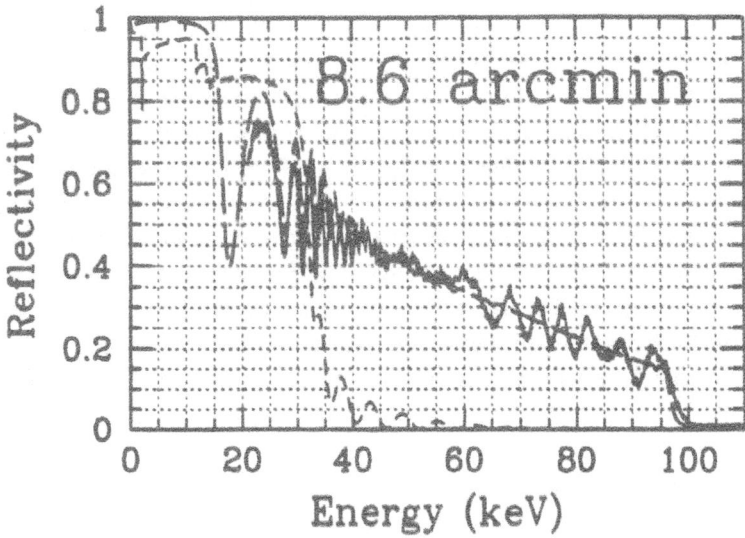

Figure 3. Measured reflectivity of a Ni/C EBB-multilayer vs. energy (solid
line). Also shown is the theoretical reflectivity of a 300 Å gold (small
dashes).

EBB multilayers reflect much better than gold above 30 keV. Figure 3
shows the reflectivity of both as a function of energy at a graze angle of 8.6
arcminutes. Scattering is not a problem with multilayer reflectors. X-rays
that would appear as a broadly distributed halo in single material reflectors
as a result of scattering are absorbed in the multilayer stack.

The effective area of an EBB coated two reflection conical mirror is shown in Figure 2 in comparison with that of a uniform period multilayer mirror. Aside from the coatings two mirror systems are identical: four modules, each with 24cm diameter and 10m focal length and reflectors that are 80 cm long.

4. Conclusions

The superior sensitivity and angular resolution of focussing instruments up to 100 keV mean that they are likely to be an important factor in future experiments. A single instrument can cover three decades of energy, because a focussing system will function down to 0.1 keV or as low as the cutoff of the interstellar medium allows. Shorter focal length instruments based upon multifocus Kirkpatrick-Baez telescopes or channel plate optics are appropriate for balloon experiments. Larger systems can be constructed by building upon the conical mirror replication technoloy being developed for XMM, modifying the dimensions to create more projected area at small graze angles, and replacing XMM's gold coatings with multilayers.

Bibliography

Briel, U.G., and Henry, J.P., 1994, *Nature* **372**, 439.

Christensen, F.E., Hornstrup, A., Westgaard, N.J., Schnopper, H.M. and Wood, K.P., 1992, *SPIE* **1546**, 160.

Elvis, M., Fabricant, D.G., and Gorenstein, P., 1988, *Applied Optics* **27**, 1481.

Gorenstein, P., 1992, *SPIE* **1546**, 91

Joensen, K.D., Christensen, F.E., Schnopper, H.W., Gorenstein, P., Susini, J., Høghøj, P., Hustache, R., Wood, J., Parker, K., 1993, *Medium-sized grazing incidence high-energy X-ray Telescopes employing continuously graded multilayers*, Proc. SPIE, **1736**, 239.

Joensen, K.D., Høghøj, P., Christensen, F., Gorenstein, P., Susini, J., Ziegler, E., Freund, A., Wood, J., 1994a, *Multilayered supermirror structures for hard X-ray synchrotron and astrophysics instrumentation*, Proc. SPIE, **2011**.

Joensen, K.D., Gorenstein, P., Høghøj, P., Christensen, F.E., 1994b, *X-ray Supermirrors: Novel multilayer structures for broad-band hard X-ray applications*, in Physics of X-ray Multilayer Structures (OSA), 159.

Joensen, K. D. , Gorenstein, P. , Wood, J. , Christensen, F.E., 1994c, *Preliminary results of a feasibility study for a hard X-ray Kirkpatrick-Baez telescope*, *Proc. SPIE*, **2279** (in press).

Herant, Pardo, Spiller and Golub, 1991, *Ap. J.* **376**, 797.

Mezei, F., 1976, *Novel Polarized neutron devices: supermirror and spin component amplifier*, Comm. on Phys., **1**, 81.

Mezei, F., Dagleish, P.A., 1977, *Corrigendum and first experimental evid on neutron supermirrors*, Comm. on Phys., **2**, 41.

Sinha, S.K., Sirota, E.B., Garoff, S., Stanley, H.B., *X-ray and Neutron Scattering from Rough Surfaces*, Phys. Rev. B.,**38**, p. 2297 (1988).

IMAGING GAS COUNTERS FOR X- AND GAMMA RAY ASTRONOMY

BRIAN D. RAMSEY

NASA/MSFC, Huntsville, Alabama 35812, USA

Abstract. Gas-filled detectors, such as proportional counters, have long been used in x-ray astronomy. They are robust, relatively easy to fabricate, and can provide large collecting areas with reasonable spatial and energy resolution. Despite coming of age in the 50's and 60's, their versatility is such that they are still planned for future missions. A vigorous development program, led mostly by the high energy physics community, has ensured continued improvements in proportional counter technology. These include multistep counters, microstrip technologies and optical avalanche chambers. High fill-gas pressures and the use of suitable converters permit operation up to 100s of GeV. The current status of imaging gas-filled detectors will be reviewed, concentrating on the lower energy region (< 100 keV) but also briefly covering higher energy applications up to the GeV region. This review is not intended to be exhaustive and draws heavily on work currently in progress at MSFC.

Key words: X-Rays – Imaging – Gas-filled Detectors – Proportional Counters

1. Introduction

Imaging proportional counters have been used extensively for x-ray / gamma ray astronomy from small-area, low-energy devices at the focus of x-ray telescopes (e.g. ROSAT [1]) to larger stand-alone devices at higher energies such as the TTM experiment on MIR [2]. In addition, they are planned for future missions such as Spectrum-X [3] and INTEGRAL [4]. The reason for their popularity is obvious - they are relatively easy to fabricate in the large areas necessary for non-focusing imaging, they are robust and reliable, and they offer reasonable performance in terms of spatial resolution, energy resolution, timing and background rejection. Fig. 1a shows the efficiency of a xenon-filled gas detector as a function of energy. Below 100 keV, large efficiencies can be obtained with modest fill gas pressures (< 5 atm) in large $(1000(s) \, cm^2)$ single-volume chambers. For higher energies, where the fill-gas pressure initially rises rapidly, standard vessels quickly become very bulky, window arrangements become difficult, and so modular approaches arraying large numbers of small, thin-walled tubular pressure vessels are typically proposed. Nevertheless, because of the high compressibility of xenon and rising pair-production cross-section, even MeV-GeV photons can be efficiently absorbed at achievable fill-gas pressures.

Fig. 1b shows the photoelectron-range-limited spatial resolution, achievable at various fill gas pressures. Since most readout schemes give a center of gravity of the primary charge cloud, this typically represents the resolution achievable on axis, especially below 100 keV where large-area imaging devices are used [5]. Typical readout schemes, particularly those utilizing

Experimental Astronomy **6**: 119–127, 1995.

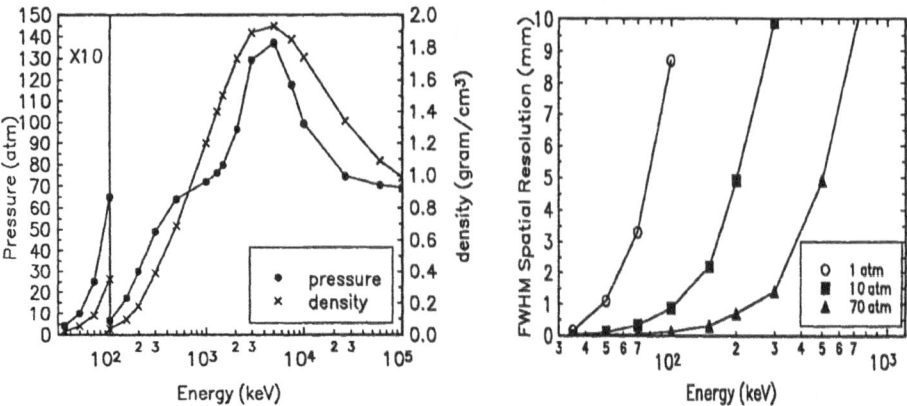

Fig. 1. a and b : (a) The pressure (left axis) and density (right axis) needed to give 50% detection efficiency in 10 cm of xenon; the data below 100 keV are expanded (y scale x 10) for clarity. (b) Photoelectron range limited spatial resolution for various fill-gas pressures. Line is intended to guide eye only.

large numbers of preamplifiers on individual, or groups of, cathode wires, have very low noise and can reconstruct the center of gravity of the charge cloud with high precision. Other schemes, such as delay line readout [3], offer similar performance. Above 100 keV, at intermediate energies where multiwire imaging chambers are not typically employed, spatial resolution is limited by other factors, such as the readout scheme or detector size for modular arrays. For very high energies (MeV-GeV), the highly extended electron tracks can be reconstructed to derive sub-mm spatial resolution once more.

2.   General techniques for improving performance of proportional counters

2.1 FLUORESCENCE GATING

Standard background-rejection uses anticoincidence and pulse-shape discrimination and routinely achieves >95% rejection below about 10 keV; these schemes are, however, less efficient at higher energies. To combat this, xenon-filled detectors can utilize fluorescence gating, which exploits the fact that most photon interactions ($\simeq$75%) above 35 keV (and below 300 keV) result in fluorescence, the reabsorption of which results in a pair of events that are a characteristic signature of true x-ray interactions. Imaging gas detectors can be designed to resolve these pairs of events with high efficiency.

For cases where the non-aperture flux dominates, fluorescence gating can provide a large background reduction. Measurements of the ratio of pair events to single events, give 1:2 for low-energy photons (<100 keV) but 1:50 for high energy photons (662 keV) and charged particles [6]. Fluorescence gating also provides a true energy measure of photoelectric interactions (no escaping photons) and improves energy resolution (by a factor of two at 40 keV) because the precise energy of one of of the two events is known. Furthermore, the ability to resolve the two events (rather than using an energy -weighted mean position), improves spatial resolution. There is, however, a small energy band (near 60 keV) where it is difficult to exploit these advantages.

2.2   PENNING GAS MIXTURES

In xenon gas, the average energy to form an electron-ion pair (21 eV) is almost twice the ionization potential (12 eV). Thus about half of the available energy is lost from the signal. Much of this "lost" energy appears in long-lived metastable states. The Penning effect uses a quench gas, with ionization potential matched to the principal metastable energy of the working gas, that becomes collisionally ionized while resonantly de-exciting the working gas. The increased number of electrons enhances energy resolution and allows a lower operating voltage. A comprehensive study of quench additives for xenon [7] found the strongest Penning effect with trimethylamine $(CH_3)_3N$, and a weaker effect with isobutylene (i-C_4H_8).

2.3   EFFECT OF ANODE DIAMETER/FIELD CONFIGURATION

The energy resolution of a gas counter is given by $2.36\,(((F+f)W)/E)^{0.5}$ where E is the energy of the x-ray, W is the mean ionization energy for the fill gas (21 eV for xenon), F is the Fano factor (0.16 for xenon) and f is the variance of gas amplification for 1 electron (typically 0.6-0.8). The theoretical work of Alkhazov [8] showed that f depends on the quantity $X = \alpha V_i/E$ where α is the Townsend coefficient, V_i is the gas ionization potential and E is the electric field strength. Higher X implies greater ionization efficiency and lower variance in the distribution of electrons in the avalanche. In general X increases with electric field, as α is a strong function of E, and thus fine anodes and close anode-to-cathode spacings are most desirable. Unfortunately, because of disruptive electrostatic forces, this is difficult to achieve in multiwire chambers. However, the microstrip proportional counter, detailed below, offers the possibility of proportional counters with extremely fine anodes and high packing density.

3.   Some new instruments

3.1   LOW ENERGY REGION ($<100~keV$)

3.1.1   *Microstrip gas detectors*
The microstrip detector replaces the usual discrete cathode and anode wire planes with conducting strips on an insulating or partially insulating substrate. Originated by Oed [9] to replace the conventional wire imaging proportional counter, it offers numerous advantages; ease of construction, more uniform response, reduction in operating voltage for a given gain, reduced charge saturation at high gain, ability to operate with little or no quench gas, better energy resolution, enhanced spatial resolution, and higher efficiency for the detection of fluorescence pairs [10]. At least two groups have been actively developing microstrips for x-ray astronomy - the Danish Space Research Institute [11], and MSFC [12]. In addition, the microstrip is being considered for the readout in a proposed X-Ray Monitor (XRM) for INTE-GRAL [4].

The microlithographic techniques used for production of microstrips permit exceptional spatial accuracy (1 μm) and uniformity, as well as extremely fine (1 μm) electrodes. These features improve gain uniformity - we have measured (1-σ) gain variations of only 1% over the area (25 cm^2) of test detectors fabricated using electron-beam technology. For comparison, in a conventional wire chamber, mechanical innacuracies and electrostatic forces typically give gain variations of 20% or more.

Microstrip technology also permits an electrode pitch much finer than the 1-to-2-mm minimum in a wire chamber and anode widths significantly less than the 15-to-25 μm practical minimum for uniform wires. The former distributes the charge signal over many anodes; the latter produces intense electric fields which effect very fast ion collection times. Together, these result in reduced charge saturation at high gain. Other benefits of fine-scale geometry include substantially reduced operating voltages and the ability to operate efficiently without a quench gas. The latter arises from the high electric fields, which channel a larger portion of the avalanche energy into ionization rather than excitation. Using little or no quench gas minimizes radiation-aging effects and improves reliability.

The high electric field also enhances energy resolution (section 2.3). Measurements as a function of decreasing wire diameter (down to a practical minimum of 12.5 μm) or width (microstrip) show gradually improving resolution, verifying the predictions of Alkhazov. To date, the best resolution achieved in xenon is just over 6% at 22 keV with 8 μm anodes and 200 μm anode/cathode separation [10]. Microstrips with anode widths down to 1μm are currently being tested. The ability to configure the electrodes at essentially any pitch also is important for imaging. Physics limited resolution (Fig 1a) is easily achievable at low energies due to the large gains available.

Practical problems in fabricating large-area microstrips have recently disappeared and now several companies have equipment capable of writing masters up to $(40 \text{ cm})^2$. A large area $(30 \text{ cm})^2$ microstrip has recently been developed for the MSFC balloon program using one of these large format devices [12].

Finally, an important issue for space applications is the stability and possible lifetime of microstrips. These are related to the choice of substrate which must be chosen so as to have enough surface conductivity to prevent charge build-up and yet keep leakage current low to minimize noise. To date most work has concentrated on ionic conduction glasses (10^6 - 10^8 ohm-cm at 250°C) which, after an initial settling time of order 1 hour after bias is applied, operate stably to very high rates, but do exhibit small gain variations with time (10% over a few weeks). Recently a new type of glass has become available having electronic rather than ionic conduction. The ohmic behavior of this glass results in predictable performance and offers the prospect of stable operation over long periods. Initial results (at CERN and elsewhere) look very encouraging and the material is currently being evaluated at MSFC. The lifetime of the microstrip should not be a problem with the very low count rates in orbit, particularly if only very tiny quantities of quench gas are used. Reports of permanent damage are typically at very high fluences ($10^9/\text{mm}^2$) or at high count rates ($10^5/\text{mm}^2$ sec). In addition, we see no signs of damage either from proton irradiation of the substrate (10 Mrad) or from alpha irradiation of the operating microstrip ($10^8/\text{cm}^2$).

3.1.2 The optical avalanche chamber

While electron multiplication in high field regions results in increased ionization, multiplication in very low fields gives large amounts of excitation. Such low-field conditions are met in the parallel-plate proportional chamber, where the avalanche extends over several mm. The addition of photosensitive quench gases to such chambers, termed optical avalanche chambers, then results in copious quantities of UV photons emitted during charge multiplication [13]. This light can then be used to encode each event. In such a chamber, a 50 keV incident x-ray photon produces 2.10^7 UV photons (peaked at 280 nm). This yield is 4 orders of magnitude greater than a NaI crystal and 3 orders of magnitude greater than the yield from a gas scintillation proportional counter (in which no charge multiplication takes place). It ensures excellent spatial resolution at low energies with a conventional Anger camera arrangement, and is also sufficient to drive a focusing system with an intensified CCD camera. Both approaches are being investigated.

A Hybrid Detector: Phoswich detectors are used extensively in hard x-ray/gamma ray astronomy, but suffer from poor performance at low energies due to the small number of visible photons produced by the NaI crystal. The hybrid instrument [14, optically couples a 40 cm x 40 cm phoswich, currently under development at HCO/CfA [15], to the avalanche chamber described above, and utilizes a common photomultiplier array to read out the light from both sections. The avalanche chamber provides low energy response, while at high energies the gas becomes transparent and the incident photons interact directly with the scintillator.

The addition of the gas detector offers significant improvements over the bare phoswich - approximately a factor of two in energy resolution and improved spatial resolution. For single events, around 1mm is achievable at 25 keV, compared with 16 mm for the bare crystal. For fluorescent pairs (section 2.1) simulations show that the very high light yield from the chamber still permits the true position of each event to be deconvolved from the overlapping light distributions despite the coarse 7 x 7 photomultiplier array [13].

With phoswich type rejection in the scintillator, escape-gating rejection in the gas detector together with risetime discrimination and the possibility of mutual rejection between the two sections, the hybrid offers broad-band high-sensitivity coverage which greatly improves the standard phoswich for little added complexity. A 1/2 scale prototype is under construction to resolve remaining design questions [16].

An Imaging Polarimeter: The copious light emission from the optical avalanche chamber permits a CCD based optics system to be utilized for imaging the photoelectron tracks of x-ray interactions. Fig. 2a shows a typical arrangement where two stages of charge amplification are used to enhance light yield and to facilitate gating. At low energies, where diffusion dominates, the spatial resolution is limited to of order 1 mm. At higher energies, where the track is more extended, software can be developed to follow the convoluted path of the electron back to its origin and the resolution becomes essentially independent of energy. Fig. 2b shows a typical photoelectron track in 2 atm of argon. Measurements of the ionization density in the track can facilitate charged particle rejection. In addition, measuring the photoelectron ejection angle leads to polarization sensitivity as the electron is emitted preferentially in one direction for polarized x-rays. A modulation factor of 30% has been measured at 60 keV in a small prototype chamber. A (balloon-borne) flight instrument is currently being developed for the 35-70 keV range [17].

3.2 MEDIUM ENERGY REGIME (100 $keV - MeV$)

3.2.1 High-Pressure Proportional Counters

Proportional counters can be successfully operated at pressures much higher than 5 atm., although the operating voltage rises significantly and the energy resolution degrades. Sakurai et al. [18] and Sood et al. [19] have taken data in xenon up to 10 atm. and 17 atm. respectively, and each demonstrate the same trends although they interpret the resolution degradation

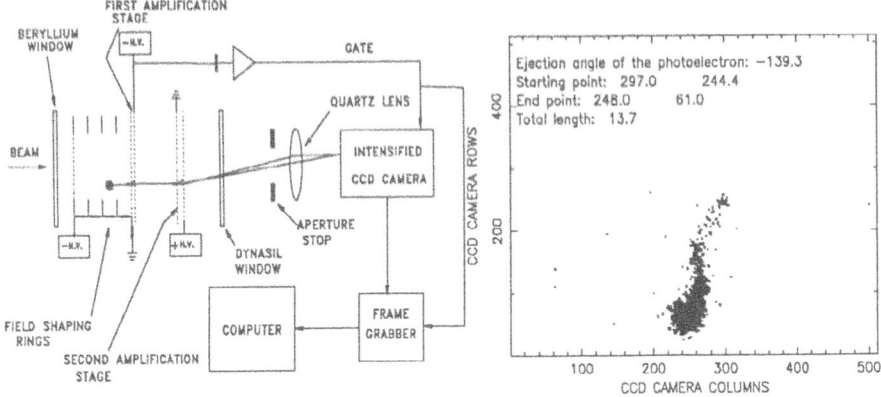

Fig 2 a and b Schematic of a CCD based x-ray imaging chamber and a 54 keV photoelectron track imaged in 2 atm of Argon + Trimethylamine

in different ways. Whatever the exact mechanism, it seems that the reduced multiplication field at higher pressures plays a key role in this reduction and that , for operating pressures above 20 atm, the resolution of a practical proportional counter would probably be comparable to that of the solid scintillator. Despite this there may still be advantages with the high pressure proportional counter over the standard NaI scintillator [19].

3.2.2 High-Pressure Ionization Chambers

As outlined in section 2.3, the energy resolution of a proportional counter is severely degraded by the charge multiplication process, particularly at high pressures. If used in the ionization mode, though, this resolution can theoretically approach the Fano limit involving only the statistics of the initial ionization process. In this case a resolution of less than 0.5% would be achieved at 1 MeV in xenon, which is within a factor of three of the theoretical resolution of germanium solid state detectors.

Several types of high-pressure ionization chambers have been investigated including parallel plate and cylindrical detectors with and without shielding meshes [20,21]. In practice the resolution is found to be limited by preamplifier noise, acoustical noise and physical processes in the gas. Fig. 3 shows the energy resolution at 570 keV as a function of xenon density [22]. Below $0.5 \, \text{gm/cm}^3$ the energy resolution is approximately constant and approaches the Fano limit. Above this density a sharp degradation of energy resolution takes place which, it is suggested, is caused by increasingly large fluctuations of charge recombination in the high-electron-density delta rays. It may be that this effect can be reduced, as with liquid xenon detectors, by doping with a photosensitive vapor having an ionization potential below the energy

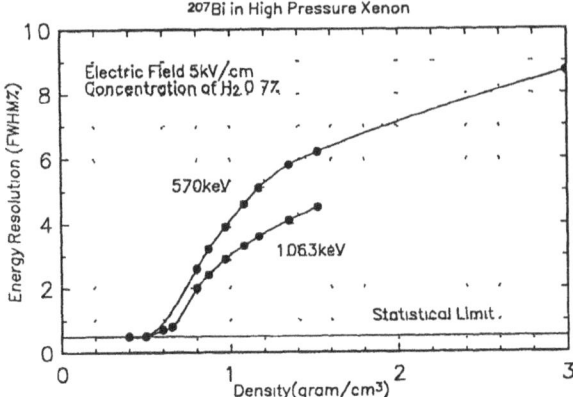

Fig. 3. Energy resolution as a function of xenon fill-gas density in a gridded ionization chamber

of the UV (xenon) photons emitted in the recombination process. This has the effect of decreasing the charge density in the delta rays and reducing the recombination effect [23]. Another explanation may be the formation of molecular clusters in the xenon which can trap electrons. The onset of this clustering is around 0.5 gm/cm^3 at room temperature [24]. If this mechanism is correct, warming the high pressure gas at densities above 0.5 gm/cm^2 could improve the resolution to that seen at lower densities.

An imaging high-pressure detector can be envisioned from an array of vertically cylindrical ionization chambers, with spatial resolution set by each tube diameter. It may further be possible to segment the collection anode, to derive an azimuthal co-ordinate within each detector and to use signal risetime to get a radial co-ordinate. The precision of such techniques, and the low-energy performance of such detectors is critically dependent upon the preamplifier noise. It may be possible to achieve around 50 electrons rms with modern (optical feedback, or no feedback) amplifiers resulting in an energy resolution of a few percent at 100 keV.

3.3 HIGH-ENERGY REGIME ($MeV - GeV$)

3.3.1 Drift Chambers and RICH detectors

When used in conjunction with a suitable converter, or sometimes without if the gas pressure is high enough, gas detectors can be used to image the charged particle tracks resulting from pair production. A three-dimensional reconstruction of these tracks permits the initial photon direction to be recovered. The proposed AGATE detector, successor to EGRET, utilizes stacks of large area (1/2 m x 1/2 m), thin (1 cm deep) drift chambers to reconstruct slices of the tracks with high precision over a proposed energy range from 20 MeV to 100 GeV [25].

The positron/electron tracks can also be reconstructed by making use of the rings of Cherenkov light emitted by the charged particles as they traverse gaseous radiators – this is the basis of the Ring Imaging Cherenkov (RICH) detector. The Cherenkov light must be focused onto a suitable detector such as a UV-windowed imaging gas detector containing a photosensitive vapor such as TMAE. Running at high gain, these detectors are sensitive to single UV photons in the Cherenkov light ring with a resolution of around 2-3 mm [26]. An alternative to this approach is to use a solid radiator as the primary converter as was proposed in [27]. Here a TMAE gas-filled UV photon detector was also used but in a very-high-gain (10^6) multistep mode to produce large quantities of visible photons which are in turn imaged by an intensified CCD camera. With this arrangement a spatial resolution of 220 μm was achieved for single UV photon detection. An ultimate angular resolution of less than 1 degree was claimed for GeV energy photons.

References

1. Pfeffermann, E. et al.: 1986, SPIE **733**, 519
2. Brinkman, A.C. et al.: 1983, Non-Thermal Processes and Very High Temperature Phenomena in X-Ray Astronomy, Publisher:Rome, 263
3. Waldron, L. et al.: 1993, SPIE **1948**, 98
4. Ubertini, P. et al.: 1994, this volume
5. Bateman, J. E: 1984, Nucl. Instr. and Meth. in Phys. Res. **221**, 131
6. Dietz, K.L. et al.: 1993, Space Programs and Technologies AIAA conference **AIAA 93**, 4254
7. Ramsey, B.D. et al.: 1989, Nucl. Instr. and Meth. in Phys. Res. **A278**, 576
8. Alkhazov, G.D.: 1970, Nucl. Instr. and Meth. **89**, 155-165
9. Oed, A.: 1988, Nucl. Instr. and Meth. in Phys. Res. **A263**, 351-359
10. Ramsey, B.D.: 1992, SPIE **1743**, 96
11. Budtz-Jorgensen, C. et al.: 1992, SPIE **1743**, 162
12. Ramsey, B.D. et al.: 1994, SPIE **2280**, 576
13. Ramsey, B.D. et al.: 1993, SPIE **2006**, 90
14. Grindlay, J.E. and Manandhar, R.P.: 1989, SPIE **1159**, 306
15. Manandhar, R.P. et al.: 1993, SPIE **2006**, 200
16. Pimperl, M.M. et al.: 1994, SPIE **2280**, 119
17. Austin, R.A. and Ramsey, B.D.: 1993, Optical Engineering **32(8)**, 1990
18. Sakurai, H. et al.: 1991, Nucl. Instr. and Meth. in Phys. Res. **A307**, 504
19. Sood, R.K. et al.: 1994, Nucl. Instr. and Meth. in Phys. Res. **A344**, 384
20. Dimitrenko et al.: 1992, SPIE **1734**, 295
21. Levin, C. et al.: 1993, Nucl. Instr. and Meth. in Phys. Res. **A332**, 206
22. Bolotnikov, A.E. et al.: 1986, Translation in Instrum. Exp. Tech. **29**, 802
23. Masuda, K. et al.: 1994, SPIE 2305
24. Bolotnikov, A.: 1994, private communication
25. Mukherjee, R. et al.: 1994, SPIE 2305
26. Swordy, S.: 1994, Nucl. Instr. and Meth. in Phys. Res. **A343**, 52
27. Charpak, G.: 1994, Nucl. Instr. and Meth. in Phys. Res. **A343**, 300

HARD X-RAY AND GAMMA-RAY IMAGING WITH SOLID STATE DETECTORS

N. GEHRELS

NASA/Goddard Space Flight Center

Abstract. Solid state detectors are used in x-ray and gamma-ray astronomy primarily for their fine spectroscopy. For some cases (e.g., gamma-ray observations with Ge detectors), the spectroscopy and sensitivity requirements drive the design of the aperture systems and only moderate-quality imaging is possible. In other cases (e.g., hard x-ray observations), the detectors can be finely segmented for high-quality imaging. The new room-temperature solid-state detectors like CdZnTe and HgI_2 are naturally well-suited for imaging. Because of their high atomic numbers, photoelectric absorption dominates over Compton scattering to >200 keV. This, combined with their high densities, allows thin detectors to be used with segmented contacts. Position resolution in the detector plane can be on ~100 μm scales giving sub-arcmin angular resolutions.

1. Introduction

Solid-state detectors have been flown in gamma-ray astronomy instruments since the late 1960's (Jacobson 1968; Womack & Overbeck 1970; Nakano et al. 1974), but until very recently (Skinner et al. 1994; Rideout et al. 1994) they have not been used with imaging systems. This is not to say that there has been no source mapping or positioning with these instruments. Even with their broad fields-of-view, the early germanium (Ge) instruments positioned sources by scanning or offset pointing. Examples include the galactic center 511 keV line source (Riegler et al. 1981; Gehrels et al. 1991) and the ^{26}Al galactic plane emission (Mahoney et al. 1984).

Recent developments in Ge segmented-electrode technology and room temperature solid-state detectors offer great promise for hard x-ray and gamma-ray imaging. In this paper, I review the history of imaging with solid-state detectors and describe the new technologies.

2. Germanium Imagers

The only true solid-state imager flown to date is the small demonstration instrument of G. Skinner and colleagues at U. Birmingham. It has a 3x3 array of close-packed Ge detectors (1 cm x 1 cm x 6 cm each) and a passive coded-aperture mask (Skinner et al. 1994). The instrument was flown as a piggyback on the GRIS balloon payload in 1993 and produced images of the Crab and Cyg X-1 (Rideout et al. 1994; Skinner et al. 1994).

Experimental Astronomy 6: 129–135, 1995.
© 1995 *Kluwer Academic Publishers.*

There have been several full-scale Ge imagers that were designed (and in one case built) but never flown (see Table 1). The GRSE and Winkler systems were not flown for lack of funding. The U.S. NAE eventually merged with the European GRASP to become the current INTEGRAL Spectrometer (see Section 5). The GRIS mask was designed and built to locate the galactic center source of 511 keV line radiation, but was not flown when GRO-OSSE data indicated that any point source in the region is currently too weak ($\lesssim 4 \times 10^{-4}$ photons cm^{-2} s^{-1}) for the GRIS imaging sensitivity.

TABLE 1
Ge Imagers Never Flown

Instrument	# Detectors	Aperture Type	Reference	Comment
GRO-GRSE	19	RMC*	Peterson et al. 1978	see Fig. 1
Winkler	9	RMC	Nakano et al. 1986	
GRIS mask	7	mask/antimask	Tueller et al. 1994	see Fig. 2
NAE	9	mask/antimask	Matteson et al. 1990	

*RMC = Rotation Modulation Collimator

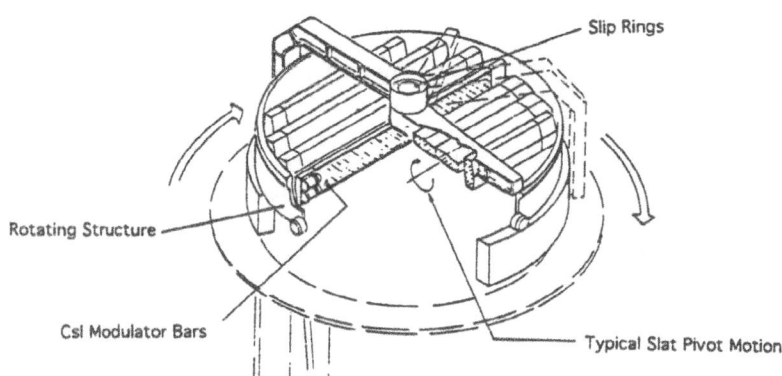

Fig. 1 - Rotation Modulation Collimator (RMC) proposed for the canceled
GRO-GRSE instrument (Peterson et al. 1978).

Since the number of detectors for the Winkler, GRIS mask and NAE concepts was small, the number of independent sky pixels would also have been small. The purpose of these imaging systems was therefore not so much to make sky maps, as to position individual sources. The mask/antimask systems of the GRIS mask and NAE would also have provided rapid and clean source modulation for background subtraction. The use and design of small-array masks is discussed by Sembay and Gehrels (1990).

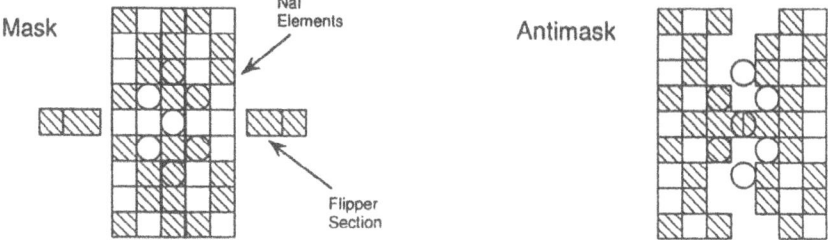

Fig. 2 - Mask/antimask system built (but never flown) for the GRIS balloon instrument (Tueller et al. 1994). The mask splits in the middle and two subsections flip in to make the antimask.

3. New Germanium Technologies

New developments in Ge detector technology are providing improved spatial resolutions which will allow finer imaging in future spectrometers. Arrays of closely-packed "finger" detectors, like the Birmingham detector (Skinner et al. 1994) described in Section 2, are becoming available. These typically have cm-sized pixels in arrays of 3x3 or 4x4. Instruments with large numbers of such arrays and with coded masks at, say, 4 m would give pixel sizes of ~10 arcmin. These are also ideal for use with crystal diffraction apertures (von Ballmoos & Smither 1994). If the detector arrays could be packed close together, they would give a better filling factor of the detector plane than current Ge instruments achieve.

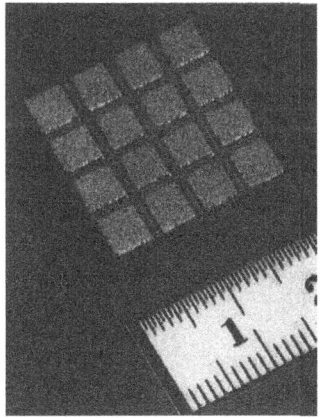

Fig. 3 - Ge pixel detector fabricated at Lawrence Berkeley Laboratories (Luke 1984).

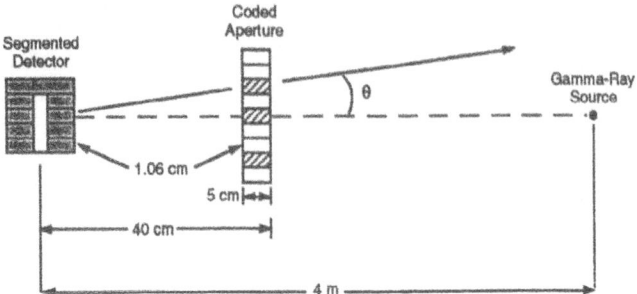

Fig. 4 - The laboratory setup for testing an imaging system using segmented Ge detectors at the Jet
Propulsion Laboratory (Mahoney et al. 1993).

Another promising area is electrode segmentation. Early work in the 1980's
produced planar detectors with 5 mm pixels (Luke 1984; see Fig. 3) and segmented
coaxials with 1 cm ring contacts (e.g., Gehrels et al. 1983; see Mahoney et al. 1993
for imaging applications as shown in Fig. 4). Recently, interest has been renewed in
planars with pixels and strip contacts as described in this volume (Mahoney 1994;
Kroeger et al. 1994; Johnson et al. 1994).

4. Room Temperature Detectors

There has long been interest in the room-temperature solid-state detectors made from
mercuric iodide (HgI_2) and cadmium telluride (CdTe) as an alternative to Ge
(properties are listed in Table 2 adapted from Gehrels et al. 1988). Their main
advantage is that they do not require cryogenic cooling as Ge does and they have
better gamma-ray stopping power. Their disadvantage is that they have poorer
energy resolution and have hole trapping problems that degrade energy resolution and
efficiency above ~200 keV.

TABLE 2
Room Temperature Semiconductor Detectors

	CdTe	HgI_2	Ge
band gap (eV)	1.5	2.1	0.7
density (g cm^3)	6.1	6.4	5.4
operating temperature	room	room	80 K
bias V (V)	100-200	100-500	3000
highest Z atomic number	52	80	32
K edge of highest Z element (keV)	32	83	11
photoelectric-Compton crossover (keV)	250	400	130
100 keV photon atten. length (mm)	1.1	0.5	3.5

In spite of the interest, there have been very few instruments flown with these materials. Only three flight instruments are known to this author, one with a small CdTe detector on ISEE-1 flown in 1977 to study gamma-ray bursts (T. Cline, priv. comm., 1994) and two HgI_2 balloon instruments flown in the early 1980's (Ricker et al. 1983; Ogawara et al. 1982). None of these had imaging systems.

Several recent developments have caused a resurgence in efforts to fly room-temperature detectors, and particularly efforts to fly large arrays for imaging (e.g., Parsons et al. 1994). These include:

1) Addition of 10%-20% zinc to make CdZnTe with higher band gaps.
2) Purer material with less trapping for both HgI_2 and CdTe.
3) Medical imaging interest which has increased the market and made large arrays financially feasible. Typical prices are now ~$50-$100 per detector of size 1 cm x 1 cm x 2 mm thick (for large orders).
4) Availability of multi-amplifier Application Specific Integrated Circuits (ASICs) with low powers of ~1 mW per channel.
5) Pulse-shaped corrections to the pulse heights to correct for energy resolution degradation due to hole trapping (e.g., Finger et al. 1984).
6) Electrode segmentation into strips and square pixels of size 100 μm - 1 mm (see Fig. 5).

With these developments it is not unreasonable to think of arrays of thousands of 1 cm detectors operating in the few keV to few 100 keV range with segmented electrodes and coded masks, giving *sub-arcminute* angular resolutions. Several papers in this volume describe work in this area (Dusi et al. 1994; Bollesteros et al. 1994; Caroli et al. 1994; Pleasants et al. 1994; Lebrun et al. 1994).

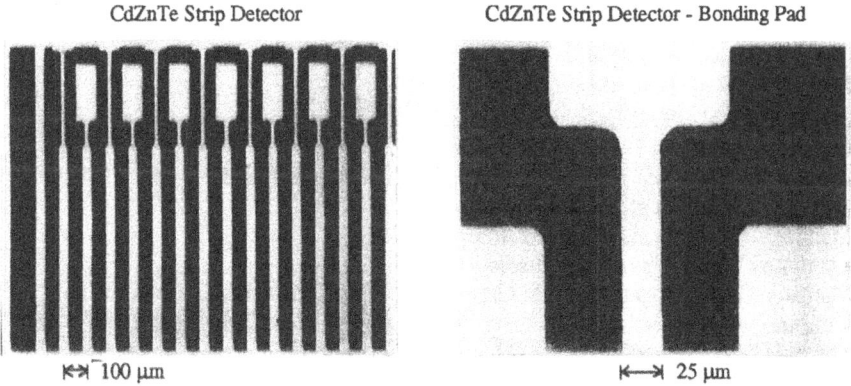

CdZnTe Strip Detector

CdZnTe Strip Detector - Bonding Pad

|←→| 100 μm |←——→| 25 μm

Fig 5 - A CdZnTe strip detector fabricated at the Goddard Space Flight Center (Parsons et al. 1994).

5. Planned Future Missions

5.1. The INTEGRAL

The first currently-funded mission to make extensive use of solid-state detectors for imaging and spectroscopy is the International Gamma Ray Astrophysics Laboratory (INTEGRAL; Winkler 1994). The baseline concept for the Spectrometer has a Ge array and coded mask to give ~1° angular resolution. The Imager may have a top layer of CdTe detectors (Lebrun et al. 1994) combined with a coded mask to give ~15' angular resolutions.

5.2. HESI

A planned, but not yet funded, solar mission called the High Energy Solar Imager (HESI) will use Ge detectors and a rotating modulation collimator for fine spectroscopy and imaging of solar-flare gamma rays (Lin et al. 1994). The imaging system uses two sets of 34 mm pitch grids separated by 1.7 m to achieve ~2 arcsec angular resolution at hard x-ray energies and ~arcmin resolution at MeV energies.

References

Bollesteros, F. J., et al. 1994, this volume

Caroli, E., et al. 1993, A&A Suppl., 97, 393

Dusi, W., et al. 1994, this volume

Finger, M., et al. 1984, IEEE Trans. on Nucl. Sci. NS-31, 348

Gehrels, N., et al. 1983, IEEE Trans. on Nucl. Sci. NS-31, 307

Gehrels, N., et al. 1988, Solar Physics, 118, 233

Gehrels, N., et al. 1991, ApJ Lett, 375, L13

Jacobson, A. S. 1968, Ph.D. thesis, UC San Diego, UCSD-SP 68-2

Johnson, W. N., et al. 1994, this volume

Kroeger, R. A., et al. 1994, this volume

Lebrun, F., et al. 1994, this volume

Lin, R. P., et al. 1994, Geophys. Monog. 84, Solar System Plasmas in Space and Time, American Geophysical Union.

Luke, P. N. 1984, IEEE Trans. on Nucl. Sci. NS-31, No. 1, 312

Mahoney, W. A., et al. 1984, ApJ, 286, 578

Mahoney, W. A., et al. 1993, A&A Suppl., 97, 385

Mahoney, W. A. 1994, this volume

Matteson, J. L., Teegarden, B. J., Gehrels, N., & Mahoney, W. A. 1990, in High Energy Astrophysics in the 21st Century, ed. P. C. Joss (AIP:New York), 343

Nakano, G. H., Imhof, W. L., & Johnson, R. G. 1974, IEEE Trans. on Nucl. Sci. NS-21, 159

Nakano, G. H. 1986, preprint

Ogawara, Y., et al. 1982, Nature, 295, 675

Parsons, A. M., et al. 1994, IEEE Trans. on Nucl. Sci., submitted

Peterson, L., et al. 1978, proposal to NASA, document UCSD-3546

Pleasants, I. B., et al. 1994, this volume
Ricker, G. R., Vallerga, J. V., & Wood, D. R. 1983, Nucl. Inst. Meth., 213, 133
Rideout, R. M., et al. 1994, this volume
Riegler, G. R., et al. 1981, ApJ, 248, L13
Sembay, S., & Gehrels, N. 1990, Nucl. Inst. Meth., A295, 477
Skinner, G. K., et al. 1994, Nucl. Inst. Meth., submitted
Tueller, J., et al. 1994, ApJ, submitted
von Ballmoos, P., & Smither, R. K. 1994, ApJ Suppl, 92
Winkler, C. 1994, this volume
Womack, E. A., & Overbeck, J. W. 1970, JGR, 75, 1811

IMAGING DESIGN OF THE WIDE FIELD X-RAY MONITOR ONBOARD THE HETE SATELLITE

J.J.M. IN 'T ZAND and E.E. FENIMORE
Los Alamos National Laboratory, Los Alamos, New Mexico, U.S.A.

N. KAWAI, A. YOSHIDA and M. MATSUOKA
The Institute of Physical and Chemical Research (RIKEN), Wako, Saitama, Japan

and

M. YAMAUCHI
Miyazaki University, Miyazaki, Japan

Abstract. The High-Energy Transient Experiment (HETE), to be launched in 1995, will study Gamma-Ray Bursts in an unprecedented wide wavelength range from Gamma- and X-ray to UV wavelengths. The X-ray range (2 to 25 keV) will be covered by 2 perpendicularly oriented 1-dimensional coded aperture cameras. These instruments cover a wide field of view of 2 sr and have a relatively large potential to locate GRBs to a fraction of a degree, which is an order of magnitude better than BATSE. The imaging design of these coded aperture cameras relates to the design of the coded apertures and the decoding algorithm. The aperture pattern is to a large extent determined by the high background in this wide field application and the low number of pattern elements (~100) in each direction. The result is a random pattern with an open fraction of 33%. An onboard decoding algorithm is dedicated to the localization of a single point source.

Key words: X-rays – gamma-ray bursts – coded aperture imaging

1 HETE, its scientific objective and instruments

In recent years the database on gamma-ray bursts (GRBs) has expanded tremendously thanks to the *Compton Gamma Ray Observatory* (Fishman et al. 1994, Meegan et al. 1994). Although this has advanced the knowledge of these phemenomena, it is generally accepted that the identification of GRBs with some known class of celestial sources in another wavelength band will provide the ultimate key to their nature. In order to obtain a unique identification between a GRB and a counterpart in another wavelength range, a good GRB location accuracy is crucial.

Instruments dedicated to the study of GRBs have until now mostly emphasized population studies rather than individual studies. As a result, accurate locations (say less than 1 degree) are rare. The best location accuracies, to 1', have so far been reached by the technique of arrival time lags as measured by gamma-ray detectors on interplanetary networks.

Finding GRB counterparts outside the gamma-ray band was the incentive for the design of the *High-Energy Transient Experiment* HETE (Ricker et al. 1992). The scientific objective of HETE is to find counterparts of high-energy transients, most notably GRBs, in other wavelength bands and perform spectral studies on these transients simultaneously over a wide range of wavelengths.

137

L. Bassani and G. di Cocco (eds.), Imaging in High Energy Astronomy, 137–142.
© 1995 Kluwer Academic Publishers.

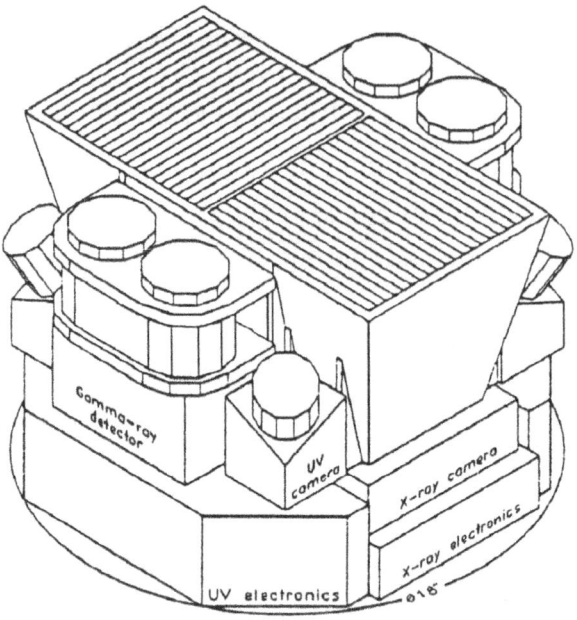

Fig. 1.   Schematic drawing of HETE's scientific payload configuration

HETE consists of three instrument packages, that all point in the same direction and have wide field of views (FOVs) of roughly a few steradians (see Fig. 1): 1) an ultraviolet transient camera array: 2) a set of four gamma-ray detectors, active between 6 keV and ~ 1 MeV: 3) a wide field X-ray monitor (WXM). Apart from the scientific capabilities of each of the three HETE instruments separately, they will in the following combined way serve the objective of fast and accurate GRB localization. The gamma-ray instrument will provide the alarm signal: the WXM will determine the sky location of the X-ray signal within 18': the UV cameras will determine the position of UV emission within at best about 10″. The best location obtained for the GRB depends on whether a significant detection occurs in the X-ray and UV instruments. Burst information will be disseminated via the VHF band to low-cost secondary ground stations which are located at several tens of radio, IR and optical observatories, where a swift follow-up study may occur.

2   WXM description

The WXM is based on the principle of coded aperture imaging (for a review, see Caroli et al. 1987). It has 4 identical 1-dimensional position-sensitive proportional counters, sensitive between 2 and 25 keV, one pair in each of two orthogonal directions. Each detector is filled with 1.4 atm xenon and contains 3 carbon fiber anode wires for event location. The position of an absorbed photon along the wires is determined from the measurement of the ratio of the pulse height at both ends of each anode wire, and the photon energy through the summation of pulse heights over both ends. Each wire is read out separately, to furnish a rough spatial resolution perpendicular to the wires.

The geometric area of each detector module is about 9 by 14 cm^2; about 70% of this area is active. The nominal position resolution of the detector is 1 mm full-width at half maximum (FWHM) at 6 keV (where the response function peaks), and improves with the square root of photon energy. The FOVs of both X- and Y-instruments are approximately 90×90 square degrees, and the overlap between the X- and Y- FOVs is about 90%. The spectral resolution is better than 20% FWHM at 6 keV.

The aperture patterns for X and Y are identical (except for small regions along the edges) and completely 1-dimensional. Since only one burst at a time will occur, a 2-dimensional aperture is not needed.

The detector data is processed in a digital signal processor. The processing can occur in 2 modes: monitor and burst mode. In the monitor mode the detector data is collected in histograms. The definition of the binning can be commanded from the ground. In the burst mode, photon-tagged data is produced and an onboard location algorithm activated.

3   The coded aperture

The aperture is about 50% larger than the detector, to ensure a large FOV within the spatial resources. A one-dimensional coded aperture is defined by four geometrical parameters: the element size, the number of elements, the fraction of open elements and the pattern of open and closed elements. We consider the optimum configuration of these four parameters for detecting GRBs with the WXM.

Mask element size. In combination with the spatial detector resolution and detector photon penetration depth, the mask element size (MES) determines the point-spread function (PSF) and thus the sensitivity, sky source location accuracy and angular resolution of the instrument. When the optimum MES with respect to these three parameters is sought, it is important to recognize that the detector resolution and the detector photon penetration are strong functions of photon energy and the photon penetration also of off-axis angle. Thus, the point-spread function (PSF) depends on the spectrum and FOV-position of the point source. To accommodate this, we used the average characteristics of GRBs that were detected by the Ginga gamma-ray detector down to a few keV (Strohmayer et al. 1994).

The angular resolution is given by the ratio of the FWHM of the PSF in the decoded image and the mask-detector distance (186 mm). For the average GRB in 1 to 20 keV, the FWHM increases monotonically with MES (see Fig. 2); at MES=2.0 mm it is 3.5 mm. The best signal-to-noise ratio S/N of a point source is achieved if it is imaged in a single pixel. When 'smearing'

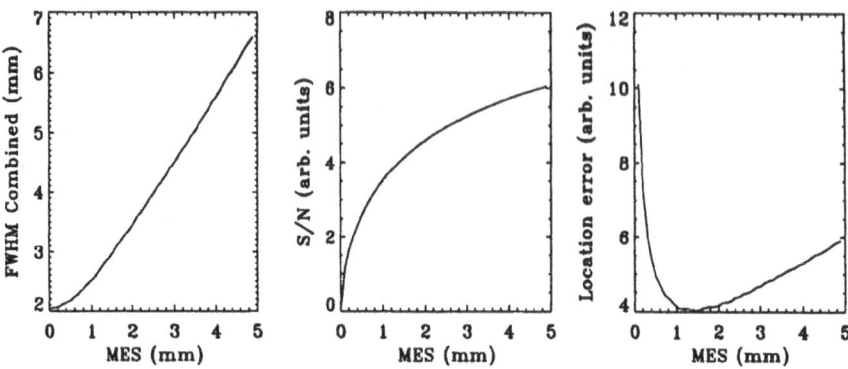

Fig. 2.   The dependencies of three point source diagnostics on the mask element size (MES). Spectral and FOV-location dependencies were averaged over the expected population of GRBs, based on Ginga results (Strohmayer et al. 1994)

causes intensity to leak to neighboring pixels, the S/N will decrease (it is typical for a decoded image that the variance varies only very slowly over the image). Obviously, the S/N increases with the MES. From MES=1 to 2 mm the relative increase is 35%. The 1-sigma source location accuracy scales to the angular resolution and to the inverse of the signal-to-noise ration S/N. Averaged over all off-axis angles and GRB characteristics, the optimum accuracy is found at MESs of roughly 1 to 2 mm.

In the burst mode, an optimum burst source location accuracy has the highest priority. Sensitivity and angular resolution are of secondary priority. Given these motives, a good compromise for the MES seems to be 2 mm; thus, the aperture patterns will measure about $N=100$ pattern elements. The accompanying angular resolution is 40' on-axis to 70' half way to the edge of the FOV, between 1-20 keV. The error box size for a burst like GB880205 is expected to be a few arc minutes when analyzed on the ground. **Fraction of open mask elements.** The open fraction of the pattern (t, the ratio of the number of open pattern elements to the total number of pattern elements, *not* the ratio of open to total area) which is optimum for the detection of point sources in a reconstructed image (we exclude the sensitivity in the count rate time profile) usually is *not* 50%. 50% only applies if the particle-induced background dominates over all other sources of events.

If the reconstruction of a GRB position and intensity is performed with a PSF fit and the particle-induced background is not important (such as in the WXM), then exercising the formulation by In 't Zand et al. (1994) results in an optimum open fraction of about 20% for a faint GRB and 39% for a bright on-axis GRB. Thus, a compromise would be about 30%. **The pattern.** The pattern is choosen to be non-repeated. Thus, there is no fully coded FOV (FCFOV) present. This choice is motivated by the desire to only need a simple onboard burst location algorithm.

URA patterns perform superior to any other type of pattern, provided that every sky position is coded with the full basic pattern (Fenimore & Cannon 1978). If the coding is not complete, due to the lack of a FCFOV, this superiority is not as strong or may even be absent. The latter may occur when for a substantial part of the FOV the coding is performed by only a few open elements, typically less than about 50, as in the WXM. For so few elements the probability is high that sizeable non-uniformities occur in the local open fraction. The reason for this is that a URA should sample all spatial scales evenly, with just a few mask elements this means heavy clustering or non-uniformity of the open fraction. With a severe non-uniformity in open fraction, a relatively large error will be made in the cross correlation for non-source positions because the cross correlation subtracts the source counts as if they were uniformly distributed. If the open fraction is strongly non-uniform, the point source counts are far from evenly distributed over the detector. Thus, the coding noise will be relatively high. The only way to minimize this effect while retaining the low number of open elements is to distribute them as uniform as possible. Since URAs do not allow this freedom while random arrays do, there may exist particular random arrays that perform better than URAs when incomplete coding is important.

To find an optimum HETE pattern, we tested patterns with $N = 103$ and $t = 0.33$. Apart from the one triadic pattern (this is a near-URA pattern, see In 't Zand et al. 1994), we tested 10^5 random patterns. The first test involved putting a point source at 3 positions (on-axis and two half way off-axis), calculating the sky images in units of standard deviations through cross correlation, determining the maximum non-source value for each of the three point source positions and adding these three values. This number is an indicator for the amount of pattern-dependent coding noise. Of the 10^5 patterns, the best 20 were selected (i.e., with the smallest coding noise indicator numbers) and tested further by extending the number of point source positions from three to the total number of binned FOV positions. Again a coding noise indicator number was evaluated which guided our final choice for the pattern (black is closed and white open):

4 Fast onboard burst location algorithm

In the burst mode, the location of the burst needs to be determined onboard fast (within ~1 s), so that the UV cameras (bin time 4 s) can reduce the amount of telemetry data. Therefore, the reconstruction of the coded detector data towards sky data needs to be as simple as possible. Fortunately, this is possible for a burst-like phenomenon.

In a coded aperture camera without a fully-coded FOV or with a random pattern, such as the WXM, coding noise is common throughout the FOV. Coding noise introduces extra crosstalk between two or more point sources, on top of the Poisson noise. However, if there is only one source in the

FOV, the single source will not be influenced by coding noise, and the coding noise peaks that this source produces elsewhere in the FOV are always smaller than the source peak itself, apart from Poisson noise, if this peak is expressed in number of Poisson standard deviations. We plan an onboard burst-locating algorithm consisting of the following points: 1) As soon as a trigger is available, a pre- and in-burst time interval is defined; 2) The pre-burst data is subtracted from the in-burst data, after renormalization with the exposure times; 3) The cross correlation algorithm is performed on this subtracted data, resulting in 4 arrays: 2 intensity arrays (in cts s^{-1} cm^{-2}), one for each X- and Y-instrument, and 2 Poisson standard deviation arrays; 4) Each intensity array is divided by the appropriate standard deviation array, resulting in intensity arrays with unit standard deviations; 5) The subtraction in step 2 ensures that the coding by all persistent sources is eliminated, provided they do not show variability on time scales equal to the time intervals used in this procedure. Therefore, only the coding of the burst source is preserved; the strongest peak in the intensity array will be the most probable location of the burst. Both X- and Y- locations are provided to the UV cameras in off-axis angle format.

5   Expected performance

With the above pattern and the definition of the reconstruction, the WXM is optimized for the detection and localization of GRBs. Given the FOV and the expected duty cycle, we expect that the WXM will detect about 24 GRBs per year for which the location can be determined to 18′ onboard and better than 10′ on the ground. With respect to persistent X-ray sources, the performance is limited by source confusion. In particular, the limit is introduced by the fact that observations, in essence, reveal separate lists of X and Y off-axis angles. The difficulty lies in combining the correct Y to each X coordinate. We expect the WXM to be able to deal with this easily as long as no more than 5 significant sources populate the FOV.

References

Caroli, E. et al. : 1987: Coded aperture imaging in X-ray and gamma-ray astronomy, *Space Science Reviews* **45**, 349
Fenimore, E.E. & Cannon, T.M.: 1978: Coded aperture imaging with uniformly redundant arrays, *Applied Optics* **17**, 337
Fishman, G.J. et al.: 1994, 'BATSE: burst performance and experiment status' in G. Fishman, J. Brainerd, K. Hurley, ed(s)., *Gamma-ray Bursts*, AIP: New York, 648
In 't Zand, J.J.M. et al.: 1994: The optimum open fraction of coded apertures, *Astronomy and Astrophysics* **288**, 665
Meegan, C. et al.: 1994, 'Two and a half years of BATSE burst observations' in G. Fishman, J. Brainerd, K. Hurley, ed(s)., *Gamma-ray Bursts*, AIP: New York, 3
Ricker, G.R. et al. : 1992, 'HETE: An international multiwavelength mission' in C. Ho, R. Epstein, E. Fenimore, ed(s)., *Gamma-ray Bursts*, CUP: Cambridge, 288
Strohmayer, T.E. et al. : 1994: X-ray spectral characteristics of GINGA gamma-ray bursts, *Astrophysical Journal* , preprint

DEVELOPMENT OF THE "CAPTIF" SOFTWARE FACILITY FOR THE SIMULATION OF CODED APERTURE TELESCOPES AND ITS APPLICATION TO "INTEGRAL" AND OTHER INSTRUMENTS.

P.H.CONNELL & G.K.SKINNER

University of Birmingham

21-September-1994

Abstract. To support the development of the next generation of coded aperture tele-scopes, a software facility has been developed to simulate the response of any mask/detector configuration and the observation of various photon sources, allowing its sensitivity and imaging properties to be evaluated. The system is suitable for any instrument which can be defined by four components: a coded mask, an array of detectors, an identical array of detector collimators and an outer shield. It allows for rotating or alternating masks, for flexibility in the choice of mask/detector shape and for off axis effects inherent in a mask of finite thickness with exponential absorption. Facilities are also provided to simulate point or extended sources and an observation program consisting of a sequence of parameters describing the pointing direction, orientation and exposure of the device. Detector count simulations and image reconstructions have been performed for the evaluation of various designs proposed for the INTEGRAL instruments and for a small, balloon carried coded aperture telescope.

1. Introduction

The need to evaluate the sensitivity and resolution of a gamma ray spectrom-eter and imager for the INTEGRAL mission has prompted the development of a general software facility to simulate the design, operation and imag-ing performance of coded aperture instruments. The software developed is suited to the simulation of an instrument comprising (Fig 1):

- A coded mask aperture (with optional alternating mask).
- A array of detectors.
- An collimating shield for each detector (optional).
- An outer shield (optional).

The system facilitates the design of models of coded masks and detector arrays and their configuration into a simulation of a complete instrument. It allows the creation of an observation program describing the sequence of pointing directions and exposures and of a list of point or extended sources comprising the field to be observed. The output of the observation simulation stage is a file of detector counts for each exposure position which then serves as input for procedures to reconstruct flux intensity images.

L. Bassani and G. di Cocco (eds.), Imaging in High Energy Astronomy, 143–147.
© *1995 Kluwer Academic Publishers.*

2.   Software structure and tasking.

The design, operation and imaging performance of an instrument are simulated by six main tasks (Fig 2):

2.1   Design and storage of masks and detector arrays.

Masks and detector arrays have a common description as regular or irregular arrays of rectangular, circular or hexagonal cells. They are stored as a sequence of parameters giving the location, shape, dimensions and a characteristic value of each cell. For masks the characteristic value is the absorption coefficient while for detectors it is the probability of absorption.

2.2   Instrument configuration and assembly.

Previously defined mask and detector arrays are selected, along with their separation distance and the shape and dimensions of the detector collimator array and outer shield. Operation of the instrument is simulated through the construction and storage of a source response function (SRF) which gives the count rate response in each detector to a unit flux source at any point in the effective field of view of the detector. The SRF is constructed by dividing each detector into small subcells, and accumulating the transmission probabilty along photon trajectory lines from each point in the FOV, through the mask, to each subcell. Storing the photon detection rate in each detector as a function of angular position in the FOV allows an accurate simulation of the response of an instrument with a fast access time, important in reducing the computation time required during image reconstruction.

2.3   Source description.

Sources to be observed are stored as a list of parameters describing the position and energy spectrum of point sources or as flux density images of extended sources for a particular energy.

2.4   Creation of a pointing or exposure program.

A pointing program divides the period of observation into a sequence of exposure intervals each of which has parameters specifying the pointing direction, exposure time, orientation and mask type of the instrument.

2.5   Simulation of source observation.

An observation is simulated by a procedure which selects a pre-calculated SRF of an instrument along with a pointing program and the sources to be observed and outputs a file of photon count expectation values for each detector and each pointing exposure interval. These are stored as decimal numbers for later addition of detector background counts and conversion to Poisson integers.

2.6   Image reconstruction.

Given the appropriate SRF, pointing programme information and detector count data, either real or simulated, procedures are provided which will reconstruct images of source emission and estimate detector background using three different methods:-

- A linear regression method, which is effectively a generalisation of the conventional 'correlation with the mask pattern' methods (see for example Skinner and Ponman, 1994). Such methods are equivalent to solving for each pixel intensity independently, while assuming that the intensity in other pixels in the field is zero. We generalize this to allow for (a) multiple pointings and, optionally, (b) mask/antimask removal of detector background from count data or (c) solution of the background rate in each detector element. The last is only possible if the number of available data points is adequate and requires multiple pointings.

- Generating deconvolved images by solving for all pixel intensities and background rates simultaneously. Again there must be enough observations with different pointings to provide sufficient independent data points. As with all deconvolution methods, solutions may be prone to instabilities and enhancement of photon noise.

- Targeting localised regions of emission using an extension of the method sometimes termed IROS (Iterative Removal of Sources Hammersley*et al.*, 1992). In the implementation used here image components are not limited to point sources but may be localised extended sources.

3.   Results

The standardized database and simulation software has proved effective and straightforward to use. It has been applied to the evaluation of a series of INTEGRAL imager and spectrometer designs (eg. Connell & Skinner, 1994). For the latter, work has concentrated on producing images of extended sources such as the clumped 1809 keV emission in the galactic plane (Diehl *et al.*, 1993) and the two component OSSE model of 511 keV emission from the galactic bulge region (Purcell *et al.*, 1993). It has also been successfully applied to the evaluation of real data obtained in observations of Cygnus X-1 and the Crab Nebula (Rideout & Skinner, 1994) during the balloon flight of a small Germanium array instrument. Future developments will include an extension of image processing methods aimed particularly at reducing instabilities in generating deconvolved images. The targeting method is of interest in searching for the most significant areas of emission to produce maps to be used as a basis for dividing a region into an irregular patchwork of pixels for a more general estimate through direct deconvolution.

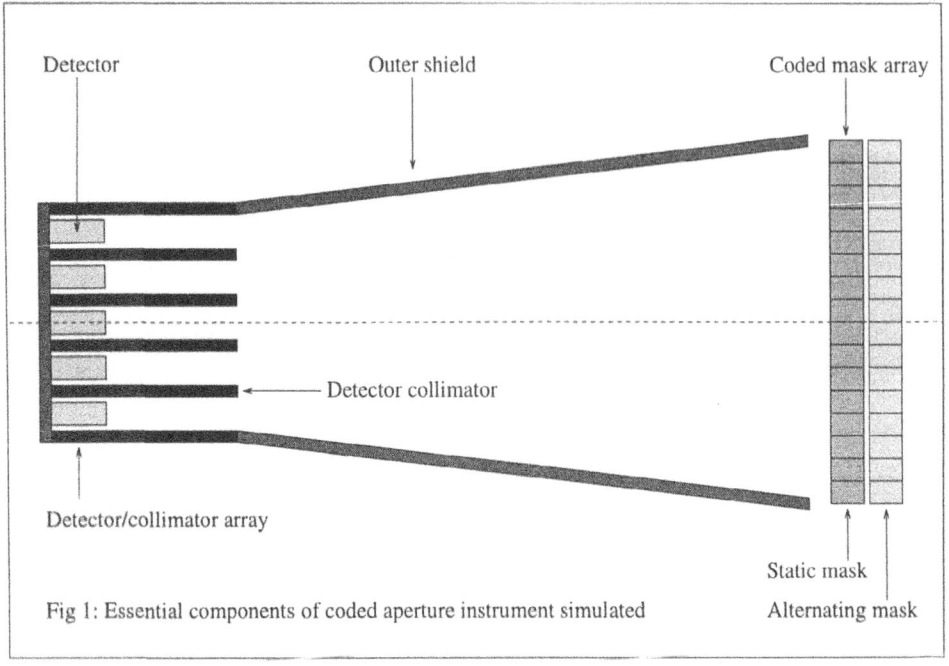

Fig 1: Essential components of coded aperture instrument simulated

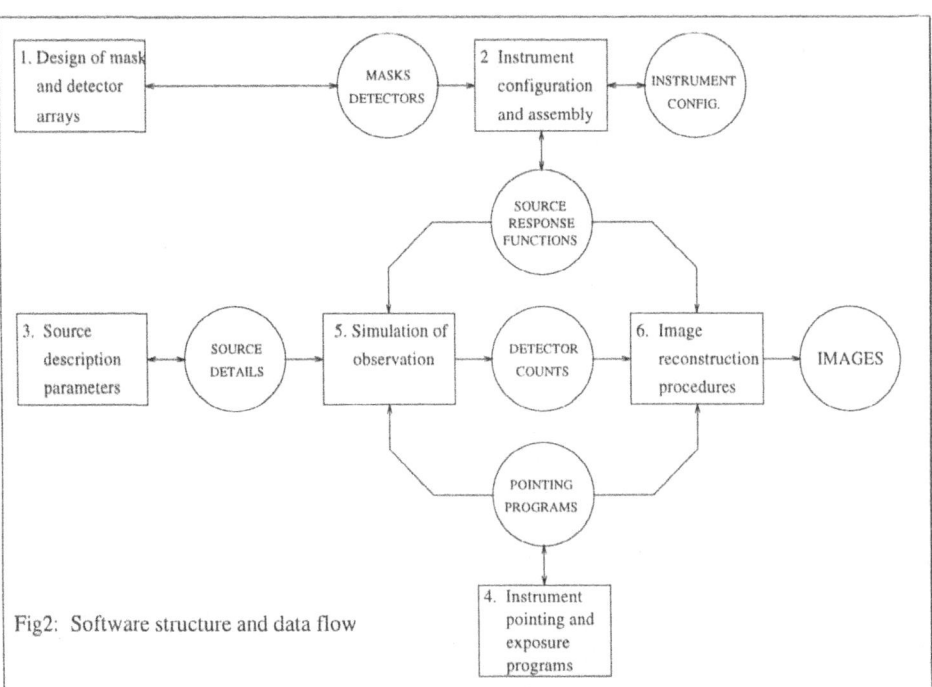

Fig2: Software structure and data flow

References

Skinner,G.K. & Ponman,T.J. *On the properties of images from coded-mask telescopes,* 1994: Mon.Not.R.Astron.Soc. 267, pp518-522.

Hammersley,A *et al. Reconstructions of images from a coded-aperture box camera,* 1992: Nuclear Instruments and Methods in Physics Research A311, pp585-594.

Connell,P.H. & Skinner,G.K. *INTEGRAL SPECTROMETER team report,* 9-Jun-1994.

Diehl *et al. Galactic Nucleosynthesis as observed through* ^{26}*Al, New insights from COMPTEL,* 1993: Proc. of the 2nd Compton Symposium, AIP Conf. Proc. 304, p.147.

Purcell *et al. The distribution of galactic 511 keV positron annihilation radiation,* 1993: Proc. of the 2nd Compton Symposium, AIP Conf. Proc. 304, p.403.

Rideout,R.M. & Skinner,G.K. : 1994, Submitted to *Proc. Workshop on Imaging in High Energy Astronomy,* Capri, to be published in *Experimental Astronomy.*

SCIENTIFIC DESIGN AND OPTIMIZATION OF
A CODED MASK HARD X-RAY TELESCOPE

L. NATALUCCI

Istituto di Astrofisica Spaziale, Frascati, Italy

and

M.E. SOGGIU

Istituto Superiore di Sanita', Roma, Italy

Abstract. The design of a coded mask X-ray instrument depends on a combination of factors: its target function (monitor, spectroscope, etc.), the field of view, the energy range and, therefore, the detection device. Recently, the high pressures attainable in gas detectors for use in space and the outcome of ultrafast electronics techniques have made proportional counters well suited for observations in the energy range from a few keV up to 200 keV. In this medium-hard X-ray range, the use of a large area detector must be coupled to an efficient shielding to reduce cosmic diffuse background.
In this paper, the criteria which have been adopted in the design of the the MART-LIME experiment on board SPECTRUM X/GAMMA are described, with emphasis on the imaging characteristics.

Key words: X-ray Astronomy – Coded Mask Imaging – Instrumentation

1. Introduction

During the past decade a number of flown space-borne experiments definitely demonstrated the ability of the coded aperture technique (Caroli et al., 1987) to perform efficient and reliable observations over the whole X-ray energy band. Two main classes of coded mask instruments have been recognized, one comprising the so called optimum or cyclic systems, in which an optimized mask design is obtained by cyclically shifted repetition of a Uniformly Redundant Array (URA) pattern (Fenimore and Cannon, 1978), and the box-camera or simple system, in which the mask and detector useful areas have the same dimensions (Brinkmann et al., 1985).
The feasibility of optimum systems is often limited by two main factors: a) presence of vignetting, which can result in a complete nine-fold ambiguity in source direction (Stephen et al., 1987); b) unfavourable ratio (~0.25) between detector useful area and mask total area. The vignetting problem can be overcome by restricting the field of view to the set of directions of the fully coded field of view (FCFOV), for example by using a passive collimator (Willmore et al., 1984). Optimum systems which do not require an unfavourable detector/mask area ratio, can be obtained for relatively narrow field of view instruments (Sunyaev et al.,1990).
An alternative configuration is obtained by the use of a URA mask slightly larger than the detector useful area, with a nearly equal space being occu-

149

L. Bassani and G. di Cocco (eds.), Imaging in High Energy Astronomy, 149–153.

pied by detector assembly and mask mechanical structure (Paul et al., 1991). This telescope is characterized by having a large partially coded field of view (PCFOV) compared to the fully coded region, but it does not critically suffer from source vignetting.

The extension of the PCFOV may be limited by the presence of a passive collimator in front of the detector. For large area hard X-ray detectors the collimator is also the most practical shielding device for cosmic diffuse background (CDB). The impact of the collimator on instrument sensitivity and on the point-spread function (PSF) will be discussed in §3.

2.   Optical design of the MART-LIME hard X-ray telescope

Hard X-rays detectors can also be divided in two broad classes: gas filled proportional counters (Ubertini, 1987) and scintillation detectors, the latter type being successfully implemented at energies higher than a few tenths of keV. Currently, the availability of space-qualified, high pressure gas cells and the outcome of the fluorescence gating technique for background rejection, have made proportional counters efficient up to ~200 keV.

An imaging detector exploiting the above capabilities has been developed at IAS for the MART-LIME experiment (Bazzano et al.,1994). MART-LIME is a coded mask instrument and is the high energy (5-150 keV) telescope on board SPECTRUM-X/GAMMA. In order to study in detail the spectral behaviour of cosmic sources and discriminate between different theorethical models, a high sensitivity ($\leq$ 1 mCrab) is necessary to resolve spectra in a typical time scale of ~1 day. This sensitivity requirement has an immediate impact on the field of view (FOV) and the detection area. The FOV cannot be too wide (mainly due to CDB at the lowest energies) and the area must be much larger than the one usually adopted in gas-filled proportional counter experiments designed to work below ~30 keV. MART-LIME has a useful detection area of 1740 cm^2 and a field of view of $6°\times6°$ FWHM.

The imaging performance of MART-LIME was described in a previous paper (Soggiu et al., 1993). The optical design of the instrument belongs to the quasi-optimum type described above. A coded mask of 105 x 105 elements is placed at a distance of 2.3 m from a high pressure (5 bars) position sensitive proportional counter. The mask design consists of a cyclically shifted URA twin prime array of 73×71 square elements, 5.8 mm in size. This ensures a FCFOV of $4.9°\times4.6°$. The detector is efficiently shielded against internal background via fluorescence gating and by standard AC technology, and features a spatial resolution less than 1 mm FWHM (Waldron et al., 1993). The total optical FOV of the instrument ($25°\times25°$ at zero response) is limited by a passive $6°\times6°$ FWHM collimator. The collimator, mainly used to reduce the CDB flux, has also the important function of supporting the detector window from the high pressure of the gas cell environment. The collimator pitch has been chosen equal to the mask element size. In this way, each mask pixel projected by a source is affected by exactly the same amount of modulation, and no systematic noise is induced in the deconvolved images. On the other hand, the collimator strongly alters the PSF on spatial scales less than the mask element size, and may have a significant impact on the off-axis source location accuracy.

3. Off-axis performance

3.1 SENSITIVITY CONSIDERATIONS

In a coded mask experiment implementing a mechanical collimator, the off-axis sensitivity is dependent on how accurately the collimator induced spatial modulation can be taken into account, that is, how good is the spatial oversampling of the detector cell defined by the collimator pitch. The aim is to select, in the deconvolution process, only the subset of each cell which is actually exposed to a given direction, hereby obtaining a significant Poisson noise reduction.

For a given celestial source distribution, the increase in signal-to-noise ratio (SNR) can be estimated as follows. Given:

A^o, A^c, the total areas coded by open and close mask elements, respectively;

f_i, the transmission factor of the collimator for the i-th direction;

$B_i' = \sum_{j \neq i} f_j S j + B$, the total intensity in counts/cm^2 due to background and other sources;

ϕ_i, fraction of B_i' within the selected area;

then the variance in the reconstructed source counts is:

$$\sigma^2[f_i S_i A^o] = B_i'(A^c + A^o) + f_i S i A^o \tag{1}$$

with no spatial selection, and

$$\sigma^2[f_i S_i A^o] = \phi_i B_i'(A^c + A^o) + f_i S i A^o \tag{2}$$

whenever spatial selection is applied. Assuming $A^c = A^o$ the increase in SNR is given by:

$$g = \sqrt{\frac{2B_i' + f_i S i}{2B_i' \phi_i + f_i S i}} \tag{3}$$

At the limit of the instrument sensitivity (where $S_i/B_i' \ll 1$ applies), the SNR is increased by $\phi_i^{-1/2}$. In the case of crowded fields with strong sources, like the Galactic Centre region, a large decrease in the minimum detectable flux can be achieved for a relatively large portion of the field of view. In

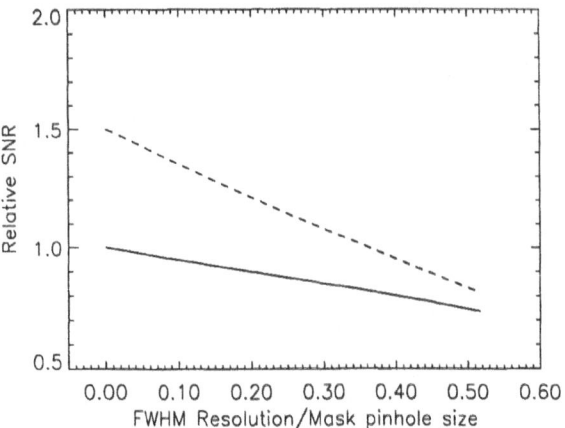

Fig. 1. The expected SNR for an off-axis source modulated by a slat collimator, as a function of spatial resolution (continuous line: no spatial selection, dashed line: spatial selection). The collimator transmission in this example is 0.44.

fact, the contribution to B_i' by a source at a given off-axis location can be eliminated or effectively reduced for the set of directions opposite to the source respect to the center of the FOV. The efficiency of Poisson noise suppression depends on how the collimator modulation is measured. The effect of the finite spatial resolution of detector is illustrated in Fig. 1 for the case in which no strong off-axis sources occur in the FOV.

3.2   SOURCE LOCATION CAPABILITY

The effect of the collimator modulation on the PSF must also be taken into account in the instrument design (see e.g. Gunson and Polychronopoulos, 1976). Without significant collimator modulation (i.e. for on-axis source) the PSF is a pyramid with a FWHM equal to the mask element size, blurred by the spatial resolution function. The capability of locating the centre of the peak distribution depends on the SNR and is related to the possibility of detecting the contours of the mask shadowgram on a spatial scale less than the mask element size. If the collimator pitch has the same size as the mask element, for a given coarse offset position there will exist a set of directions for which the contours of the mask elements are hidden in those areas of detector which are obscured by the collimator walls. In this case the positional accuracy is limited to the value d/h, where d is the linear size of the obscured area within each collimator cell. If indeed the mask element contours are visible, the PSF will assume a "ghost-like" shape with two visible shoulders and a narrow peak in between (Fig. 2), the relative heigth of the two shoulders being correlated with the fine position of the source.

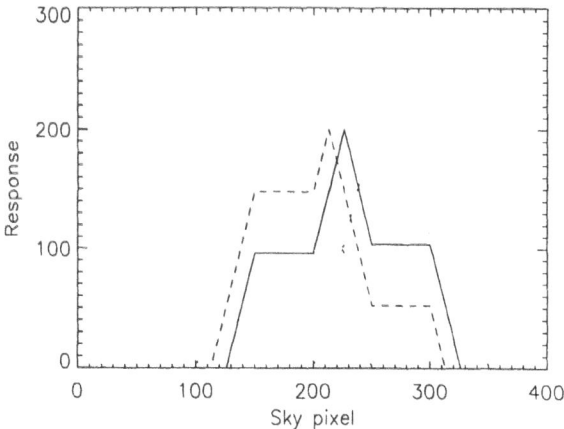

Fig. 2.   One-dimensional PSF of a source having an offset angle of half the collimator FOV FWHM. The mask element size is divided in 100 subpixels. Dotted shapes correspond to directions in which the mask contours are hidden, while continuous and dashed lines represent the PSF for two finely spaced positions in which the contours are visible.

Therefore the shape of the distribution will be variable on the arcmin scale. The degradation of source location accuracy caused by loss of the shadowgram contours could be reduced by choosing a collimator pitch larger than the mask element. In this case, however, off-axis sources may induce artifacts in the image, the spatial distribution of which depends on the particular mask design chosen. The normalized rms, associated to this systematic noise, increases with both the ratio between the sizes of the collimator cell and the mask element, and the source offset angle.

References

Bazzano,A., Amoretti,M., La Padula,C., Natalucci,L., Soggiu,E., Ubertini,P. and Waldron,L.: 1994, *Proc. SPIE* **2279**, 446

Brinkmann,A.C., Dam,J., Mels,W.A., Skinner,G.K. and Willmore,A.P.: 1985, *Non Thermal and Very High Temperature Phenomena in X-ray Astronomy*, G.C. Perola and M. Salvati: Roma, 263

Caroli E., Stephen,J.B., Di Cocco,G., Natalucci,L. and Spizzichino, A.: 1987, *Space Sci. Rev.* **45**, 349

Fenimore, E.E. and Cannon, T.M.: 1991, *Appl. Opt.* **17**, 337

Paul, J. et al.: 1991, *Adv. Space Res.* **11**, 289

Stephen,J.B., Caroli,E., Di Cocco,G., Maggioli,P.P., Natalucci,L. and Spizzichino,A.: 1987, *Astron. Astrophys.* **185**, 343

Soggiu,E., Bazzano,A., Ubertini,P. and Waldron,L.: 1993, *Proc.SPIE* **1948**, 109

Sunyaev et al.: 1990, *Adv. Space Res.* **10**, 233

Ubertini, P.: 1987, *Space Sci.Rev.* **46**, 1

Waldron, L., Ubertini, P., Bazzano, A., Hall, C.J., Lewis, R.A., Soggiu, M.E.,: 1993, *Proc. SPIE* **1948**, 20

Willmore,A.P., Skinner,G.K., Eyles,C.J. and Ramsey,B.: 1984, *Nucl.Instr.Meth.* **221**, 284

X-RAY IMAGING WITH ART-P/GRANAT

Application of the Wavelet Transform

S.A. GREBENEV, M.N. PAVLINSKY and R.A. SUNYAEV

Space Research Institute, Russian Academy of Sciences,
84/32 Profsoyuznaya, 117810, Moscow, Russia

February 15, 1995

Abstract. We discuss the use of a wavelet transform for the subtraction of nonuniform background and for filtering images obtained with the coded-mask X-ray telescope ART-P aboard the Granat spacecraft.

Key words: X-ray images – coded-mask telescopes – wavelet transform

1. Introduction

The wavelet transform is a new method for multi-scale analysis. It already has been successfully applied to the study of fractal structures in physics (Argoul et al., 1989), clustering and subclustering of galaxy distribution in astronomy (Slezak et al., 1990; Escalera et al., 1992). Recently, Grebenev et al. (1995) applied this method to the spatial analysis of the ROSAT PSPC and Einstein HRI images of the cluster A1367. They have found it to be a powerful tool for the subtraction of the cluster diffuse emission and for the detection and characterization of point sources and small-scale extended features embedded within the intracluster medium. Futhermore, they have developed a new technique based on the wavelet transform to quantitatively measure angular extent of these sources.

Describing below very briefly the results of the wavelet transform application to the Granat ART-P data, we are willing to demonstrate that this method can be also beneficially used for the analysis of X-ray images obtained with coded-mask telescopes.

2. The Coded-Mask Telescope ART-P

The telescope ART-P aboard Granat was designed for imaging and spectroscopy in the 2.5-60 keV energy band (Sunyaev et al., 1990). It consists of a position sensitive detector (a multiwire proportional chamber with a window area of 625 cm^2) equipped with a URA coded mask and a collimator.

In 1990-1992 more than 400 sky fields were observed with the instrument. A good $\sim 5'$ FWHM angular resolution made it especially useful for studying the fields containing several nearby sources, e.g. the Galactic

155

L. Bassani and G. di Cocco (eds.), Imaging in High Energy Astronomy, 155–158.

center field. Due to rather a narrow $3°4 \times 3°6$ field-of-view of ART-P, bright sources surrounding the Galactic center contributed nothing to Poisson noise of reconstructed images that provided the achievement of a high ~ 1 mCrab sensitivity level during the observations (Pavlinsky et al., 1992; 1994).

One of the major problems of the image restoring with ART-P was connected with the difficulty to correctly estimate the local background. The instrument background (produced by charge particles and variable on a time scale of days) was nonuniformly distributed over the detector area being higher at the edge than near the center. Such nonuniformity led to the appearance of false large-scale features in reconstructed images. To take the actual background distribution into account we applied a complex technique based on using the differences in temporal (reflecting changes in the spacecraft orientation) behaviour of the background photons and those coming through the coded mask. The wavelet transform allows the problem to be resolved in much easier way.

3.   Wavelet Transform Analysis

The wavelet transform can be defined as a correlation of the sky image $s(x, y)$ with the analyzing wavelet $g(x, y, a)$. Due to special properties of the wavelet (it must be a localized function having a zero mean), the sky image is decomposed into a set of the wavelet coefficient images $w(x, y, a)$ with different resolutions corresponding to a complete set of scales (i.e. a, $2a$, $4a$, ...). Following Grebenev et al. (1995) we use the so-called radial Mexican Hat wavelet given by

$$g\left(\frac{x}{a}, \frac{y}{a}\right) = \left(2 - \frac{x^2 + y^2}{a^2}\right) e^{-(x^2+y^2)/2a^2} \tag{1}$$

The wavelet coefficient image corresponding to the wavelet transform of scale a of the discrete sky image s_{ij} is computed in each pixel (m, n) as follows

$$w_{mn}(a) = \frac{1}{a} \sum_{i,j} s_{ij} g\left(\frac{x_i - x_m}{a}, \frac{y_j - y_n}{a}\right) \tag{2}$$

If the value s_{ij} in any pixel of the sky image is statistically independent of that in any other pixel, the variance of $w_{mn}(a)$ also can be computed

$$q_{mn}(a) = \frac{1}{a^2} \sum_{i,j} d_{ij} g^2\left(\frac{x_i - x_m}{a}, \frac{y_j - y_n}{a}\right) \tag{3}$$

Here d_{ij} is the variance of the value s_{ij}. Each mask pixel of ART-P covers the detector area of 4×4 'fine' detector pixels. For the wavelet analysis, we use the delta decoded sky image (Fenimore and Cannon, 1981), because in this

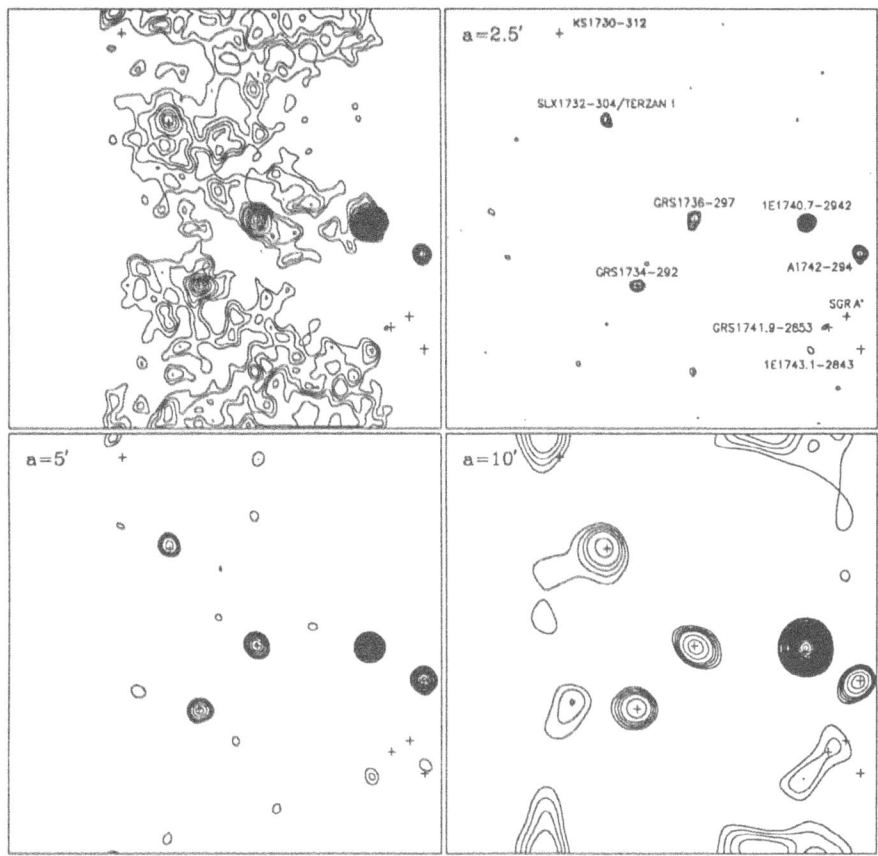

Fig. 1. Contour maps for the original ART-P image of the Galactic center field (smoothed with the instrument PSF) and those computed in wavelet space using different values of the wavelet scale, $a = 2.5'$, $5'$ and $10'$. The contours are shown at the levels of 3, 4, 5, 6, 8, 10, ... standard deviations.

case the s_{ij} values in different 'fine' pixels remain statistically independent and having the same variance ($d_{ij} = N/16$, where N is the total number of detected photons). The wavelet coefficients also have the same variance, $q_{mn}(a) \simeq \pi N/8$, in different pixels. The left upper panel of Figure 1 shows the signal-to-noise (S/N) ratio map for the Galactic center image in 3-17 keV X-rays obtained with ART-P on October 6, 1990. The image was previously smoothed with the instrument point spread function (PSF) – a Gaussian of $\sigma \simeq 2.5'$. The other panels show S/N maps for the wavelet coefficient images with different scales ($a = 2$, 4 and 8 pixels corresponding to $2.5'$, $5'$ and $10'$). The S/N ratio of the wavelet coefficients is equal here to $w_{mn}(a)/\sqrt{q_{mn}(a)}$ and referred to hereafter as measured in standard deviations (or σ), although the coefficient noise distribution is slightly different from the Gaussian one.

There are several bright spots of extended background emission in the original ART-P image. They appear due to nonuniformity of the local background (no corrections to the image similar to the above mentioned have been applied). Some of the spots are present also in the wavelet image with scale $a = 10'$. In the images with smaller scales, $a = 2.5'$ and $5'$, only point-like sources are observed. The significance of each source varies with the scale of the analyzing wavelet, reaching a maximum on the scale comparable with the source angular extent, i.e. on the scale approximately equal to the width of the instrument PSF ($a_m = \sqrt{3}\,\sigma \simeq 0.74$ FWHM, see Grebenev et al., 1995). In the case of ART-P, $a_m \simeq 4.3'$, so the image with scale $a = 5'$ is most appropriate for point source detection. Two of the sources visible in Figure 1, GRS1734-292 and GRS1736-297, were first discovered with ART-P (Pavlinsky et al., 1992). On the other hand, the well known sources Sgr A* and 1E1743.1-2843 were not detected during this observation. They were at the edge of ART-P field-of-view in the region of low collimator transparency.

The wavelet coefficients $w_{mn}(a)$ in nearby pixels are correlated and their statistical behaviour is complex (Grebenev et al., 1995). To set the correspondence between the computed S/N ratio of the coefficients and actual levels of significance we performed Monte-Carlo simulations. The probability that one of the sources, detected at the 4σ S/N threshold in the 164×172 pixel wavelet image with scale $a = 5'$, is spurious was found to be only 20%.

The results of the wavelet transform application to the ART-P data confirm its high ability to subtract from the image any flat or extended background components, whenever their characteristic scale exceeds that of the analyzing wavelet. In addition, the wavelet transform, like some other techniques such as convolution of a sky image with the instrument PSF, its smoothing with a Gaussian filter or just its binning, is found very effective to improve the statistical confidence of source detection.

Acknowledgements

We would like to thank W.Forman for useful comments and stimulating discussion during the course of this research.

References

Argoul, F., Arneodo, A., Elezgaray, J., et al.: 1989, *Phys. Letters* **135**, 327.
Escalera, E., Slezak, E., and Mazure, A.: 1992, *Astron. Astrophys.* **264**, 379.
Fenimore, E.E., and Cannon, T.M.: 1981, *Appl. Opt.* **20**, 1858.
Grebenev, S.A., Forman, W., Jones, C., and Murray, S.: 1995, *Astrophys.J.* **444**, in press (*SAO preprint* 4010).
Pavlinsky, M.N., Grebenev, S.A., and Sunyaev, R.A.: 1992, *Sov. Astron. Lett.* **18**, 217.
Pavlinsky, M.N., Grebenev, S.A., and Sunyaev, R.A.: 1994, *Astrophys. J.* **425**, 110.
Slezak, E., Bijaoui, A. & Mars, G.: 1990, *Astron. Astrophys.* **227**, 301.
Sunyaev, R.A., Babichenko, S.I., Goganov, D.A. et al.: 1990, *Adv.Space Res.* **10**, 2(233).

ABRIXAS

A BRoad-band Imaging X-ray All-sky Survey

G. RICHTER, G. HASINGER, P. FRIEDRICH and K. FRITZE

AIP
An der Sternwarte 16
14482 Potsdam

J. TRÜMPER, H. BRÄUNINGER and P. PREDEHL

MPE
Karl-Schwarzschild-Str. 1
85748 Garching

and

R. STAUBERT and E. KENDZIORRA

AIT Univ. Tübingen
Waldhäuserstr. 64
72076 Tübingen

Abstract. In a cooperation between Astr. Inst. Potsdam, MPE Garching and Astr. Inst. Univ.- Tübingen we have proposed to DARA to build a small X-ray satellite for an all-sky survey in the 0.5-10 keV band in order to observe an expected population of AGN, absorbed by a gas and dust torus.

Key words: satellite – X-ray telescope – AGN

1. Astrophysical Background

There is a longstanding debate on the nature of the extragalactic X-ray background between 1 and 100 keV: whether it is the integrated effect of faint discrete sources, or whether there is a truly diffuse radiator, e.g. a hot intergalactic gas. The latter would fit the observed spectrum quite well, but would have a substantial effect on the cosmic microwave background, which has already been excluded by the recent COBE measurements (Mather et al. 1990).

In the past the discrete source model, however, seemed unattractive because of the so-called "spectral paradoxon", i.e. the fact that no known class of sources has a spectrum similar to the background.

Recently, deep ROSAT pointed observations (Hasinger et al. 1993) have shown that a substantial fraction (about 60%) of the background in the 1-2 keV band can be resolved into discrete sources at a flux level of about $3 \times 10^{-15} erg/cm^2 s$. A fluctuation analysis of the remaining background is consistent with all the soft background being due to discrete objects, and restricts a possible diffuse process to less than 25%.

L. Bassani and G. di Cocco (eds.), Imaging in High Energy Astronomy, 159–162.
© 1995 *Kluwer Academic Publishers.*

The vast majority of optical identifications in the deep ROSAT fields are active galactic nuclei (AGN), in particular quasars of medium to high z (Shanks et al. 1991). The logN-logS function shows a distinct flattening at the faint end, consistent with the same evolutionary models as for QSOs in the optical waveband (Boyle et al. 1993). Thus, it seems likely that the majority of the soft (1-2 keV) X-ray background is due to unresolved AGN.

However, what about the higher energy background, which increases (in $E * I(E)$) from about 3 keV to higher energies and peaks around 40 keV, and is a factor of at least 10 higher than the extrapolated flux of the above-mentioned soft AGN population?

If one accepts the unified scheme of AGN (e.g. Antonucci 1993), which assumes that the active nucleus is surrounded by a thick obscuring torus, then we receive unobscured soft X-rays only in the relatively rare cases where we see the torus almost face on. With increasing aspect angle the soft part of the spectra becomes more and more absorbed, resulting in a population of hard spectrum AGN. Comastri et al. 1995 succeeded to model the hard background and the related constraints by an appropriate mixture of such obscured AGN. There is also evidence, that the gas in the torus is partially ionized (e.g. Nandra & Pounds 1992) so that a variety of spectral features like absorption edges and flourescence lines are expected. Hence, the energy range above 2 keV seems to comprise a very interesting, up to now almost unknown population of extragalactic X-ray sources, which contain crucial information on structure and mechanism of AGN.

The main reason for this strange ignorance in just the classical X-ray domain 2-10 keV is the lack of imaging instruments with appropriate resolution to identify the fainter sources optically. The UHURU and HEAO-1 all-sky surveys have detected about 800 objects, but only a small number of them are identified with AGN, mostly local Seyfert galaxies (Piccinotti et al. 1981). On the other hand, by the year 2000, X-ray observatories with good imaging capabilities and bandpass up to 10 keV will be available (e.g. XMM and AXAF) and will be able to measure directly a broad distribution of AGN source spectra and absorption at all redshifts. However, they are not survey instruments and need precise coordinate inputs of already known AGN, and they cannot provide the large, statistically complete and unbiased source samples needed to study the evolution, etc. of the AGN population. In principle, the ROSAT all-sky survey with its approximately 60000 new sources (Voges et al. 1992) could serve as a pathfinder. But because of its soft energy band which is extremely sensitive to the absorption, it provides only a subsample of the AGN population.

An all-sky survey in a wide and hard band (say 1-20 keV), ideally fore-running AXAF and XMM missions, would therefore be most desirable and scientifically very valuable. Its main aim would be to study the expected absorbed AGN population. Additional goals might be the observation of the soft end of γ-bursts, the spectrum and isotropy of the X-ray background and the longterm monitoring of bright X-ray sources.

2. Mission Concept

To meet the scientific goals defined above, we need on the order of 1000 optically identified sources. Taking into account the expected large scatter in optical-to-X-ray luminosity ratio, the satellite should be able to detect at least one order of magnitude more X-ray sources, say 10000, to get an unbiased sample. This means a sensitivity well below 0.1 mCrab.

In order to make reliable optical identifications with a reasonable observational effort (for all objects in the error box spectra are needed!) the angular resolution should be better than 1 arcmin.

We have designed a satellite to be a pure survey instrument without pointing capability. This makes the attitude control very easy and inexpensive, because the active control can be very crude and the attitude parameters must be known to better than 1 arcmin only post factum. The rotation of the craft will be bound to the orbital motion in such a way, that the field of view never sees the Earth nor the Sun. Magnetocoils and one reaction wheel will control it. The orbital plane will rotate by 360° in one year, i.e. every source is observed every 6 months. The design life time is 1.5 years, and the expected one 3 years.

Because of the very short time scale until the scheduled launch (1997/98) and due to the tight financial bounderies, only already well known and proven technologies and components will be used. As the focal instrument, the CCD camera MAXI developed at MPE Garching and AIT for XMM EPIC (Bräuninger et al. 1993) will be used. It contains a pn-CCD with a deep depletion layer, 64×120 pixels of $150 \times 150 \mu^2$ each, and an efficiency of about 1 in the range 0.5-10 keV. Twelve such CCDs will be produced together as an array on a Si wafer of roughly $6 \times 6 cm^2$.

3. Telescope Alternatives

We have studied and simulated the complete orbital behaviour of two alternative designs for the X-ray telescope, both matching the scientific requirements.

The first one is a coded mask telescope of 1.5m length. The mask would have a size of $70 \times 70 cm^2$ and elements of $1 \times 1 mm^2$. A detector of $35 \times 35 cm^2$ would be needed, consisting of 36 XMM CCD wafers, to provide a field of view of 24°. In order to use it for continuous scanning, a new reconstruction method has been developed (direct backprojection of each photon) which very effectively smoothes the fixed pattern noise. The scientific requirements are met very well, in particular at the high energy part of the band, for X-ray background measurement, variability monitoring, and γ-bursts. Problems arise from the high telemetry rate and huge computational expense of reconstruction, but fatal for this design is the requirement of passive cooling of so many CCDs in a low earth orbit.

The second design is a battery of 7 identical grazing incidence telescopes of 1.6m focal length, each consisting of 27 nested biconical mirrors of 30cm length each, and 16cm diameter of the outer one. The 7 focal planes of 2cm diameter each are sharing one $6 \times 6cm^2$ CCD chip. Thus, the optical axes diverge and the 7 fields of view, each of 40 arcmin diameter, are about $6°$ apart of each other on the sky. Nevertheless, the maximum of 4 arcmin shift between two successive orbits provides a continuous coverage of the sky. DARA has completed a phase-1 study in competition between two independent companies. The conclusion is, that the ABRIXAS mission is feasible under the preset narrow budged and time constraints.

References

Antonucci, R.: 1993, *Ann. Rev. Astron. Astrophys.* **31**, 473

Boyle, B., Griffith, R., Shanks, T., Stewart, G., Georgantopoulos, I.: 1993, *Mon. Not. R. astr. Soc.* **260**, 49

Bräuninger, H., et al.: 1993, *Nucl. Instr. Meth.* **A326**, 129

Comastri, A., Setti, G., Zamorani, G., Hasinger, G.: 1995, *Astron. Astrophys.* , in press

Hasinger, G., Burg, R., Giacconi, R., Hartner, G., Schmidt, M., Trümper, J., Zamorani, G.: 1993, *Astron. Astrophys.* **275**, 1

Mather, E., et al.: 1990, *Astrophys. J. Lett.* **354**, L37

Nandra, K., Pounds, K.: 1992, *Nature* **359**, 215

Piccinotti, G., Mushotski, R., Boldt, A., Holt, F., Marshall, F., Serlemitsos, P., Shafer, R.: 1982, *Astrophys. J.* **253**, 485

Shanks, T., Georgantopoulos, I., Stewart, G., Pounds, K., Boyle, B., Griffith, R.: 1991, *Nature* **353**, 315

Voges, W.: 1992, *ESA ISY-3* , 9

THE MART-LIME HIGH-ENERGY TELESCOPE

L. WALDRON, M. AMORETTI, L. BOCCACCINI, M. FEDERICI,
M. FRUTTI, G. GIANNI, C. LA PADULA, R. PATRIARCA,
G. SABATINO, P. UBERTINI and U. ZANNONI
Istituto di Astrofisica Spaziale, Frascati, Italy

Abstract. MART-LIME is a coded mask imaging hard X-Ray telescope to be flown on-board the international SPECTRUM RG observatory in early 1996. This instrument, the heart of which is a high-pressure proportional counter sensitive to the 5–150 keV energy range, will be characterised by a limiting sensitivity of about 1 milliCrab for a 10^5 s observation period. The imaging capability of MART-LIME is provided by a coded-mask aperture system, used in conjunction with the position sensitive detector. The basic pattern of the square coded aperture is a 71×73 URA mask (twin prime). This is surrounded by an outer frame comprising 17 and 16 pixels, respectively, on adjacent sides, giving a mask of 105×105 pixels in total. With this configuration, a fully coded field-of-view of $5.2^\circ \times 5.2^\circ$ is obtained, whilst a partially coded field-of-view is achieved up to 6°, the acceptance angle of the ferrite collimator. The basic concept of the MART-LIME telescope is presented.

Key words: X-ray Astronomy – Coded Mask Imaging – Instrumentation

1. Introduction

SPECTRUM RG (SRG) is a russian X-Ray Observatory class astronomical satellite scheduled for launch in late 1996, with an intended operational lifetime of three years. This satellite comprises a dozen instruments, two of which utilise grazing-incidence optics to attain a sensitivity up to ~ 15 keV. The void at higher energies will be fulfilled by MART-LIME, a telescope currently under joint construction by the Istituto di Astrofisica Spaziale, Frascati, Italy, and the Space Research Institute (IKI), Russian Academy of Science, Moscow, Russia. MART-LIME is a a broadband ($5 \sim 150$ keV) hard X-Ray telescope having sub-milliCrab sensitivity, high spectral resolution ($E/\triangle E \geq 20$), a sub-arcminute point-source location capability, and temporal resolutions of 1 ms and 80 μs in "imaging mode" and "science mode" respectively. In addition to the obvious scientific merits associated with extending the spectroscopic, positioning and timing capabilities of SRG to over 4 decades of energy (Bazzano et al, 1994), MART-LIME is a useful stand-alone instrument in its own right. It will provide a probe of X-Ray binary sources and associated cyclotron lines, the location of hard X-Ray emission from complex regions, AGN spectral morphology, spectra (energy-resolved) of clusters of galaxies and hence any associated non-thermal emission components, and the X-Ray background in the hard X-Ray range.

L. Bassani and G. di Cocco (eds.), Imaging in High Energy Astronomy, 163–166.

2.   The Experimental Configuration

The MART-LIME telescope comprises the following basic systems:
- optical bench (structural chassis);
- photon detection system;
- electronics; and
- Thermal Balance System (TBS), Multi-Layer Insulation (MLI) and thermal screens.

2.1 THE OPTICAL BENCH

This structural chassis of box-like architecture is the framework designed to accomodate all MART-LIME subsystems. It is fabricated from a carbon fibre laminate of high elastic modulus and low coefficient of thermal expansion, internal to which is a conducting mesh for both electromagnetic insulation and grounding.

2.2 THE PHOTON-DETECTION SYSTEM

The photon-detection system is a high-pressure (5-Bar Xenon) titanium Multi-Wire Proportional Counter (MWPC) behind a coded mask system. The MWPC has an active area of 1800 cm^2 and a detection efficiency of $> 10\%$ over the $5 \sim 150$ keV range of sensitivity. The detector will be used in the fluorescence gated coincidence mode with active veto (Waldron et al., 1993) so as to both reduce inherent background noise and obtain sub-millimeter spatial resolution. Bi-directional spatial information is extracted by implementing the continuous delay line concept.

The passive stainless steel collimator, an integral support structure of the MWPC entrance window, limits the FOV of the detector to (6×6) square degrees (FWHM), thereby reducing the instrument background due to diffuse X-Rays and low-energy cosmic-rays at large offset angles. The imaging capability of the instrument is is provided by a self-supporting (105×105) pixel coded mask $(71 \times 73$ URA, twin-prime) aperture system, coplanar with and 2.3 m infront of the detector entrance window. This mask comprises a tungsten sheet laminated with a thin layer of carbon fibre, together with a carbon-fibre slat support structure. The tungsten sheet is laser-etched with a pattern of holes such that there is $\sim 50\%$ overall transparency to X-Rays. The result is a fully coded field of view of (5.2×5.2) square degrees.

As a monitor of any drift in gain or degradation of energy resolution, inflight spectroscopic calibration will be accomplished by illuminating one pixel of the detector image plane with a low count-rate calibration source embedded within the collimator. The linearity of the detector image plane will be confirmed or re-calibrated by observing a portion of the sky containing several known sources which are expected to be visible.

2.3   Electronics

MART-LIME Electronics comprise four subsystems:
- Front-End Electronics (FEE)
- High Voltage Units (HVU)
- Data Processing Electronics (DPE)
- Interface with the spacecraft on-board computer (BIUS Interface)

2.3.1   FEE
The FEE (analogue, e.g. pre-amplifiers etc.) reside in the base of the detector pressure vessel. FEE are devoted to the processing of signals generated by the absorption of X-Rays in the active volume of the MWPC. They comprise a fast spectroscopic amplifier connected in unison to anode wires, together with four cathode pre-amplifiers (Two for each cathode plane). The FEE extract the requisite positional, spectral and temporal information from the MWPC readout system. Moreover, implementation of the fast delay-line readout concept enables a nanosecond discrimination capability.

2.3.2   HVU
The HVU reside in the base of the detector pressure vessel. They comprise three digitally programmable power supplies which are used to power the anodes, drift region and anticoincidence region of the detector, respectively.

2.3.3   DPE
The DPE (digital), together with the interfaces to spacecraft, resides within a dedicated box that is isolated from the detector. The tasks to be performed by this multi-functional micro-computer controlled system include event selection and validation, background rejection, and collection and formatting of scientific data. It also provides the interface to the spacecraft, command and control interface, detector control, the collection of housekeeping data, power conversion and conditioning, and an overall experiment "watch-dog".

2.3.4   BIUS Interface
This unit interfaces MART-LIME with the SRG on-board computer. It is essential for the transmission of packed scientific/housekeeping telemetry data and communication of telecommands to the experiment.

2.3.5   TBS, MLI and Thermal Screens
The TBS utilises both a passive radiator cooling system and a semi-active system comprising heaters and thermal sensors that are controlled and managed by the Thermal Control Unit (TCU). MLI is used to thermally decouple the experiment from deep space and solar effects. Thermal screens consisting of Kapton foil (X-Ray transparent) are interposed in the telescopes field of

view, thereby providing protection against the impinging sun and cooling effects due to deep space exposure.

The predicted in-flight performance parameters of MART-LIME are presented in Table 1.

Table 1: Predicted Performance Parameters of MART-LIME

Energy Range:	5–150 keV (10% efficiency)
Sensitive Area:	1800 cm^2
Spectral Resolution	5%, $E > 35$ keV; $< 20\%$, $E < 35$ keV
Field of View	$6.0^o \times 6.0^o$ degree FWHM
Angular Resolution	8.7×8.7 arcminute
Typical Source Location	20 arcsec (5-σ source)
Temporal Resolution	1 ms imaging mode; 80 μs timing mode
Continuum Sensitivity	
3-σ in 10^5 s	1.5×10^{-6} ph cm^{-2} s^{-1} keV^{-1}
3-σ in 10^6 s	5×10^{-7} ph cm^{-2} s^{-1} keV^{-1}
Line Detection Sensitivity	(50 keV, narrow line); line width $\simeq 2$ keV
3-σ in 10^5 s	1×10^{-5} ph cm^{-2} s^{-1}
3-σ in 10^6 s	3×10^{-6} ph cm^{-2} s^{-1}

3.   Conclusions

The basic concept of the MART-LIME telescope has been presented. The scientific target of the telescope is the generation of X-Ray sky images with sub arcminute angular resolution, good spectral resolution ($E/\triangle E \geq 20$) and milliCrab sensitivity during a typical one-day observation period. The Deep-sky surveys thus obtained over a wide field of view will herald a breakthrough in high-energy Astrophysics with the provision of a complete hard X-Ray catalogue at the milliCrab sensitivity level.

4.   References

Bazzano, A. et al., *SPIE Proceedings*, **Vol. 2279**, 446 (1994)
Waldron, L. et al., *SPIE Proceedings*, **Vol. 1948**, 20, (1993)

5.   Acknowledgements

L. Waldron wishes to acknowledge the continued support of Dr G. Babalian, the MART-LIME representative of IKI, Moscow.

THE CODED MASK DESIGN AND TEST FLIGHT RESULTS
OF A SMALL ARRAY GERMANIUM TELESCOPE

R.M. RIDEOUT & G.K. SKINNER
Space Research Group
University of Birmingham
Birmingham, UK.

Abstract. A small germanium array detector had been incorporated into a coded mask telescope and was flown as a piggy-back experiment on the NASA/GRIS high altitude balloon payload. The flight, on the 24th September 1993 from Fort Sumner, New Mexico, successfully demonstrated the use of a compact 3×3 germanium array detector to achieve good imaging of X-ray and γ-ray sources. This paper describes the techniques used to select a suitable coded mask pattern for the telescope as well giving a brief overview of flight results. Comparisons are made of the imaging ability of several different classes of coded mask patterns. By applying constraints to the design of the otherwise 'random' coded mask pattern, the performance in terms of point source resolution is seen to be better than that of masks based on cyclic difference sets and 'constraint-less' random masks.

Key words: Coded mask – Germanium detector – X-ray Telescope

1. Introduction

Coded mask techniques have been widely investigated and used in a variety of experiments to image photon source energies above ≈ 10 keV where the effectiveness of grazing incidence optics breaks down. Previously, efforts have concentrated on achieving good image resolution using position sensitive detectors with excellent spatial sensitivity but poor energy resolution. Here we will discuss the coded aperture design considerations and flight results for a new type of coded mask telescope, flown on the NASA/GRIS (Tueller *et al.*, 1994) balloon gondola, which successfully combined the high energy resolution property of cooled Germanium (Ge) detectors with the spatial resolution obtained by positioning the detectors into a compact array. This allowed the use of coded mask techniques to image the targeted sources, Cygnus X-1 and the Crab Nebula.

Laboratory tests of the compact $45 \times 45 \times 50$ mm, 3×3 Ge array detector which was incorporated into the experimental coded mask telescope, have already been discussed in Skinner *et al.* (1993a). Here we concentrate on the coded mask properties, and use a few of the flights results as examples. More extensive results and conclusions from the flight data can be found in Skinner *et al.* (1994).

L. Bassani and G. di Cocco (eds.), Imaging in High Energy Astronomy, 167–170.

2. Coded mask design

The characteristics of the instrument considered important here are:
- The coded mask elements had the same size and pitch as the Ge array, ie. 15.0 mm and 15.2 mm respectively.
- The mask was a 21×7 element array. A point source casts a 3×3 element *sub-mask* shadow at the detectors.
- Targeted sources were able to drift through the field of view (FOV), casting a constantly changing sub-mask shadow onto the detectors.

Image reconstruction of coded data is usually done by performing a correlation of the recorded shadow with some variation on the mask pattern itself. The search for unambiguous image reconstruction has generally lead to the use of cyclic mask patterns with particular artifact-free properties, in particular the URA's (Fenimore and Canon, 1978). The point source function (PSF) for such masks consists of a central peak sitting on a perfectly flat pedestal which eliminates ambiguities within one cycle of the base pattern. Unfortunately, URA's do not exist in a 3×3 configuration and so were not an option here. A variation of URA's, the 'Modified URA' (MURA) (Gottesman and Fenimore, 1989) which has the same PSF as the URA's, does exist in a 3×3 array. Repeating a 3×3 MURA in the 21×7 coded aperture was considered but simulations showed unacceptable smeared ghost images in the reconstruction every $\approx 3°$. It was decided that the telescope mask pattern would be random in nature but would be selected following a search procedure, constraining the mask pattern to have certain desirable characteristics.

The properties of the selected coded mask pattern were as follows.
1. Each sub-mask had either 4 or 5 opaque elements.
2. All 95 sub-masks were different by at least one element.
3. Any two sub-masks separated by 1 or 2 elements along either mask axis were different in at least 3 elements, but averaged 5.4 element difference.

In making all the individual sub-mask patterns different from each other (item 2 above), the amount of coded noise in the reconstruction is kept to a reasonable level and ghost images are reduced in intensity. Property 3 implies that the image reconstruction is free from ambiguities in pixels locally surrounding point sources in the FOV. The comparison of the expected detector counts from a nearby sky pixel, through a neighboring sub-mask pattern, with the actual recorded detector counts, will correlate badly and so will be assigned a low intensity value.

3. Results

We can consider the expected imaging performance of a system by examining the PSF derived by convolving the autocorrelation function (ACF) of the

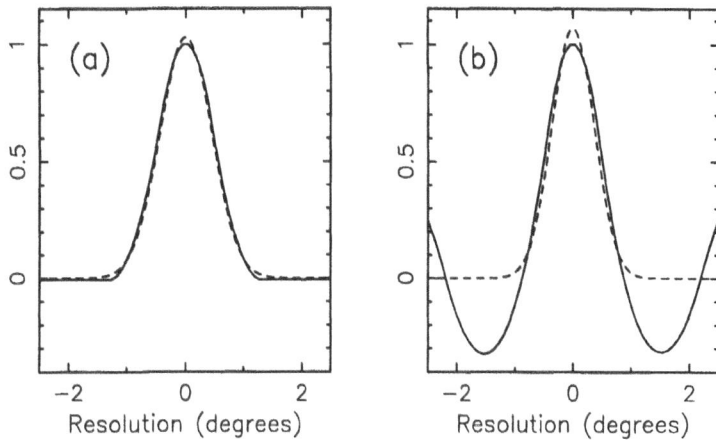

Fig. 1. Normalised slice through the PSF for (a) an (M)URA or pinhole, and (b) the coded mask pattern of the experimental telescope. The dashed line is a best fit 1-d gaussian on a mean level plateau for the PSF. FWHM's are 0.92° and 0.73° respectively.

mask with a square element, of size 0.9° in the case of the experimental telescope, to represent the angular resolution of a single Ge element. This PSF indicates the best imaging ability possible, obtained after a raster of pointings, recording data through all possible sub-masks in a cyclic fashion. We can then use this property to compare the resolution of the selected mask with that of a URA or similar ideal mask when used with the same detectors. Figure 1 shows a 1-d slice through the PSF of the experimental telescope mask and an artifact-free mask, such as a URA, for comparison.

Property 3(§2) of the mask is responsible for the large negative side-lobes either side of the central peak in the PSF (figure 2(b)). These are seen to improve the apparent imaging ability as measured by the full width at half maximum (FWHM) of a gaussian fitted to the peak, on top of the flat, mean PSF value. The results of fitting shows the chosen mask to have a resolution of 0.73° FWHM, compared with the fit to the ideal mask pattern of 0.92° FWHM. Investigating further shows that a typical 'constraint-less' random mask has worse resolution still.

Figure 2 shows reconstructions obtained by using the CAPTIF software (Connell and Skinner, 1994) on the Cygnus X-1 flight data, and a simulation of a 'noiseless' source at the Cygnus X-1 position and flux strength. The FWHM of best fitted 2-d gaussians are 1.11° and 1.05°, respectively. These values compare very well and indicate that the telescope performed as expected in terms of imaging ability. It is not possible to compare, directly, the FWHM of the reconstructed images and the PSF, as the latter utilises

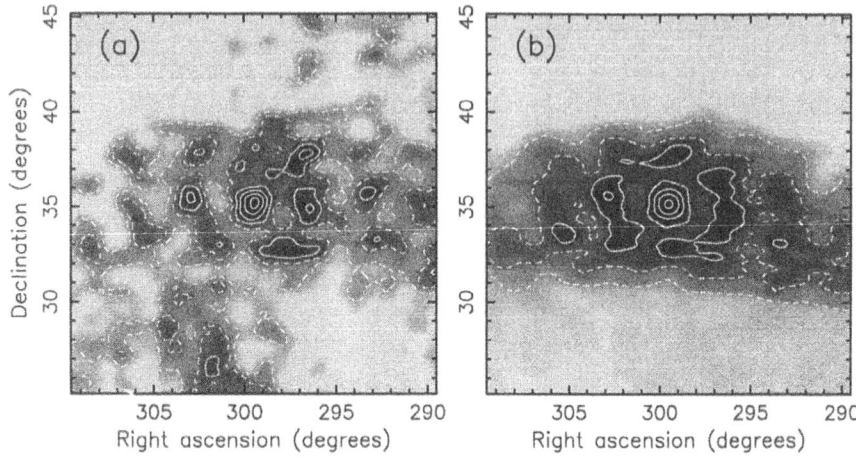

Fig. 2. Reconstructed images between 30-80 keV of (a) flight data during the Cygnus X-1 observation, and (b) simulated data of a 'noiseless' source at the Cygnus X-1 position and flux. Images are significance plots with contours at 1σ levels (dashed contours are at 1σ and 2σ).

many more pointings in its derivation. However, it is apparent that the constraints applied to the chosen mask pattern are responsible for the reduction of coded noise in the reconstructed images, local to the source position.

4. Conclusions

Since it is not possible to utilise the artifact-free properties of (M)URA's with the experimental telescope, it has been shown that by selectively designing the 'random' pattern, it is possible to improve the resolution or 'clarity' with which sources can be imaged. Also, the coded noise which results from a 'non-ideal' mask pattern is reduced by utilising the movement and rotation of the sky field within the mask FOV, a method intended to be employed by the INTEGRAL mission (Skinner et al., 1993b).

References

Connell, P.H. and Skinner, G.K. : 1994, Submitted to *Proc. Workshop on Imaging in High Energy Astronomy*, Capri, to be published in *Experimental Astronomy*.
Fenimore, E.E. and Canon, T.M. : 1978, *App. Op.* **17**, 337.
Gottesman, S.R. and Fenimore, E.E. : 1989, *App. Op.* **28**, 4344.
Skinner et al. : 1993a, *SPIE* **1945**, 465.
Skinner et al. : 1993b, *SPIE* **1945**, 112.
Skinner et al. : 1994, Accepted for publication in *Nucl. Instr. and Meth.*
Tueller et al. : 1994, Submitted to *Ap. J.*

HEXIT : HIGH ENERGY X-RAY IMAGING TELESCOPE

R. K. MANCHANDA

Tata Institute of Fundamental Research, Colaba, Bombay-400 005, India.

Abstract. A balloon-borne High Energy X-ray Imaging Telescope is presently being developed for imaging studies in 20 keV to 1 MeV energy region. The payload consists of a 40 cm diameter 'Phoswich Anger Camera' made of NaI(Tl) and CsI(Na) scintillation crystals viewed by 13 phototubes and a URA-based coded mask, mounted on a servo-stabilized alta- azimuth platform. This paper describes the design characteristics and laboratory performance of the detector system.

Key words: X-ray imaging; X-ray astronomy

1. Introduction

High resolution spatial and spectral mapping in the hard X-ray region has been lacking mainly due to fundamental limitation of the sensitivity of the detector systems and imaging capability of the instruments. The past sky surveys, except for a few balloon-borne observations, in the 20-200 keV range have relied on satellite-borne relatively small ($100\text{-}500\ cm^2$) detectors using a few square degree collimators. Imaging capability with good angular resolution is essential for localizing the hard X-ray sources. Also an imager allows the simultaneous measurement of background and the source, which is crucial at low flux levels.

The main performance factor for a high energy X-ray imager is the low background large area detector with a suitable choice of imaging technique. The best imaging technique currently in use in the hard X-ray and gamma ray region is the use of coded aperture mask along with a position sensitive detector.

At energies above 30 keV, the only available detector system is the scintillation counters. The use of position sensitive proportional counters beyond 30 keV, suffers on two counts. First, the detection efficiency is extremely small. Second, the observed energy loss spectra does not directly map into incident photon spectrum due to large K-escape probability (85% in xenon). Monte-Carlo method used to deconvolve the incoming spectrum, critically depends on the knowledge of systematic effects. In contrast, scintillation detectors do have high detection efficiency but no spatial sensitivity. A light division technique can be used to determine the point of interaction[1]. Satellite and balloon-borne imaging payloads employing single crystal scintillators are currently in use[2,3,4]. Large area CdTe and HgI_2 array detectors for astronomical imaging are currently in the development stages[5].

The inherent background produced in a scintillator due to cosmic ray interaction and the Compton scattering of the high energy photons dominates the aperture flux and therefore, the flux sensitivity does not directly scale with the geometric area. The most favoured arrangement for reducing the detector background is the use of phoswich techniques using NaI(Tl) and CsI(Na), which provides a shielding factor of about 4-6 in cosmic environment.

HEXIT is a new balloon-borne hard X-ray imaging telescope, which combines a large area phoswich anger camera with a coded aperture URA mask.

L. Bassani and G. di Cocco (eds.), Imaging in High Energy Astronomy, 171–175.

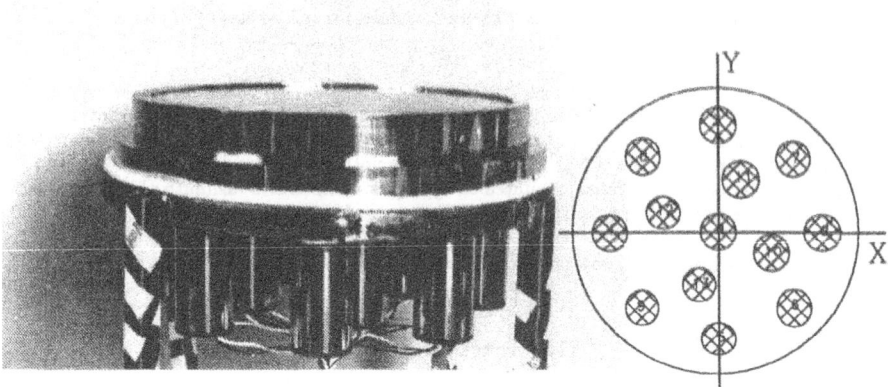

Fig. 1. Detector Assembly and schematic of the phototubes

2. Detector system

The HEXIT uses a new type of detector which we have been developing for the past few years[6,7], in order to build a high sensitivity imaging telescope in the hard X-ray range of 20-500 keV. The detector is a large area phoswich assembly having 40 cm dia and made of NaI(Tl) and CsI(Na) scintillator crystals. Since the optimum thickness of the two scintillators in a phoswich detector depends on the targeted energy range, HEXIT assembly consists of 12 mm NaI(Tl) as the prime detector coupled to a 40 mm thick CsI(Na) crystal. The combination is vacuum sealed with a 0.127 mm aluminum enterence window and a 10 mm thick Pyrax back plate. Thirteen 76 mm phototubes are directly coupled to the optical window in a quadrant symmetric fashion. The large area detector and the arrangement of the photomultiplier tubes is shown in the figure 1. A passive slat collimator made from lead (.5 mm) laminated with tin (.25 mm) and copper (.15mm) on both sides is placed above the detector to limit the field of view to $5^o \times 5^o$. The electronics associated with the phoswich anger camera is presently being fabricated and consists of individual pre-amplifier and amplification units for each of the thirteen phototubes and a common rise-time-discrimination (RTD) circuit. For each event, the relative amplitudes from all the 13 signal are digitized using 8 bit flash ADC. The pulse height data along with the ID flag for each signal is further processed by an 386/SX embedded PC, on-board computer to evaluate the position of interaction for the accepted event. The X-ray energy is determined from the summed amplitude from all the photomultipliers.

2.1 PERFORMANCE

Since the signal separation from the prime and rear detector primarily depends on the pulse shape discrimination and the expected event rate in the shield detector is much larger both due to large area and the higher detection efficiency, an ultra-fast RTD system is essential for a very low dead-time. We have developed

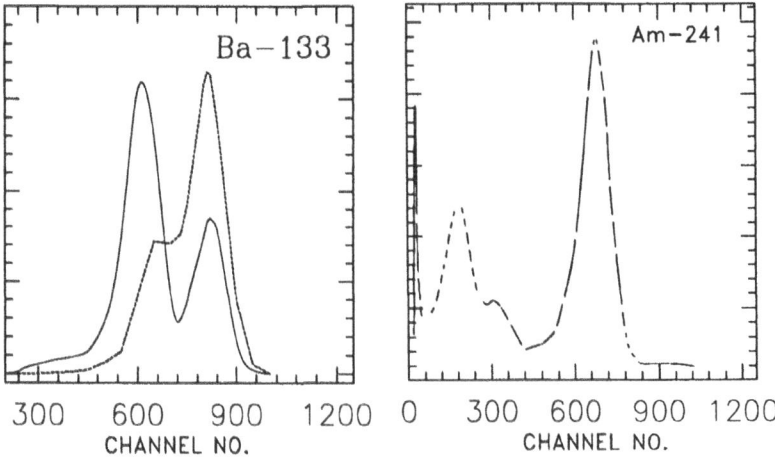

Fig. 2. **a**. Rise time distribution for 80 and 350 keV photons from Ba-133. *solid line* : NaI(Tl); *dotted line* : CsI(Na). **b**. Energy resolution of the Anger Camera for 60 keV photons

a new RTD circuit which can handle up to 10^5 counts/sec. A sample data of the rise time discrimination of the signals from the two detectors using a collimated Ba^{133} radioactive source from above and below is shown in figure 2a. The signal corresponding to 350 keV photons reaching the second detector is clearly seen in the figure.

2.1.1 Energy Resolution

The pulse height spectrum of the HEXIT detector obtained for the 60 keV line emission from Am^{241} is shown in figure 2b. The observed energy resolution for an uncollimated source is 16%. Even though the phototubes are relatively widely spaced in our geometry, the summed light output is uniform within 5% over the entire area of the detector. Partial collection of the signal from less number of tubes for any event does lead to poorer energy resolution.

2.1.2 Spatial Resolution

Variation of the signal for the central and a side tube for different positions of Am^{241} source is shown in figure 3a. A linearized radial response of a phototube is shown in figure 3b. To obtain the location of the photon interaction over the surface, all signals from 13 phototubes are individually pulse height analyzed. The event coordinates will be calculated on-board and transmitted to ground along with relevant pulse height information and time-tag. We are planning to create a look-up table for quick computation of the event co-ordinates. From the laboratory tests, using χ^2 minimization of 13 normalized data points for every event having two free parameters (x,y), we estimate the position resolution to be about 7mm even at 30 keV, within 90% confidence limit.

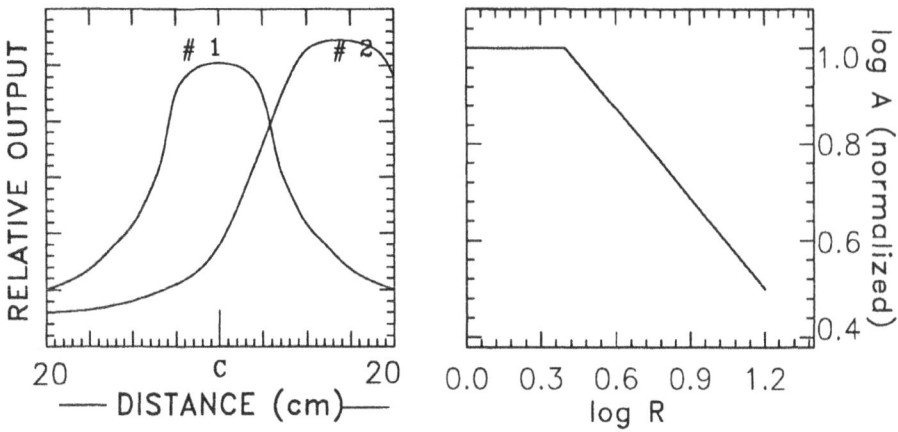

Fig. 3. **a**. Signal variation along X-axis for the central and side phototube **b**. Linearized radial dependence of the phototube signal

3. Payload gondola

Imaging in the HEXIT experiment will be performed using a coded aperture mask located at 1.5 meter in front of the detector. A rectangular URA mask is planned at present. With the mask element size of 15 × 15 mm the expected angular resolution is ∼40 arcmin. We are also studying the use of zone-plates for imaging via shadowcasting.

The HEXIT gondola is fully automatic and is controlled by an on-board microprocessor based star tracker for payload orientation and source tracking. The front-end analysis will be performed by an on-board 386/SX computer and the data transmitted on a PCM link. A real-time analysis of the housekeeping and quick-look data is performed on ground using 486/AT computer and will provide the status of the payload during the flight. All the parameters of the instrument can also be commanded from ground if required, using manual-mode override command.

From our earlier experiments from Hyderabad, using phoswich detectors, the observed background counting rates at 4 gm cm^{-2} float altitude are 2.5, 1.7 and 4.8 ct sec^{-1} in the energy range 20-40, 40-60 and 60-150 keV respectively, for a detector of 100 cm^2 area and FOV of 11° × 11°. The estimated 3σ sensitivity of the HEXIT detector for a source observation of 10^4 sec at 4 gm.cm^{-2} is thus < 10^{-5} $ph\ cm^{-2}s^{-1}keV^{-1}$.

4. Acknowledgments

It is pleasure to thank sarvashri J.A.R. D'Silva, P.P. Madhwani, N.V. Bhagat and A.B. Ghodke for the electronics and mechanical fabrication of the payload.

References

1. Anger H. O. : 1958, *Rev Sci Instr* **29**, 27.
2. Laurent P. et al.: 1993, *Astr Astrophy* **278**, 444.
3. Palmer et al.: 1993, *Ap J* **412**, 203.
4. Grindley et al.: 1992, *Astrophys Sp Sci* **272**, 733.
5. Dusi et al.: 1994, *This conference*.
6. Manchanda R.K and Sood R.K.: 1991, *Ind J Radio & Sp Phys* **20**, 268.
7. Sood R.K., Manchanda R.K. and Staubert R.: 1994, *Adv Sp Res* , in print

A COMPENDIUM OF CODED MASK DESIGNS

G.K. SKINNER & R.M. RIDEOUT

University of Birmingham, Edgbaston, Birmingham B152TT, UK

May 30, 1995

Abstract. Many different patterns have been proposed for coded masks, from Fresnel zone plates, through scatter-hole designs, to various patterns which are 'optimal' in different senses. We present a compendium of such patterns, with examples and their autocorrelation functions, together with a bibliography of associated literature.

Key words: X-rays – Gamma-rays – Imaging – Coded-masks

1.   Introduction

Since Mertz and Young (1961) introduced the concept of indirect imaging using shadows cast by Fresnel Zone plates, a wide range of patterns has been proposed for what has become known as 'coded mask imaging'. Visually some of these patterns are strikingly different, while in many cases their performance, encapsulated in the autocorrelation function (ACF) of the mask pattern, is identical to that of other patterns. We present a compilation of patterns which have been proposed.

The patterns illustrated in Fig. 1 are accompanied by the Point Spread Function (PSF) when reconstruction is by correlation with the mask pattern. In most cases this is simply the cyclic ACF. Exceptions are marked by ‡ in the notes below. The patterns in (k)–(q) all have ACFs (or PSFs) which are essentially the same. Some patterns allow reconstruction of an image with flat sidelobes in the PSF by correlation with a function similar to, but not identical with, the mask pattern. These are indicated by $\mathcal{G}$ in the notes.

2.   Notes

The following notes are in approximate chronological sequence.

(a) Fresnel zone plate. ‡ As proposed by Mertz and Young (1961), who suggested an optical method for image reconstruction. An infinite pattern with true $Cos(r^2)$ transmission would give a δ-function.

(b,c) Random scatter holes. True random positions or regular grid (Ables, 1968; Dicke, 1968).

(k) 'Optimum' patterns of Gunson and Polychronopulos (1976). These authors pointed out that any cyclic difference set can be used as the basis of a mask pattern with optimum properties. The 2-d case was mentioned but not described explicitly until Proctor *et al.*, (1979). The example shown is based on an $m = 2$ Singer cyclic difference set.

L. Bassani and G. di Cocco (eds.), Imaging in High Energy Astronomy, 177–182.
© 1995 *Kluwer Academic Publishers.*

(l) **Miyamoto (1977).** ‡ These designs are equivalent to the $m = 2$ Singer sets of Gunson and Polychronopulos, but with a different mapping onto 2-dimensions which results in a pattern which is not simply cyclic. The mask area corresponding to 4 times the area of one cyle is shown, alongside the cross-correlation of this with a zero-extended cycle.

(m) **URA.** This term was introduced by Fenimore and Cannon (1978) to describe patterns like those of Gunson and Polychronopulos based on twin-prime cyclic difference sets.

(d) **Geometric coded apertures** ($\mathcal{G}$) (Gourlay and Stephen, 1983) are generally square and self supporting. Correlation with a three-valued array closely related to the mask pattern gives an ideal PSF.

(n) **Giles (1981)** ‡ found patterns based on $m = 2$ Singer sets tesselated in the way proposed by Miyamoto, which could be made self-supporting. The area of mask shown and type of correlation are as in (l).

(o) **HURA.** Finger and Prince (1985) showed how to generate Hexagonal URAs by mapping cyclic difference sets onto a hexagonal grid.

(e) **Modified URA (MURA)** ($\mathcal{G}$). These were introduced by Gottesman and Fenimore (1989). Reconstruction by correlation with a two-valued array differing from the mask pattern by one element achieves flat side-lobes in the PSF.

(f) **Pseudo-noise products (PNPs)** ($\mathcal{G}$). (Gottesman and Schneid, 1986) are produced by multiplying two 1-dimensional patterns of type (k) or (l)

(p) **Perfect binary arrays (PBAs)** have been discussed by many authors (*e.g.* Bomer and Antweiler, 1987; Jedwab and Mitchell, 1988; Luke *et al.* 1989), but only recently considered for mask designs. They can be made square and have flat side-lobed ACFs.

(g) **M-P and M-M products** ($\mathcal{G}$) (Byard, 1992) are generalisations of PNP's (see (f)) where one or both of the 1-dimensional arrays are MURA's.

(h) **Two spatial scale** masks have been proposed in which one scale dominates at high energies and another at low (Skinner and Grindlay, 1993).

(i) **Triadic residue** ($\mathcal{G}$) arrays have been proposed by in't Zand *et al* (1994) as mask patterns with transparencies $\sim \frac{1}{3}$, which can offer advantages. Their properties are, however, only close to ideal, especially as complete sequences cannot be used as 2-d arrays.

(q) **Two dimensional cyclic difference** sets of Kopilovich and Sodin (1991, 1994) are effectively the same as the PBA's (p).

(j) **Generalised two dimensional cyclic difference sets** ($\mathcal{G}$) of Kopilovich and Sodin (1991, 1994) have 4-level autocorrelation functions.

References

Ables, J.G.: 1968, *Proc. Astron. Soc. Austr.* **1**, 172.

Byard, K.: 1992, Nucl. Instr. Meth. **A322**, 97.

Dicke, R.H.: 1968, *Ap. J.* **153**, L101.

Fenimore, E.E. and Cannon, T.M.: 1978, *Appl. Opt.* **17**, 337.

Finger, M.H. and Prince, T.A.: 1985, *19th Cos. Ray Conf. (La Jolla)* **OG 9.2-1**, 295.

Giles, A.B.: 1981, *Appl. Opt.* **20**, 3068.

Gottesman, S.R., and Fenimore, E.E.: 1989, *Appl. Opt.* **28**, 4344.

Gottesman, S.R., and Schneid, E.J.: 1986, *I.E.E.E. Trans. Nucl. Sci.* **NS 33**, 745.

Gourlay, A.R., and Stephen, J.B.: 1983, *Appl. Opt.* **22**, 4042.

Gunson, J., and Polychronopulos, B.: 1976, *Mon. Not. R. Astr. Soc.* **177**, 485.

in't Zand, J.J.M., Heise, J. and Jager, R.: 1994, *Astron. & Astrophys.* **288**, 665.

Jedwab, J. and Mitchell, C.: 1988, *Electronics Lett.* **24**, 650.

Kopilovich, L.E. and Sodin, L.G.: 1991, *Proc. I.E.E. -H* **138**, 233.

Kopilovich, L.E. and Sodin, L.G.: 1994, *Mon. Not. R. Astr. Soc.* **266**, 357.

Luke, H.D., Bömer, L. and Antweiler, M.: 1989, *Signal Processing* **17**, 69.

Mertz, L. and Young, N.O.: 1961, *Proc. Int. Conf. on Optical Instr. Tech*, Chapman and Hall, London, 305.

Miyamoto, S.: 1977, *Space Sci. Instr.* **3**, 473.

Proctor, R.J., Skinner, G.K. and Willmore, A.P.: 1979, *Mon. Not. R. Astr. Soc.* **187**, 633.

Skinner, G.K. and Grindlay, J.E.: 1993, *Astron. Astrophys.* **276**, 673.

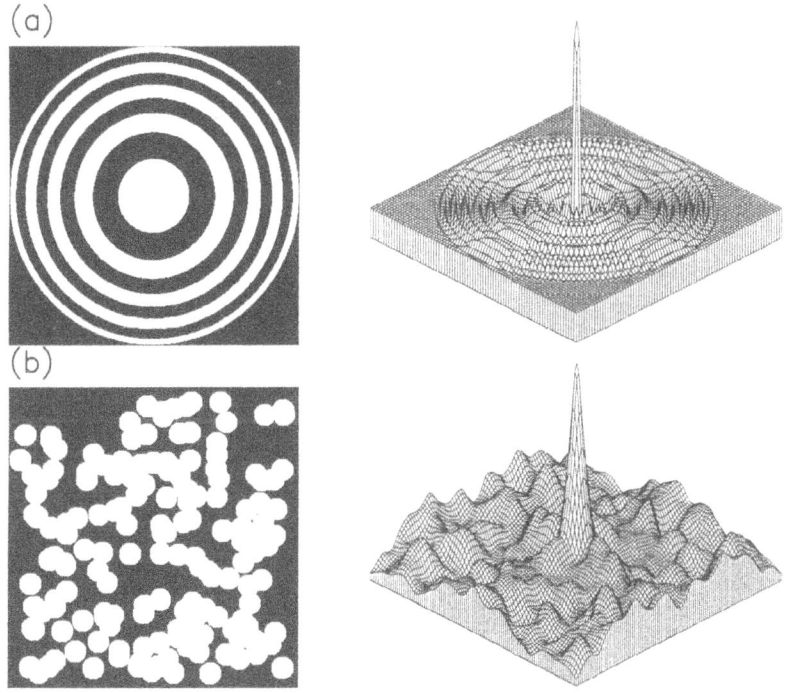

Figure 1. Patterns which have been proposed for coded mask telescopes. (a–i) Patterns with non-flat ACFs; (k–q) patterns with flat ACFs or PSFs.

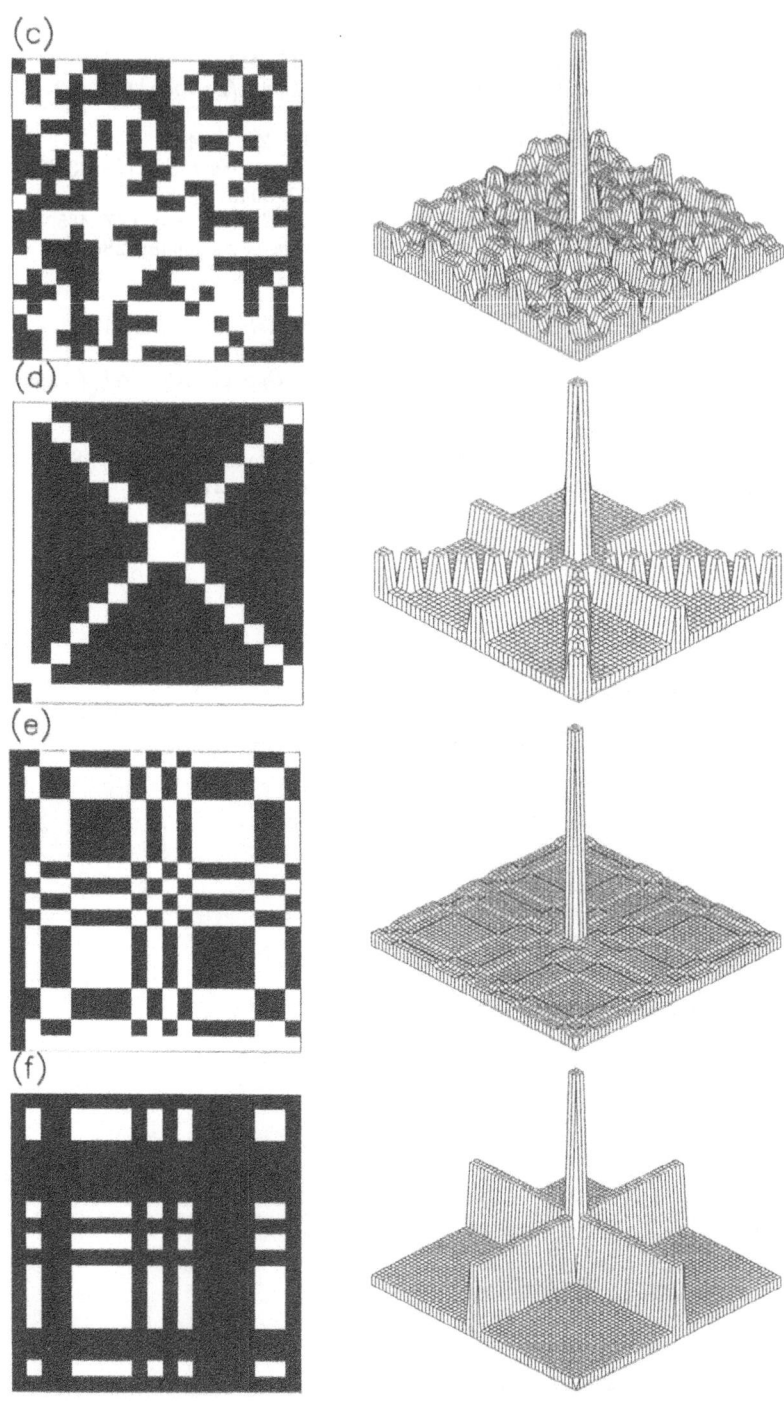

Figure 1. Continued

Figure 1. Continued

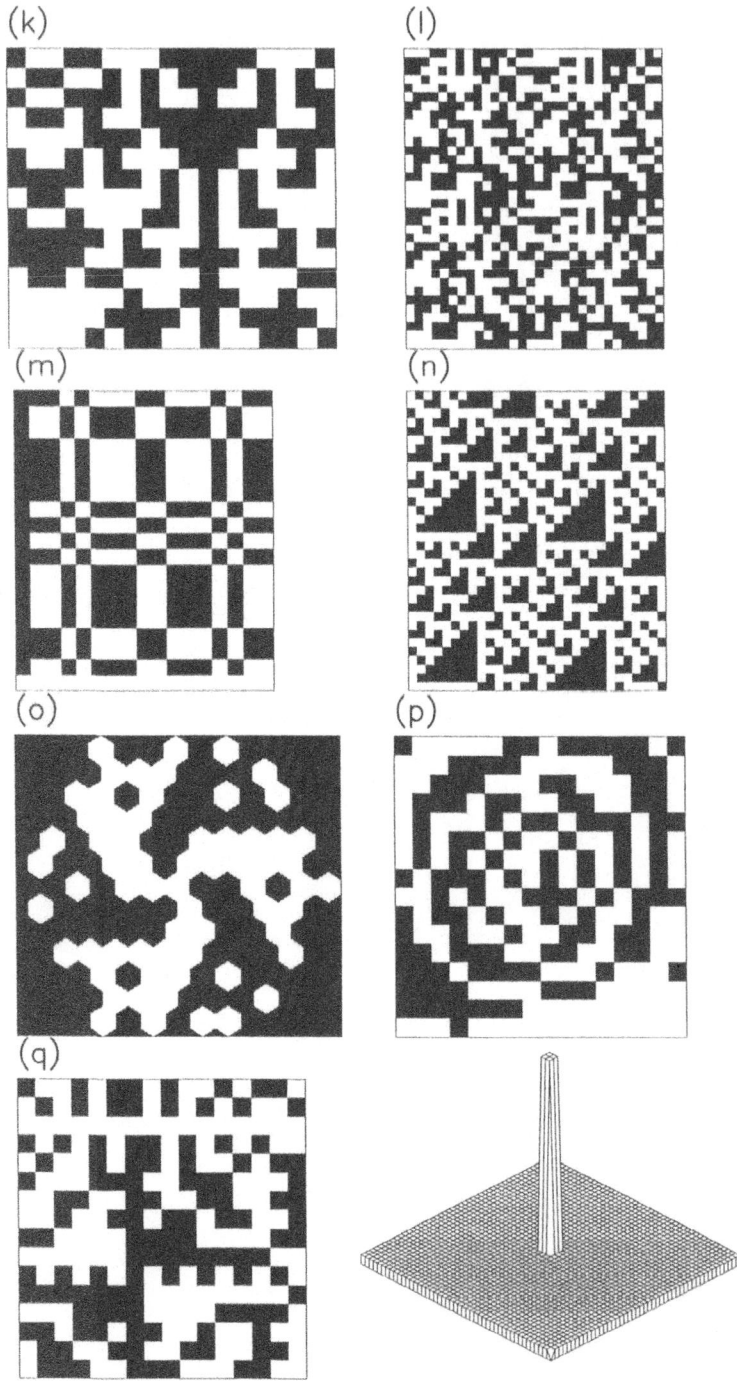

Figure 1. Continued

LOW ENERGY GAMMA RAY IMAGER (LEGRI)

BALLESTEROS F., BERNABEU G., GIMÉNEZ A., FABREGAT J., PÉREZ F., and
PORRAS E., REGLERO V., REIG P., ROBERT A., SÁENZ A., SÁNCHEZ F.
Universidad de Valencia

DÍAZ E., MARTÍN J.A., MÁS J.M., REINA M., SÁNCHEZ A., TORRALBO I.
Instituto Nacional de Técnicas Aeroespaciales "Esteban Terradas" (INTA)

GONZÁLEZ M., MARÍN J., MARTÍNEZ L., OLMOS P., PÉREZ J.M.
Centro de Investigaciones Energéticas Medioambientales y Tecnológicas (CIEMAT)

CRUISE M., SWINYARD B.M.
Rutherford Appleton Laboratory (RAL)

EYLES C., SKINNER G.
University of Birmingham

and

BIRD A.J., DEAN A.J., GHUATAURE H.S.
University of Southampton

May 10, 1995

Abstract. The Low Energy Gamma Ray Imager (LEGRI) will be one of the three instruments carried by the first MINISAT mission. LEGRI aims to demonstrate the technological feasibility of a new generation of low energy gamma-ray telescopes with imaging, medium resolution spectroscopy and high continuum sensitivity in the 20-200 keV spectral region, based on HgI_2 solid state detector technology.

The data supplied by LEGRI will allow us to investigate HgI_2 detectors in astrophysics by examining semiconductor stability, radiation damage and survival capability under space conditions. Furthermore, LEGRI will provide a good scientific output in the very interesting low energy γ-ray/hard X-ray domain.

LEGRI was accepted by the MINISAT consortium in March 1993. The delivery is scheduled by September 1995 and the MINISAT-01 launch at the mid of 1996 with a nominal life of 2 years.

1. Scientific and technological background

Imaging combined with high continuum sensitivity is a prime requirement for high energy astronomy. Accurate positioning of the gamma-ray point sources and imaging for the extended sources have been identified as key issues for the GRO and SIGMA successors.

Precise location of the gamma-ray emitters is crucial to identify X-ray, optical, infrared and radio counterparts, while mapping diffuse Galactic emission and extended sources will be crucial in understanding the interstellar processes working in space and the injection of heavier elements by the supernova explosions. Extragalactic gamma-ray astronomy is also potentially a very rewarding discipline. Accurate measurements of gamma-ray

183

L. Bassani and G. di Cocco (eds.), Imaging in High Energy Astronomy, 183–187.
© 1995 *Kluwer Academic Publishers.*

AGN's and their relation with X-ray extragalatic emitters is a priority for future high energy missions. The extension of AGN spectra from the X-ray range to higher energies is fundamental to the understanding of the nature of powerful mechanisms acting on these objects.

In this framework the 20-200 keV spectral region has a particular importance. Unique astrophysical information regarding nuclear excitation process, radioactivity, cyclotron emission and line formation is contained in this region of the electromagnetic spectrum. For many astronomical objects it is very important to determine the end of their X-ray tails and the possible extension of their spectra to gamma-ray energies. Despite the importance of this energy range, a significant data gap actually appears at the 20-200 keV region. Efficiencies of currently available and new generation CCD X-ray detectors drop abruptly at 20 keV and many scintillator based instruments have been limited to operation above 100 keV. LEGRI aims to fill in this gap performing high sensitivity measurements and continuous monitoring of hard X-ray/low energy gamma emitters.

HgI_2 detectors will be used to fill in this vital energy gap. HgI_2 presents a greater resistance to radiation damage than other semiconductors which can be used in this spectral range such as Ge, Si and CdTe. They have been tested under fluences of 40 keV photons with counting rates up to 10^{12} ph cm^{-2} sec^{-1} without any significant deterioration [1]. It has also been demonstrated that HgI_2 detectors can endure doses of 10^{14} neutrons cm^{-2} and 10^{12} protons cm^{-2} of 10 MeV without suffering any damage. The above referred resistance to proton radiation damage is particularly important for space borne gamma-ray instrumentation which may have to pass through regions of extreme high energy particle fluxes.

Detectors based on HgI_2 technology do not show evidence of internal worsening over periods of seven years, working efficiently at room temperature. They provide medium energy resolution with a spectroscopic performance which is intermediate between Ge and scintillation devices while avoiding the complexity of the cooling systems necessary for Ge. For all these reasons, low energy gamma-ray detectors based on this technology are extremely promising devices to develope low gamma-ray astronomy in the space.

The technology and expertise required to fabricate HgI_2 crystals are not extensively available. A few laboratories in the world can produce suitable sensors, mainly for laboratory purposes. Over the last five years, the Centro de Investigaciones Energéticas, Medioambientales y Tecnológicas (CIEMAT) in Spain, has carried out a development program on this material with successful results and a production line is now available.

2.   The Low Energy Gamma Ray Imager

2.1   LEGRI PERFORMANCE

LEGRI on board of the MINISAT-01 (INTA) mission has been partially conceived as a technology demonstration model for a new generation of gamma-ray telescopes based on the use of solid state HgI_2 detectors.

·LEGRI is a gamma-ray imager with moderate resolution (2.2°) and good point source location capability (20'), providing medium resolution spectroscopy (4 keV at 30 keV) and high continuum and broad line sensitivity at 20-200 keV. The estimated continuum sensitivity of 3 mCrab at 30 keV will allow survey to be made of selected areas and deep measurements of specific fields. A summary of LEGRI key performance parameters can be found in Table 1. In Figure 1 we present the estimated continuos LEGRI sensitivity by comparison with the currently operational space missions SIGMA and OSSE [2,3].

ENERGY RANGE	20-200 keV
CONTINUUM SENSITIVITY	3 mCrab at 30 keV and 10^5 s (3σ)
SPECTRAL RESOLUTION	E/ΔE = 8% at 30 keV
ANGULAR RESOLUTION	2.2°
POINT SOURCE LOCATION CAPABILITY (PSLC)	20'
FULLY CODED FOV	11°
DETECTORS	10x10 1cm^2 HgI$_2$ crystals array 0.5 mm thick
MASS	30 kg.
POWER	20 W

TABLE 1. LEGRI key performance parameters.

Space does not permit a more detailed discussion of the expected LEGRI scientific scope but the wide range of astronomical objects that can be studied includes pulsars, black hole candidates, X-ray, binaries, supernovae and AGN.

2.2   LEGRI CONSTRUCTION

The LEGRI system will consist of the following units:

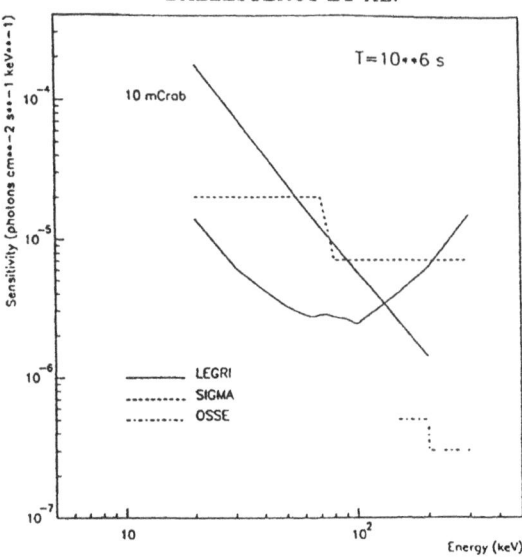

Fig. 1. LEGRI sensitivity at 3σ. Comparison with SIGMA (GRANAT) and OSSE (GRO).

- Detector Unit: This is a position-sensitive gamma-ray detector consisting of an array of 100 HgI_2 detection elements, each 0.5 mm thick, arranged on a 12x12 mm^2 square grid; the associated Front End Electronics, based on a 16 channel low-noise preamplifier chip, using gate array technology and a commercial ADC; a mechanical collimator made of tantalum; and a main mechanical assembly with passive shielding.

- Mask Unit: located at 540 mm from the detector plane and parallel to it, consisting of a static coded aperture mask made from 24x24 mm^2 tungsten elements, 1 mm thick, within a carbon fibre honeycomb plate and support structure.

- Power Supply and Digital Processing Units: which will provide interfaces to the PLM for power, commands and data transmission.

- High Voltage Unit: This will supply the high voltage (500 V) needed to operate the detectors.

- Star Sensor: which will be used to determine with sufficient precision the satellite attitude , allowing for the reconstruction of the gamma-ray images on the ground without spatial blurring caused by platform drift or jitter.

- A Science Operation Center which will be located at Valencia.

In Fig. 2 we show LEGRI's location on the MINISAT-01 payload module, together with the other instruments which will complete this first MINISAT mission.

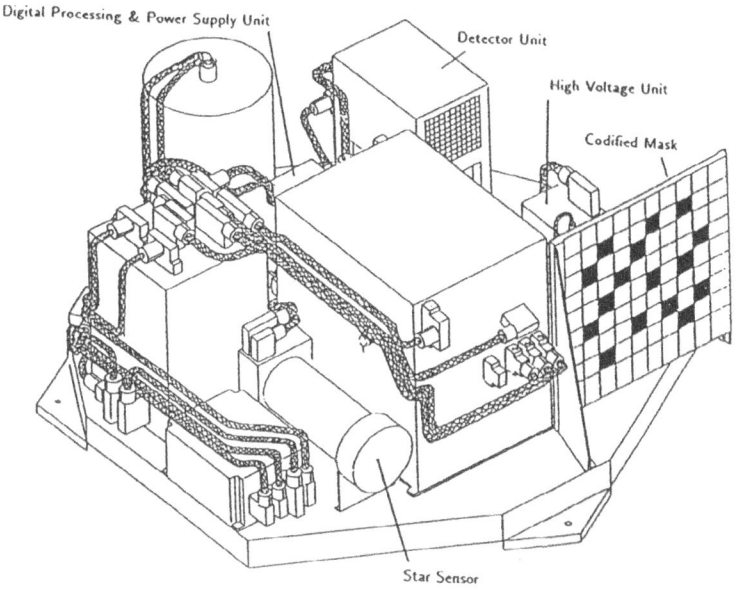

Fig. 2.   LEGRI's location on the MINISAT PLM

2.3   LEGRI COLLABORATION

LEGRI is being designed and built by a consortium including the following Universities and Laboratories:

University of Valencia/CSIC/University of Alicante
    Detector Mechanical Assembly and Integration, Coded Mask Assembly, Science Operation Center (SOC)

CIEMAT: HgI_2 Manufacturing and Testing, Detector Matrix Assembly

INTA: Management, Thermal Control, Integration and Test

RAL: Front End Electronics, High Voltage Unit, Star Sensor

University of Birmingham: Digital Processing Unit, Power Supply Unit

University of Southampton: Electrical Ground Support Equipement

This work has been partially supported by the IMPIVA.

References

[1] Pérez J.M.: 1990, 'Ph D Thesis', *CIEMAT. Madrid* ,
[2] Bouchet L.: 1992, 'Ph D Thesis', *Université Paul Sabatier. Toulouse* ,
[3] Bergeson-Willis, S. et al.: 1993, 'INTEGRAL Phase-A Report', *ESA SCI(93)I* ,

ADVANCED TECHNIQUES OF IMAGE CORRECTION

FOR THE CODED MASK TELESCOPE SIGMA

M-C. Schmitz-Fraysse and J. B. Stephen*
*Centre d'Etude Spatiale des Rayonnements,
BP 4346, 31029 Toulouse Cedex, France*

and

J. Ballet
*Service d'Astrophysique, Centre d'Etudes de Saclay,
91191 Gif-sur-yvette Cedex, France*

Abstract. The position sensitive detector of the SIGMA telescope, a variety of the Anger camera, suffers from intrinsic non uniformity defects. In order to maximize the signal to noise ratio of the observed gamma ray sources, the removal of the systematic defects should be nearly perfect.

In this paper we describe and give the achieved performances of some correction methods. In particular we describe an adaptative filtering method which is automatically adjusted for the best possible signal to noise ratio.

1. Introduction

The position sensitive detector, PSD, based on the Anger-type camera of the SIGMA coded aperture imaging telescope [1], [2], suffers from a position dependence of the linearity of the detector. The size of the pixel is variable and produces a variation in the pixel count rate of about 10% of the average value in the case of a uniform exposure [3]. The image reconstruction using a cross correlation method amplifies this variation. This level of non-uniformity implies that some kind of image correction is necessary before performing the deconvolution.

Herein, we present a review of the methods employed to correct them. Those methods may be used for other space experiments incorporating coded aperture imaging systems with γ-camera type PSD's.

---

* on leave from ITESRE/CNR via P. Gobetti 101, 40129 Bologna Italy

L. Bassani and G. di Cocco (eds.), Imaging in High Energy Astronomy, 189–193.

2. Necessity of Correction and Correction Method

The detector records an image P (Fig.1a) which is the superposition of all the mask shadows cast by the sources in the field of view (FOV). Denoting N_P the total number of counts in matrix P and N the number of pixels, the variance of counts in a flat image would be : $(\Delta P)^2 = \dfrac{N_P}{N}$. Since P is an image with a Poissonian distribution.

If the image P suffers from systematic effects, the variance becomes :

$$\left(\Delta P\right)^2 = \left(\Delta P_{stat}\right)^2 + \left(\Delta P_{sys}\right)^2 = \frac{N_P}{N}\left(1 + \alpha^2\frac{N_P}{N}\right) \quad (1)$$

where α is the systematic effect factor (α x 100 is the average percentage non-uniformity). If $\alpha = 0.1$ and the mean number of counts per pixel is 100, then the variance of the image P is multiplied by a factor of 2 with respect to the ideal Poissonian case.

According to Laudet and Roques (1987) [4], the variance of the reconstructed image $\tilde{W}$ of the observed sky, is still multiplied by a factor due to the non-uniformity effects. It is necessary to correct $\tilde{W}$ for these systematic effects in order to obtain the most uniform image.

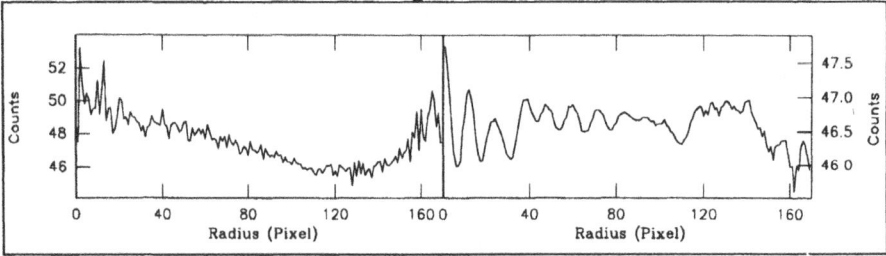

Fig.1: a) (left panel) the centre to edge profile of the image detector P; b) (right panel) application of the low-pass filter to the corrected image Pc.

The effects of non uniformity of the SIGMA telescope are periodically measured by observing sky regions without any source. A calibration matrix D, representing the state of the detector uniformity at a given time, is built and included in a database used to correct images recorded at anytime. The calibration matrix to use is chosen by the proximity with the image P to correct.

The corrected image, P_C is given by : $P_C(i,j) = P(i,j)\dfrac{N_C/N}{C(i,j)}\phi \quad (2)$

where C is an adaptation of the matrix D to P. The matrix D is smoothed by a low-pass filter with gaussian cut-off in the frequency domain [4] and convolved by the attitude drifts of P. N_C/N is the average number of counts and ϕ is a normalisation factor to ensure the flux conservation.

The application of the low-pass filter to the corrected image Pc to remove the statistical effects shows the residual systematic defects in the centre to edge profile of this matrix (Fig.1b).
To better correct these effects we must employ different filters. A test on many linear filters are described below.

3. The Filters

Two methods of deconvolution can be employed : the standard deconvolution called DCV where the image Pc is convolved by the matrix M which describes the aperture transmission, the deconvolution called PSF to which we add a convolution of a gaussian of which the sigma is adapted to the spatial resolution (2.61 pixels SI in 40-75 keV energy band). This new convolution allows to obtain the most significant flux and to eliminate the high frequency noise [5].
We have used the normalized variance of the reconstructed image as indicator. To compare the variance values, one must normalize the methods to obtain the same value at the theoretical source position.
Denoting Vth, Vfilter the theoretical variance (number of counts in the image Pc) and the variance after filtering respectively and F, Ffilter the flux of the theoretical source in the deconvolved image and the flux of this source after filtering respectively, the normalized variance V is then :

$$V = \frac{V_{filter} \times \left(\frac{F}{F_{filter}} \right)^2}{V_{th}} \qquad (3)$$

In the theoretical image case with a poissonian background only (no systematic effects), if the theoretical variance is 1.000 in the DCV image, it goes down to 0.465 in the PSF image with the spatial resolution in 40-75 keV energy band. In the ideal case, the use of filters always produces an increase of the variance in relation to the theoretical variance for DCV and PSF processes (see Fig.2 dashed line). But in the case of a real image with systematic effects the action of filters can produce a gain in the variance especially with the PSF process (see Fig.2 solid line). This difference is due to the convolution by the spatial resolution which enhances the systematic effects. And the filters compensate this effect.
We have investigated a range of filters which are listed in table I, as eight examples from 4 classes of linear high-pass filter. Their variable parameter, the cut-off frequency, allowes a better variance, thus obtaining a stable quality of the images.

TABLE I

Filter	Class	Order	Form
a	High Pass		1;w>wc, 0;w<wc
b	Triangular		1;w>wc, w/wc; w<wc
c, d, e	Exponential	1, 2, 3	$1-\exp(-(w/wc)^{order})$
f, g, h	Butterworth	1, 2, 3	$1-1/(1+(w/wc)^{order})$

There is a very little difference between the best variance produced by the various filtering techniques. Figure 2 shows the improvement in sigma (standard deviation equal to the square root of variance) for one particular image Pc. This adaptative filtering technique allows to work with a better sensibility level in the images.

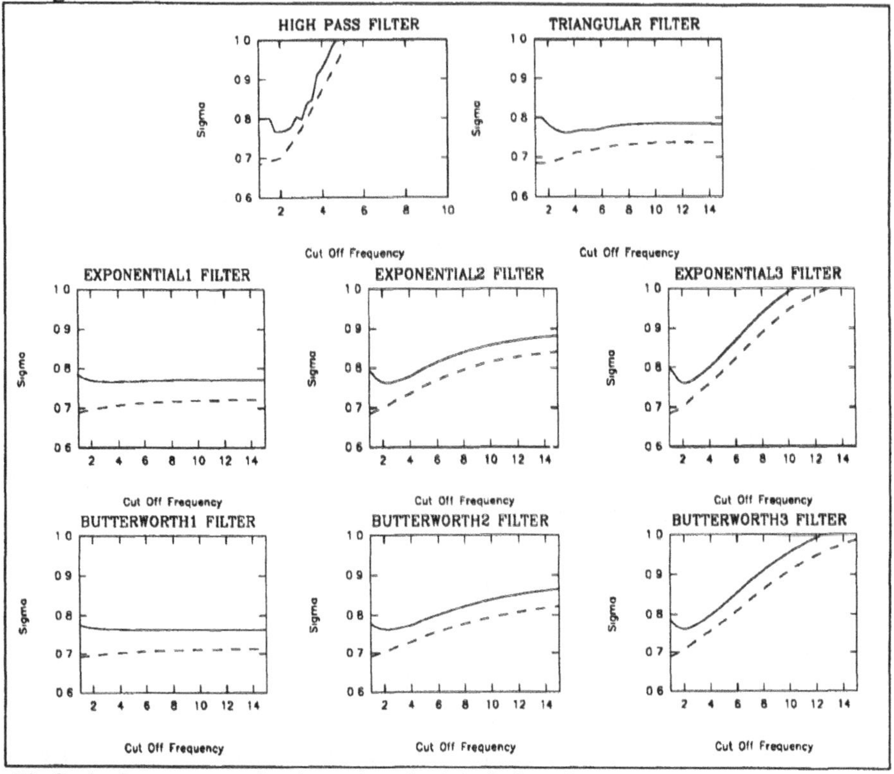

Fig.2: the improvement in sigma (standard deviation) for one particular image with the different filters (solid line) and the sigma for theoretical image without systematic effects (dashed line).

4. Results and Conclusion

In the detector image P the ratio of the systematic effects is around 3-6 %. After the uniformity correction this ratio is now of 1-2 %. For the case of the detector image P which undergoes the deconvolution process directly, the ratio is very important around 25-45 %. The action of the correction itself with the deconvolution process reduces it to 5-6 %.

We can conclude that the smallest loss of the ratio of the systematic effects in the image P implies a strong decrease of this ratio after the deconvolution process.

This fact is illustrated on figure 3 where the centre to edge profiles of the filter applied before the deconvolution process and after, respectively are represented .

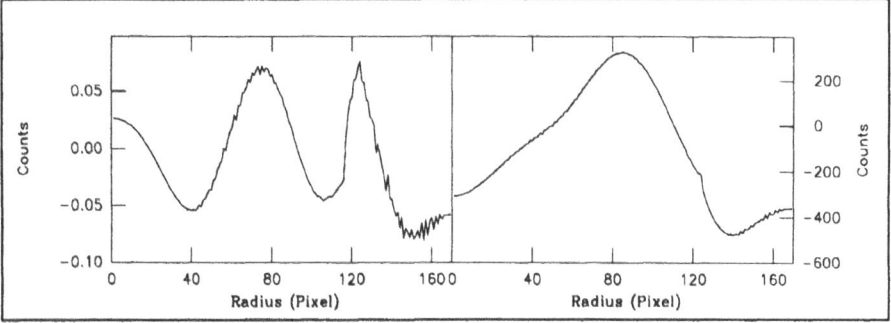

Fig.3: the centre to edge profile of the systematic effects suppressed in the image Pc with the filter exponential order 3 (left panel) and after PSF with the same filter (right panel)

We have tested the improvement of the adaptative filtering techniques on the SIGMA images. Two examples are described in following.

For the case of the session N°0154 with the source 4U1700-377, the value of the signal to noise, SNR, for the PSF process only is $1.58 \cdot 10^{-4}$. This ratio SNR becomes equal to $2.01 \cdot 10^{-4}$ with the use of the exponential order 3 filter corresponding to a gain of 27 % in the SNR. In this case the detection level of the source goes to 17.52σ, with the PSF process only, to 22.46σ, with the exponential order 3 filter.

For the case of a session with a very weak source, e.g. the session N°0529 with the source LMCX-3, the value of the SNR for the PSF process only is $1.56 \cdot 10^{-4}$. The best SNR for the exponential order 3 filter is $1.82 \cdot 10^{-4}$, 17 % more.

The variable linear filters give the best improvement with the PSF process on the image with or without a source. The application of the filters enables the research of the very weak sources where the systematic effects are very present. Thus, the adaptative technique allows to increase the level of detection in the image.

This work will be developped in a future paper [6].

5. References

[1] J.Paul et al., Adv. Space. Res. vol.11, N°8, 289-302, (1991).

[2] P.Mandrou et al., in : P.Durouchoux, N.Prantzos (Eds), Proc. Int. Symp. on Gamma line Astrophysics, Saclay, AIP Conf. Proc. 232, New York, p492, (1991).

[3] M.D.Short, Nucl. Instr. and Meth. 221, 142, (1984).

[4] P.Laudet and J.P.Roques, Nucl. Instr. and Meth. A267, 212-217, (1988).

[5] W.R. Cook et al., IEEE Tr. Nu. Sc., Vol. NS-31, N°1, (1984)

[6] M-C. Schmitz-Fraysse et al., in preparation.

THE SAX MISSION FOR WIDE BAND X-RAY ASTRONOMY

G. DI COCCO*

Istituto di Studio e Tecnologie sulle Radiazioni Extraterrestri

SAX Consortium

Istituto di Studio e Tecnologie sulle Radiazioni Extraterrestri, CNR, Bologna (ITeSRE)
Istituto Astrofisica Spaziale, CNR, Frascati, (IAS)
Istituto Fisica Cosmica e Tecnologie Relative, CNR IIFCTR) ed Unità GIFCO, Milano
Istituto di Fisica Cosmica ed Appl. Informatica (IFCAI) ed Unità GIFCO, Palermo
Istituto Osservatorio Astronomico, Università di Roma "La Sapienza"
Space Reasearch Institute of Utrecht/SRON (SRU), The Netherlands
Space Science Departments (SSD), ESA, Noordwijk, The Netherlands

Abstract. In the framework of past and future X-ray missions, the SAX satellite, to be launched in February 1996, stands out for its very wide spectral coverage from 0.1 to 200 keV, with well balanced performances of the low energy and high energy instrumentation. The sensitivity of the scientific payload will allow the exploitation of the full band of SAX also for weak sources (1/20 of 3C273), opening new perspectives in the study of spectral shape and variability of several classes of objects. SAX will produce X-ray imaging in the 0.1-10 keV energy range having on board concentrator-spectrometer system and in the 2-30 keV energy range via a coded mask proportional counter system. In this paper we describe the main aspects of the mission, the instruments and the scientific objectives.

Key words: X-Ray Telescope - High Energy Astrophysics

1. Outline of the mission

The study of the spectral behaviour of celestial sources over a wide range of energies is of primary importance in understanding comprehensively the emission mechanisms that, in several instances, produce spectral features localized in different regions of the electromagnetic spectrum.

The Italian-Dutch X-ray Satellite SAX will be the first X-ray mission that will have the capability of observing sources over more than three decades of energy - from 0.1 to 200 keV - with a relatively large area, a good energy resolution, associated with imaging capabilities (resolution of about 1') in the range of 0.1-10 keV.

The instrument complement dedicated to such purpose (fig 1) is composed of a medium energy (1-10 keV) concentrator optics/spectrometer, MECS, consisting of three units, a low energy (0.1-10 keV) concentrator

* On behalf of the SAX Consortium

L. Bassani and G. di Cocco (eds.), Imaging in High Energy Astronomy, 195–202.
© 1995 *Kluwer Academic Publishers.*

optics/spectrometer, LECS, a high pressure gas scintillation proportional counter (3-120 keV), HPGSPC, and a phoswich detector system (15-300 keV), PDS, all of which have narrow fields and point in the same direction (Narrow Field Instruments, NFI).

The other characterization of the mission is its capability of monitoring large regions of the sky with a resolution of 5' in the range 2-30 keV to study long term variability of sources down to 1 mCrab and to detect X-ray transient phenomena. This is realized by means of two coded mask proportional counters (Wide Field Cameras, WFC) pointing in diametrically opposed directions perpendicular to the NFI (fig. 1). Finally, the anticoincidence scintillator shields of the PDS will be used as a gamma-ray burst monitor in the range 60-600 keV.

SAX will be launched by an ATLAS G-Centaur into a 600 km orbit at 3 degrees inclination in February-March 1996. During each orbit up to 450 Mbits of data will be stored onboard and relayed to ground during station passage. The average data rate available to instruments will be about 60 kbits/s, but peak rates of up to 100 kbits/s can be retained. The satellite will pass above the ground station, placed near the equator in Malindi, every orbit. This will allow a prompt operation and control the satellite.

The SAX mission is a major joint program of the Italian Space Agency (ASI) and the Netherlands Agency for Space Programs (NIVR). Prime contractors for space and ground segments are Alenia and Telespazio respectively. The mission development is supported by a consortium of institutes in Italy together with institutes in The Netherlands and the Space Department of ESA. A collaboration with the Max Planck Institute for Extraterrestrial Physics also exists for X-ray mirror testing and the calibration of the concentrator/spectrometer system.

2.   The instruments

The main characteristics of the SAX instruments, (see also Butler and Scarsi 1990 and references therein) are given in table 1 and figure 2.

3.   The Concentrator/Spectrometer system

The concentrator/spectrometer system consists of four separate concentrator mirror assemblies with a focal length of 185 cm, each one with a position sensitive, Xenon filled, GSPC in the focal plane, three covering the 1-10 keV range (ME), the fourth extending the range down to 0.1 keV (LE).

The mirror assemblies are composed of 30 nested gold coated, confocal double cone approximation to a Woltjer 1 configuration. X-ray tests on the prototype mirrors (Citterio et al. 1990, Conti et al. 1993, 1995) and on the flight unit in Panter have shown a resolution of 40" (HPR), better than the design goal of 1 arcmin and confirmed the prediction on the effective area (fig 2). The concentrators are designed to maximize their effective area around 7 keV for studies of iron line.

The four focal plane detectors are imaging Gas Scintillation Proportional Counters. Three of the GSPC's have 50 micron beryllium windows, thus being sensitive in the range of 1-10 keV (ME). The fourth (LE) GSPC developed by SSD/ESTEC (Parmar et al. 1990) is similar to the ME, except that a thin window (Bavdaz et al. 1994) allows transmission of low-energy X-rays into a detector.

4.   The high energy instruments

The high energy instruments onboard SAX are respectively a 5 atm Xenon filled Gas Scintillation Proportional Counter (High Pressure GSPC) and a NaI/CsI phoswich scintillator (Phoswich Detector System, PDS).

Great care has been taken over the design so to ensure the suppression of the main sources of systematic effects. Both instruments adopt the technique of rocking collimators, continuously pointing on and off source (with a period of the order of a minute) to monitor the background.

The High Pressure Gas Scintillation Proportional Counter is a new development of a GSPC filled with 5 atm. Xenon/He mixture, which allows the detector to be sensitive up to 120 keV. The cell of the HPGSPC is seen by an Anger camera arrangement of seven PMT. The event position is used to correct the event energy with an overall improvement of a factor of two in energy resolution compared to proportional counters (Giarrusso et al 1989).

The Phoswich Detector System (PDS) consists of a square array of four independent NaI(Tl)/CsI(Na) phoswich scintillation detectors. Each of the four detectors is made of two crystals of NaI(Tl) and CsI(Na) optically coupled and forming what is known as PHOSWICH (acronym of PHOsphor and sandWICH). The scintillation light produced in each phoswich is viewed, through a light guide of quartz, by a photomultiplier tube (PMT). The NaI(Tl) acts as X-ray detector, while the CsI(Na) scintillator acts as an active shield.

In order to further reduce the phoswhich background level, both a lateral and a top anticoincidence shields are provided. The lateral shielding system is made of 4 CsI(Na) scintillators 10 mm thick. The top shield is made of an organic scintillator 1 mm thick.

The collimators can be independently rocked back and forth to allow the simultaneous monitoring of the source and background. Finally a particle monitor provides information on high environmental particle fluxes and triggers the action to prevent the damage of the detectors PMT's (Frontera et al. 1991).

5.   The Wide Field Cameras

The two identical Wide Field Cameras (Jager et al. 1989) consist of two position sensitive proportional counters filled with 2 atm of Xenon, which view the sky through a coded mask perpendicularly to the axis of the NFI (fig 1) and 180 degrees away from each other. They combine a large field of view ($20° \times 20°$ per unit) with an angular resolution of 5 arcmin (FWHM). The operative energy band ranges from 1.8 to 30 keV with a resolution of 18% at 6 keV. The effective area of one camera is shown in figure 2. The mask pattern is square, with 256×256 elements each of 1 mm^2. The open fraction of the pattern is 33% (effective 26.4%), that increases the S/N ratio for faint sources.

The sensitivity depends mainly on the pointing direction in the sky, because each source in the FOV contributes to the overall background. Towards high-galactic latitude the Cosmic Diffuse X-Ray Background is the main contributor to the background. In this case the sensitivity is of the order of a few mCrab in 10^4 s.

This will allow the monitoring of faint sources like AGN, along with their primary objective, the survey of the galactic plane and the search for X-ray transients for follow-up studies with the narrow field instruments.

On the basis of the logN-logS distribution of gamma-ray bursts and assuming fx about 1/100 of fγ we expect to detect about 3 X-ray counterparts to gamma-ray bursts, thus positioning the events within 5' and gathering broad band information with the simultaneous observation of the gamma-ray burst monitor.

6.   Scientific Program

6.1   SCIENTIFIC OBJECTIVES

The primary characteristics of SAX in comparison with past and future missions is its very wide spectral coverage, and the well balanced performances between the low-medium (0.1-10 keV) and medium-high (1-200 keV) energy bands. In this respect the example shown in figure 3 is very enlightening. It shows the complex spectrum of a Seyfert 1 galaxy, namely $MCG-6-30-15$, observed by SAX-NFI in 40000 s with all the spectral component and features detected by several satellites in the past. Starting from the low energy part there is: a soft excess observed by EXOSAT (Pounds et al. 1986), a OVII edge around 0.8 keV observed by ROSAT (Nandra and Pounds 1992) and ASCA (Fabian et al. 1994), an iron line at 6.4 keV and a high energy bump above 10 keV detected by GINGA (Matsuoka et al. 1990). All those components can be measured with good accuracy by SAX in a single shot for the first time.

Given the instrumental capabilities over the wide energy range described in the previous sections, SAX can provide a significant contribution (and unique contribution for science involving the exploitation of the wide band) in several areas of X-ray astronomy such as compact galactic sources, active galactic nuclei, clusters of galaxies, supernova remnants, normal galaxies, stars, gamma-ray bursts.

With its minimum lifetime of two years (extendible to four years), SAX will be able to perform more than 2000 pointings. While the NFI will be the prime instruments most of the time, the WFC will be periodically used to scan the galactic plane to monitor the temporal behaviour of sources above 1 mCrab and to detect transient phenomena. Thanks to their large field of view, the WFC will be operated to monitor selected objects when the NFI perform their sequence of pointed observations. We expect to detect about 10-20 bright X-ray transients per year during the WFC observations.

Acknowledgements

The SAX program is the result of the work of many people and it is impossible to thank them all here. A particular thought to the memory of the project manager Massimo Casciola. Part of this paper was based on the contributions to the SAX Observers Handbook by L. Piro, C. Butler, L. Chiappetti, G. Conti, E. Costa, D. Dal Fiume, F. Frontera, S. Giarrusso, R. Jager, C. Maccarone, G. Manzo, G. Matt, T. Mineo, A. Parmar, G.C. Perola, S. Re, B. Sacco, L. Salotti, A. Santangelo. Finally I wish to acknowledge the contributions to the mission of J. Bleeker, G. Boella, B. Taylor and L. Scarsi; F. Favata and M. Trifoglio for the Ground Segment activities; G. Manarini and B. Negri of ASI.

References

Bavdaz, M. et al.: 1994, *Nucl. Instr. & Methods* **345**, 549
Butler, C., and Scarsi, L.: 1990, *SPIE proceedings* **1344**, 464
Citterio, O., et al.: 1990, *Il Nuovo Cimento* **13C** 2, 375
Conti, G., et al.: 1993, *SPIE proceedings* **2011**, 118
Conti, G., et al.: 1995, *SPIE proceedings*, in press
Fabian, A. et al.: 1994, *PASJ* **46**, L59
Frontera, F., et al.: 1991, *Adv in Space Res.* **11**, 281
Giarrusso, S., et al. 1989, *SPIE proceedings* **1159**, 514
Jager, R., et al.: 1989, *SPIE proceedings* **1159**, 2
Matsuoka, M., Piro, L., Yamauchi, M., Makishima, M.: 1990, *ApJ* **361**, 440
Nandra, P., and Pounds, K.: 1992, *Nature* **359**, 215
Parmar, A., Smith, A., Bavdaz, M.: 1990, in *Observatories in Earth orbit and Beyond*, p. 456, Y. Kondo ed., Kluwer
Pounds, et al.: 1986, *MNRAS* **221**, 7P

TABLE I

SAX Instruments

Instrument	Range (keV)	FOV (°FWHM)	Angul. Res. (arcmin)	Area (cm^2)	Energy Res. (% FWHM)
Concentrators					
1 LECS	0.1-10	0.5	$1.5\frac{E}{6}^{-0.5}$	20 @0.25 keV	$8\frac{E}{6}^{-0.5}$
3 MECS	1-10	"		150 @6 keV	
HPGSPC	3-120	1(coll.)	-	450 (geom.)	$4\frac{E}{60}^{-0.5}$
PDS	15-300	1.5(coll.)	-	800 (geom.)	$18\frac{E}{60}^{-0.5}$
2 WFC	2-30	$20 \times 20^{\dagger}$	5	250† (through mask)	$18\frac{E}{6}^{-0.5}$

TABLE II

Optics characteristics (per unit)

Configuration	Double cone
Focal	185 cm
geometrical area	124 cm^2
n. mirrors	30
mirror length	30 cm
grazing angle external grazing angle inner	0.62° 0.23°
micro roughness Resolution (HPR)	$\leq 10\text{Å}$ 1 arcmin (design) 40" FM (measured)
Effective area measure on FM	84 cm^2 @ 0.3 keV 81 cm^2 @ 0.9 keV 57 cm^2 @ 6.4 keV 38 cm^2 @ 8 keV

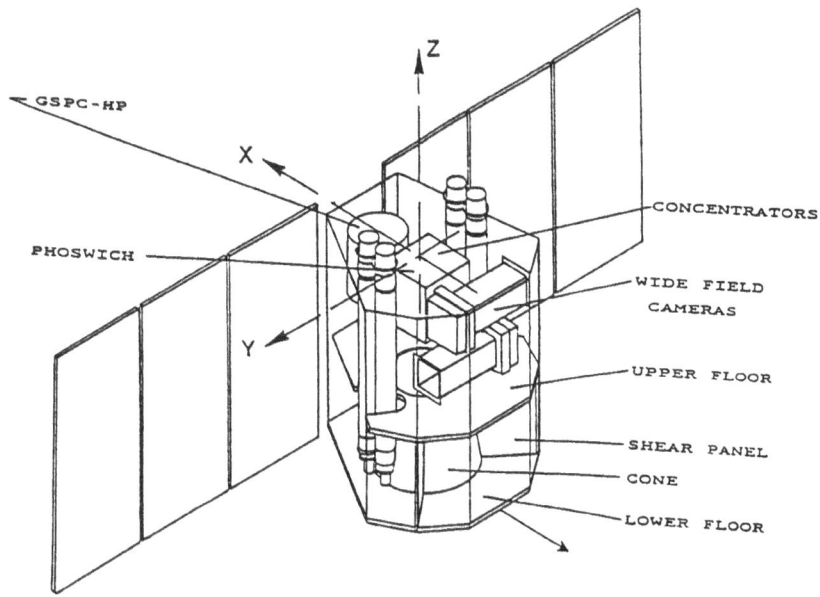

Figure 1: SAX P/L accomodation.

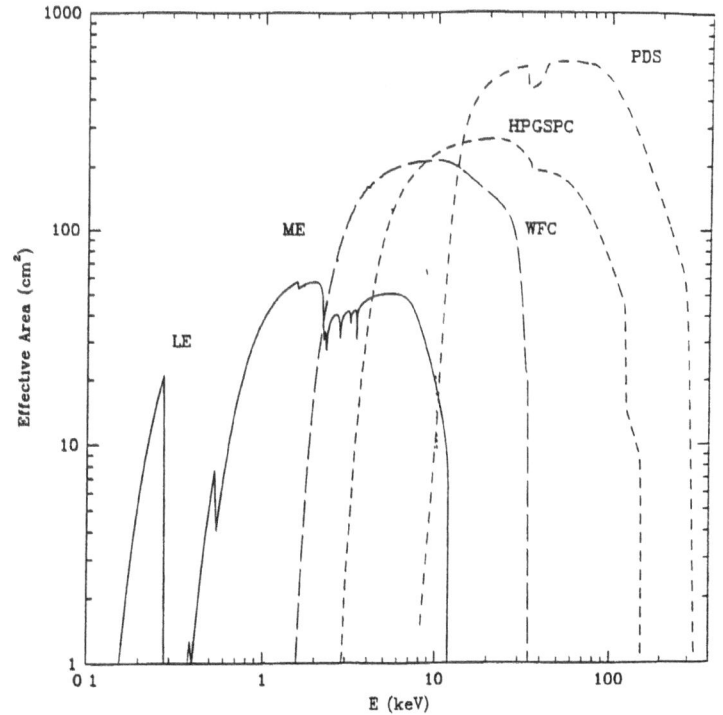

Figure 2 Effective area of the SAX instruments

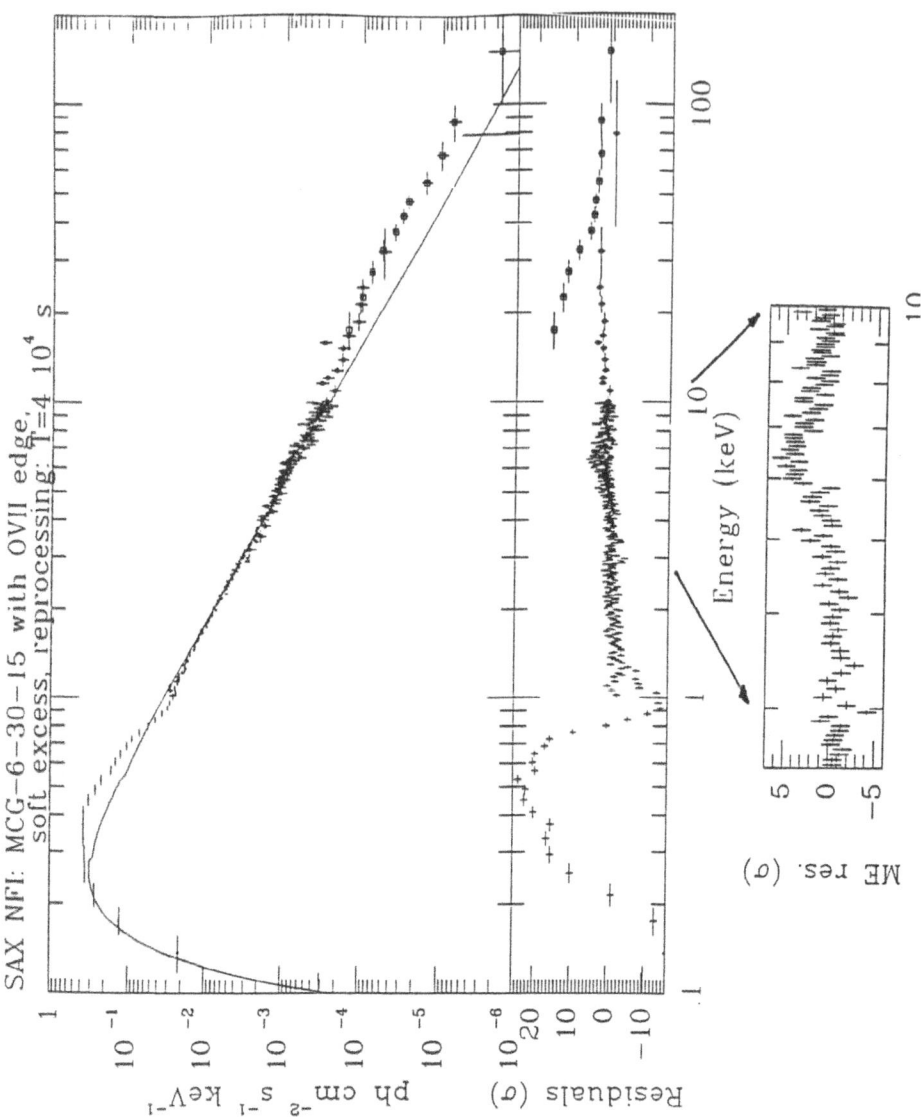

Figure 3: Simulation of an observation by SAX-NFI of a Seyfert 1 galaxy (MCG-6-30-16) with several of the features observed in the last 10 years by different satellites: a soft excess below 1 keV (EXOSAT), an edge of ionized oxygen at 0.8 keV (ROSAT, ASCA), a broad iron line (FWHM = 0.7 keV) at 6.4 keV and a high energy bump between 10 and 200 keV (GINGA). The spectrum is fitted with a simple power law and the residuals clearly show all those components. In the blow-up in the bottom of the figure the residuals of the MECS around the broad iron line are plotted, showing that the line is well resolved by the detector.

OPTIMISED SAMPLING FOR HEXAGONAL ARRAY
CODED MASK TELESCOPES

I. D. JUPP, A. R. GREEN and A. J. DEAN

Astronomy Group, Physics Department, University of Southampton,
Southampton, SO17 1BJ ENGLAND

Abstract. The sensitivity of hexagonal geometry, gamma-ray coded aperture telescopes has been studied in order to investigate the trade-off between the nominal angular resolution and the mean reconstructed image signal to noise ratio (SNR) that occurs when a pixellated detector array is used and images are retrieved by a simple correlation analysis. Without high fine sampling of the coded mask by the detector plane, the image SNR is seriously compromised if the source under observation lies close to a sky pixel boundary. Increasing the integer sampling partially restores the image SNR, but at the expense of the angular resolution. A method of improving the image SNR across sky pixel boundaries has been investigated, resulting in sky images that are free from the SNR modulation encountered with integer fine sampling of the shadowgram, and without any significant loss in angular resolution.

Key words: γ-ray imaging, coded masks

1. Introduction

The position sensitive detectors used for astronomical imaging in the γ-ray domain can take on a number of forms, including the Anger camera, scintillation crystal bars, and multiwire gas proportional counters. A recent modification of the scintillation bar arrangement is the detector that employs a stacked array of scintillation crystal-photodiode pixels [1]. It is the sampling configuration of this type of detector array that is investigated in the following work. The emphasis of the study is the quantification of the imaging performance available when using such a detector with reference to the variations in image quality when different sampling configurations are employed. It should be noted that although the correlation analysis used here is not necessarily the optimum method of image construction, it does have the advantage of requiring very little CPU time, and for the coded masks implemented here it is a reliable method of obtaining the first approximation to the true sky distribution prior to more rigorous analysis. The optimisation of the available angular resolution of a coded mask telescope is dependent on several factors one of which is the positional resolution of the detector *i.e.* the spatial sampling frequency. At low incident photon energies where single-site photoelectric events are the dominant interactions, the pixelated detector provides sampling of the shadowgram with a positional resolution equal to the pixel size. At higher energies as Compton scattering becomes increasingly dominant, multiple scattering of the incident photons increases the probablity of incorrect incident pixel reconstruction, and hence the

203

L. Bassani and G. di Cocco (eds.), Imaging in High Energy Astronomy, 203–207.
© *1995 Kluwer Academic Publishers.*

effective sampling ratio varies with source photon energy. The effects of this variation in sampling size on the instrument sensitivity are presented else-where.

2.   Problems Arising from Non-Optimised Sampling

It has been demonstrated [2],[3] for square geometries that as a result of both a poor sampling configuration, and a low mask element to detector pixel area ratio, phasing errors arise at any source position where the result-ing shadowgram falls across detector pixel boundaries. Since the source flux is binned into pixel sized samples, upon image reconstruction the source count is blurred between several sky pixels. For a single point source in the fully coded field of view (FCFOV) both the total flux reconstructed and the pixel to pixel statistical variations remain unchanged with source position. Consequently sources may be 'lost' if they happen to lie at unfavourable sky pixel positions because of the degraded peak height. For integer fine sampling configurations where an exact number of detector pixels sample each mask element, this phasing error can be supressed and a more uniform response achieved by increasing the sampling. Problems arise however if there are restrictions on the minimum detector pixel size - employing a higher sam-pling ratio results in a loss in angular resolution. Similarly it has been shown [4] that the use of hexagonal element uniformly redundant arrays (HURAs) with discrete pixel detectors introduces an intrinsic systematic coding noise producing non-zero sidelobes in the system point spread function (SPSF) unless specific sampling configurations are employed. The specific sampling orientations that were identified are equivalent to the integer fine sampling configurations used by [2] with square geometries. Although these orienta-tions enable the fast finely sampled cross-correlation algorithm to be used, and produce images free from the coding noise inherent at other mask orien-tations, the severe degradation in SNR towards sky pixel boundaries is still present. The specific mask orientations described by [4] also suffer from a second loss in sensitivity irregardless of the source position resulting from a degraded peak height. It should at this point be noted that the latter sensi-tivity loss can easily be rectified by tesselating the mask pattern onto a unit element that is divisible by a detector pixel. The properties of the HURA are retained, as are the features of the SPSF, which retains the central spike and flat side-lobes, and in particular the angular resolution remains unchanged. Figure 1. shows an example of a mask pattern with the improved tesselation elements. As fig. 2 suggests however, unfavourable sky pixel positions remain a potential loss in sensitivity.   Outlined now is a method of improving the sampling configuration in this situation, providing a more uniform response to all sky positions without significant loss in angular resolution.

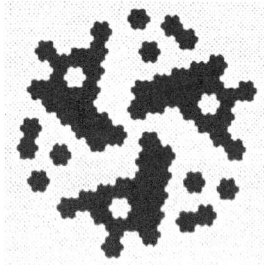

Fig. 1.   An example mask pattern with improved element tesselation

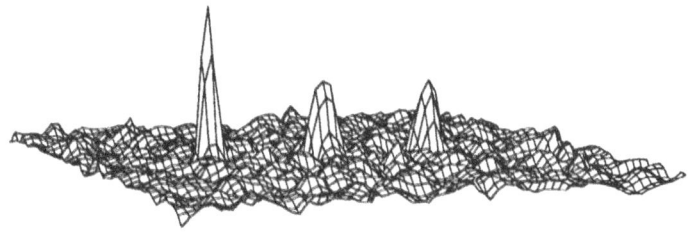

Fig. 2.   The result of phasing errors in the correlation image

3.   Non-Integer Fine Sampling

A non-integer fine sampling arrangement where the detector pixel repetition is out of phase by a fractional value with the mask element repetition has been investigated on the assumption that such an arrangement will supress any particular phasing error. For any given source location, only a small proportion of the mask element shadows are unfavourably matched with the detector pixels, whereas for integer fine sampling the shadowgram is either completely in phase with the detector sampling, or out of phase at all shadowgram boundaries. Although it has been highlighted that for hexagonal geometries coding noise is unavoidable for all sampling configurations bar a few specific orientations, with the aid of commonly used preprocessing techniques such as plateau removal [6] and mask-antimask imaging, the detrimental effect of coding noise on the re-constructed image can be successfully supressed. A variation on the previously proposed 'weighted deconvolution' [6] has been used to retrieve the source distribution which involves rebinning both the deconvolution and detector arrays into 'virtual' pixels that are compatable for performing cross-correlation. Coding noise in the FCFOV region of the SPSF has been restricted to less than 0.1% of the peak. As fig. 3 shows, there are now no unfavourable source positions.

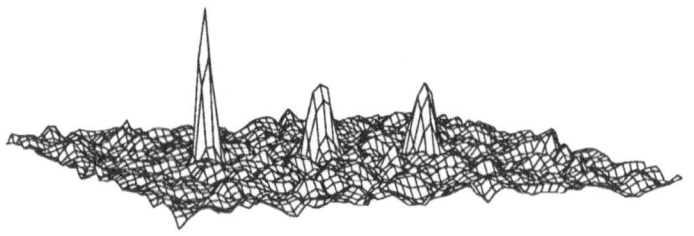

Fig. 3.   Supression of phasing errors with non-integer fine sampling.

Although the image SNR remains modulated with source position, the variation is negligable in comparison to the likely background fluctuations thus providing a smoother instrument response across all sky pixel positions than for integer sampling. Since the dominant parameter governing the sensitivity for a pixellated detector configuration is the mask element to detector pixel area ratio, non-integer fine sampling will not improve the overall sensitivity of an instrument, it will however ensure that the correlation map provides a most likely (in the least squares sense) 'a priori' estimation of the source distribution without the potential 'source loss' found with integer sampling.

To verify that the sensitivity of hexagonal geometry telescopes is related to

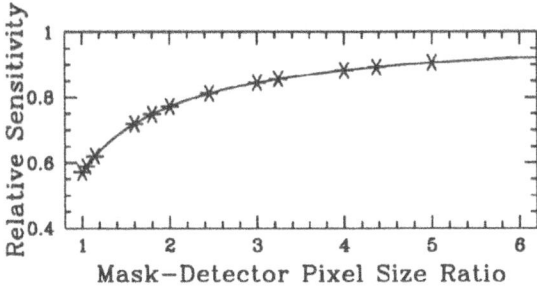

Fig. 4.   The sensitivity dependence (on the sampling ratio) of the correlation map for hexagonal geometries

the sampling ratio in much the same way as for square geometries [3],[7], a model has been derived analytically to determine the mean sensitivity over all source positions for any hexagonal sampling configuration. The following equation was derived initially for integer hexagonal sampling configurations, but will also hold for non-integer sampling values:

$$SNR_m = SNR_T \left(0.054 \left(\frac{D_{af}}{M_{af}} \right)^2 - 0.483 \left(\frac{D_{af}}{M_{af}} \right) + 1 \right)$$

where SNR_T is the maximum theoretical SNR [5] which is measured as the peak pixel height above the statistical fluctuations of the surrounding plateau. D_{af} and M_{af} are the detector pixel and mask element size respectively. Figure 4. shows a verification of the accuracy of the model both numerically and with simulations of telescopes employing hexagonal sampling configurations. It should be noted that the coefficients describing the equivalent equation for square geometries [3] are 0.0625 and 0.5 (compared with 0.054 and 0.483 above).

4.   Conclusions

The trade-off between angular resolution and sensitivity has serious implications for the reconstructed image SNR of a source which lies close to sky pixel boundaries, lowering the mean SNR within a sky pixel when the angular resolution is increased. However, the sensitivity across sky pixel boundaries can be restored to a uniformly high level without drastically compromising the telescope angular resolution with the use of non-integer fine sampling. Calculations have shown that the boundary SNR can be increased by a factor of 1.75 with a corresponding loss in angular resolution of only 1%.

References

[1] *INTEGRAL* : International Gamma-Ray Astrophysical Laboratory, ESA - NASA, SCI(93)1 (1993).

[2] Fenimore E.E., Weston G.S.: 1981, 'Fast delta Hadamard transform', *Appl. Opt.* **20** 3058-3067.

[3] Jupp I.D., Byard K., Dean A.J.: 1994, 'An improved sampling configuration for a coded aperture telescope', *Nucl. Instr. Meths. Phys. Res.* **345** 576-584.

[4] Byard K., Dean A.J., Goldwurm A., Hall C.J., Harding J.S.J., Lei F.: 1990, 'Imaging using HURA coded apertures with discrete pixel detector arrays', *Astron. Astrophys.* **227** 634-639.

[5] Fenimore E.E.: 1978, 'Predicted performance of URAs', *Appl. Opt.* **17** 3562-3570.

[6] Goldwurm A., Byard K., Dean A.J., Hall C.J., Harding J.S.J.: 1990, 'Laboratory images with HURA coded apertures', *Astron. Astrophys.* **227** 640-648.

[7] Skinner G.K.: 1995, 'Coding (and decoding) coded mask telescopes', This volume.

IMAGING WITH TWO ANGULAR SCALES USING CODED MASK TECHNIQUES

I. D. JUPP, A. R. GREEN and A. J. DEAN

Astronomy Group, Physics Department, University of Southampton,
Southampton, SO17 1BJ ENGLAND

Abstract. Coded aperture telescopes employing a high angular resolution specifically for accurate point source imaging, are subject to a severe loss in sensitivity with respect to the extended source imaging capability. However, by choosing a mask with a larger element size the telescope becomes sensitive to more extended regions, but consequently there will be considerable source confusion in crowded fields. This paper describes two systems which simultaneously take images of point sources and extended regions over the same energy range.

Key words: coded mask telescopes, dual scale imaging

1. Introduction

Coded mask imaging techniques are now generally accepted as the optimum method of producing high angular resolution images of astronomical sources at hard X-ray and γ-ray energies. However, from a scientific viewpoint coded mask telescopes have a serious failing which is their poor sensitivity to a wide range of angular scales simultaneously. For optimised sensitivity, the angular extent of the mask elements (relative to the detector plane) of a coded aperture telescope, must be similar to the angular extent of the proposed target sources. The following work is based on an investigation into finding viable options for imaging on more than one angular scale simultaneously.

2. Scientific Requirements

Before designing a coded aperture telescope that enables imaging on more than one angular scale, the extended source distribution throughout the sky must be quantified. Listed below are the regions within our galaxy in which extended emission is thought to occur.

1. The galactic 1.806 MeV emission from ^{26}Al. COMPTEL on the Compton Gamma Ray Observatory has mapped the distribution of ^{26}Al within the galactic plane. Although it is possible that the ^{26}Al is from a number of point sources it seems more likely that the emission is extended, and has structure on scales of $< 4°$.

2. Cosmic-ray induced γ-rays from interstellar matter. This emission has been mapped by COMPTEL in the 1 MeV – 10 MeV range, showing angular structure on scales larger than the angular resolution of COMPTEL.

L. Bassani and G. di Cocco (eds.), Imaging in High Energy Astronomy, 209–212.
© *1995 Kluwer Academic Publishers.*

3. <u>The Orion Nebula.</u> COMPTEL has detected emission from the Orion complex with energies consitent with the ^{12}C and ^{16}O nuclear de-excitation lines which should follow the distribution of CO in Orion, and so structure should be detected at spatial scales of $\sim 1°$.

For this investigation the designs have been optimised for imaging on two angular scales - point sources and extended regions of the order of $\sim 1°$.

3.   DESIGN 1 - Adjacent Masks

To image on two angular scales simultaneously the simplist solution would be to have two co-aligned telescopes mounted side by side, requiring an increase in detector area to retain sensitivity. Alternatively a single detector plane and two rotating masks could be used. a fine mask directly above the detector plane and a coarse mask adjacent to the fine mask (see Fig. 1(a)). For the configuration shown in Fig. 1(a) the two shadowgrams are simulta-

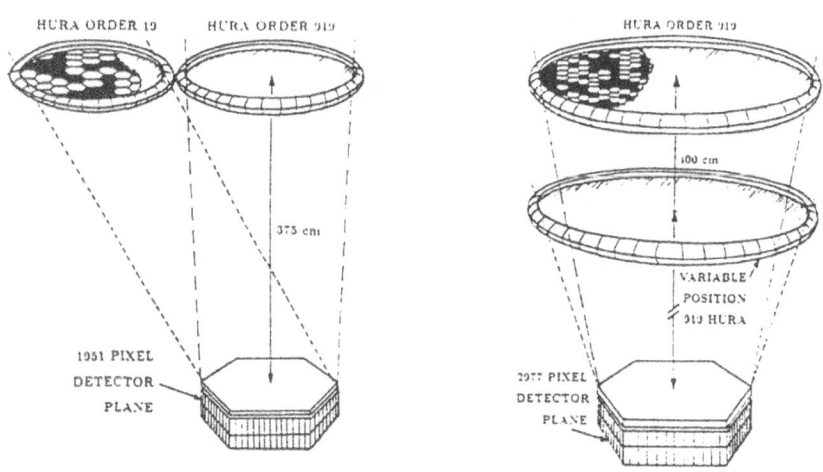

Fig. 1(a)   Fig. 1(b)

neously collected on the same detector plane and mask-antimask techniques are used to separate information on both angular scales. For simulations of this design, hexagonal URAs (antisymmetric on 60°) have been used with a hexagonal detector plane consisting of 1951 hexagonal pixels, each of dimension 9mm across flats (A/F) and with a centre-centre spacing of 10mm A/F, giving a total effective area of 1368cm². The size of the mask pixel was determined by the required angular resolution and the mask-detector separation. In this study a mask detector separation of 375 cm is used. A HURA of order 919 was used for the fine scale mask with a pixel size of 14.6mm A/F giving a nominal angular resolution of 13.4'. For the coarse mask a HURA of order 19 was chosen with pixel size of 102mm A/F giving an angular resolution of 1.56°.

3.1 Simulated Performance

An observational scenario consisting of 3 point sources of flux 2.7, 2.1, & 1.4 $\times 10^{-3}$ ph cm^{-2} s^{-1}, within an extended region of radius 2° and flux 1.7 $\times 10^{-3}$ ph cm^{-2} s^{-1} deg^{-2} has been modelled. The observation time was set as 10^5s, and a background count rate of 0.5 cts cm^{-2} s^{-1} was simulated. Fig. 2(a) shows the image collected from the fine mask. The three point sources can clearly be seen, however the extended region is barely visible. On the contrary, for the image collected using the coarse mask the unresolved point sources distort the reconstructed distribution of the extended region. However, the position and flux information retrieved from the fine scale image enables iterative removal of the point sources in the coarse resolution image and thus enables reconstruction of the extended region only (Fig. 2(b)). The failing of this configuration however is the inability to

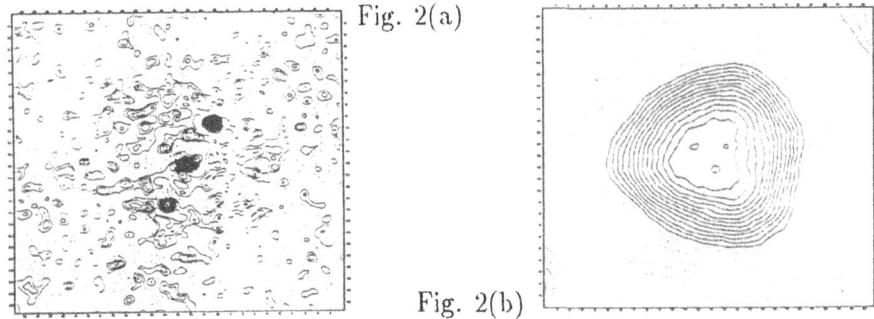

Fig. 2(a)

Fig. 2(b)

observe *the same region of sky* on two angular scales simultaneously over the same energy range. A configuration is now described that enables this.

4. DESIGN 2 - Overlaping Masks

The design shown in Fig. 1(b) employs a fine resolution mask dedicated to point source imaging, beneath which a second identical mask is positioned with an adjustable mask-detector separation to vary the angular resolution. By using identical masks for both angular scales, the two can be placed together enabling imaging on a single scale only. For dual scale imaging the masks are separated and a mask-antimask sequence is used to remove cross-talk between the two convolved datasets. The penalty of acheiving simultaneity in image collection is that only 25% of the detector area is exposed, however in situations were the telescope background is dominated by diffuse cosmic sources, this may not be such a disadvantage. The telescope modelled in these simulations (see Fig. 1(b)) used two 919 HURAs and a hexagonal pixellated detector plane that consisted of 2975 hexagonal pixels, with an across flats pixel size of 1.175 cm giving a total sensitive area of 2977 cm^2. The mask elements were 2.115 cm A/F which for the fine mask fixed at a separation of 400 cm gave an angular resolution of 18 arcminutes, and for the coarse mask at 100 cm an angular resolution of 1.21°.

4.1 SIMULATED PERFORMANCE

The previous configuration demonstrated that it is possible to retrieve source information on both angular scales from data that has been collected simultaneously. It is also possible to image on angular scales far removed from the angular sensitivity range of each mask, without detriment to either image. Consequently the simulated region consists of three Crab-like point sources each of flux 1×10^{-3} ph cm^2 s^{-1}. within an elliptical extended region 12° long and 2° wide and with a flux 1×10^{-3} ph cm^2 s^{-1} deg^2, over the energy range 100 keV - 10 MeV. A total background count rate of 0.7 cts cm^{-2} s^{-1} has been simulated and observations were made over a 2×10^5 second period. The fine scale shadowgram is initially deconvolved, giving the first approximation of the point source positions and fluxes. On deconvolving the coarse resolution image, since the system point spread function (SPSF) will be known accurately, the effect of the point sources can be removed. Similarly the effect of the extended region is removed from the fine resolution image. This is an iterative process when performed correctly, but for the images shown here, the sources are sufficiently bright for a direct subtraction using the known SPSFs and estimates of source flux. Fig. 3(a) shows the fine resolution image where only the point sources are visible. The simultaneously collected coarse scale data, when reconstructed (Fig. 3(b)) reveals the extended structure surrounding the point sources.

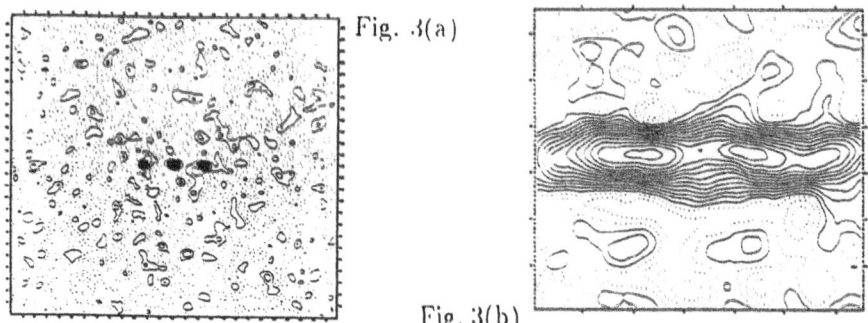

Fig. 3(a)

Fig. 3(b)

5. Summary

The telescope designs presented here, in particular the second, require mechanically complex mask support structures to achieve simultaneous imaging. For future missions, such engineering requirements will be technologically viable. Two possible methods of simultaneous imaging on more than one angular scale over the same energy range have been successfully demonstrated, although a range of solutions to this problem and investigations into the scientific requirements *i.e.* extended source distribution throughout the sky, are ongoing and will be presented elsewhere.

NEW WAVELET METHODS FOR FLATFIELDING CODED APERTURE IMAGES

J.E. GRINDLAY, D. BARRET,* K.S.K. LUM, R.P. MANANDHAR,
B. ROBBASON and S. VANCE**

Harvard-Smithsonian Center for Astrophysics, Cambridge, MA

Abstract. We describe preliminary results from a new investigation we have undertaken for the processing of "flatfield" images used in coded aperture imaging analysis to remove spatial variations in detector background. We have explored wavelet methods for multiresolution smoothing of detector background images which preserve both high and low spatial frequency components of the detector background. Whereas original EXITE images, from the EXITE1 detector and telescope, were processed with a high-statistics average detector background image for a subtractive flatfield, any temporal variations in background shape would be better removed using the observations themselves, appropriately smoothed, to define the background shape. Wavelet smoothing, combined with a pointing program including small dither offsets (planned for EXITE2), appears to be a promising technique. Preliminary results of actual EXITE1 image analysis and simulations are presented.

1. INTRODUCTION

Coded aperture imaging requires, for optimum sensitivity and resolution, that the detector image be "flat". If spatial variations in the detector background exist, these will contribute noise to the reconstructed image. This problem has long been recognized for coded aperture imaging (cf. review by Caroli et al 1987), and it has led to the development of "flatfielding" methods (cf. Laudet and Roques 1988, Covault et al 1991). An alternative approach to remove spatial and temporal variations in detector background is to use a rotating HURA mask (cf. Althouse et al 1985), although this introduces the complexity of a mechanically rotating mask.

The Energetic X-ray Imaging Telescope Experiment (EXITE) (Grindlay et al 1986, Covault 1991), which has now been extended to a much more sensitive and larger area imaging phoswich detector, EXITE2 (cf. Manandhar et al 1993, Lum et al 1994), employs a fixed coded aperture mask (4 × 3.5 cycles of a 13 × 11 URA). With EXITE1, our first detector/telescope combination, we have employed "subtractive flatfielding" (cf. Covault et al 1991) to remove the spatial variations in the detector background. The detector background shape was derived from a high-statistics average of many "background" detector images recorded during the flight while strong sources were not being observed. This has allowed detector spatial variations

---

* Also, Centre d'Etude Spatiale des Rayonnements, CESR-CNRS-UPS, 9 Av. du Colonel Roche, 31029 Toulouse, FRANCE
** Also, Department of Physics, Andrews University, Berrien Springs, MI

L. Bassani and G. di Cocco (eds.), Imaging in High Energy Astronomy, 213–220.
© *1995 Kluwer Academic Publishers.*

to contribute no more than ~10-20% additional noise above that expected from Poisson statistics. In an effort to further reduce this by allowing for any temporal variations in background shape, we have explored "self-flatfielding" methods: the detector images recorded during the observation of a source are used to define the background shape as well as being used to reconstruct the sky image. We emphasize that self-flatfielding techniques are well suited for balloon-borne coded-mask experiments with limited exposure time because background and source data can be then be acquired simultaneously. They are also relevant for a space-borne mission for which the background shape may be changing rapidly along the satellite orbit.

This paper presents initial results, employing real and simulated data for the EXITE1 detector from a Crab observation in the 1989 Alice Springs flight (Covault, 1991), of a first implementation of self-flatfielding as planned for the EXITE2 detector on its upcoming first science flight.

2. SELF-FLATFIELDING METHODS

Self-flatfielding, whereby the source pointing itself is used without aspect correction to define the detector background shape, requires the aspect to shift during the observation. (Note: This presumes the internal detector and diffuse gamma-ray shield leakage and not the aperture flux, dominate the background, as is indeed the case even with the low background of the EXITE2 phoswich system with its 4.6° (FWHM) field of view. In this case, the detector background shape is *independent* of telescope aspect, at least over small angles.) A "perfect" aspect pointing, without shifts of the source shadow on the detector, would result in loss of signal and thus sensitivity even if a smoothed image of this detector "background" were subtracted for the flatfield. Thus dithered offsets, or smooth scans, are required for optimum self-flatfields with fixed-mask detectors.

For EXITE2 we plan a "dithered" pointing offset scheme to enable optimized self-flatfielding. A semi-automated pointing program has been developed whereby instead of fixed pointing on target, typically to within the ~ 3 arcmin stability achieved with EXITE1, a mini-raster scan of offset pointings can be carried out. The basic raster contemplated is a 3 × 3 grid, with each element being offset by 1 imaging resolution element (i.e. 1 mask pixel, or 23 arcmin) from the central on-axis raster point. With only ~10 sec needed to offset to each point, and an integration time of perhaps 3 min per point, the dead time is negligible and the background sampling in both space and time is optimized. The same modulation of source vs. background regions on the detector could be achieved with a smooth scan, or coarse pointing (within limits). However, the simplicity of known offsets for aspect makes this the best operational choice. Nevertheless, we are simulating (see below) both fixed offsets and limited-continuous scans to study their effects

on self-flatfielding. In this paper we do not consider the additional improvement (over self-flatfielding) that could be achieved by also incorporating
deconvolution techniques: the dithered offset introduces negative sidelobes
in the system PSF (Manandhar 1995). We shall report in a forthcoming
paper an optimized analysis scheme which incorporates self-flatfielding with
wavelet smoothing and correlation analysis (as discussed here) followed by
deconvolution analysis with a bipolar PSF.

3. WAVELET SMOOTHING

Subtractive flatfielding using the raw detector image, uncorrected for aspect
(as with self-flatfielding), might be expected to reduce the final signal to
noise ratio (SNR) in the image by a factor $\sqrt{2}$ because of the additional statistical fluctuations introduced by subtracting the flatfield. Thus it is
desirable to smooth the detector image, to be used for the flatfield, to minimize the statistical fluctuations. Normal linear smoothing (e.g. convolution
with a gaussian) is not optimum because the real background shape is likely
to vary on several spatial scales (e.g. a smooth interior detector background
surrounded by a sharp ring, as in the EXITE1 detector background shown
below).

Therefore we have explored wavelet smoothing techniques and have adopted the multiresolution filtering techniques employing wavelet transforms as
described by Starck, Murtagh and Bijaoui (1994). This wavelet transform
algorithm was chosen as it preserves statistically significant features in the
image (the flatfield background) while allowing smoothing down to a prede-
termined scale. This allows smoothing over a wide range of spatial frequencies, as shown in Starck et al (1994).

In Figure 1 below, we demonstrate the result of applying this package,
which we have adapted for EXITE image file analysis, to the detector background image recorded in flight with the EXITE1 detector in "band 5"
(82-111 keV). This background image was constructed from 68 files (i.e. 232
minutes exposure) obtained from pointings on several weak source fields in
our May 1989 flight from Australia (cf. Covault 1991). The bright source
at the edge of the detector (in the enhanced background ring) is the ^{109}Cd
source (88 keV line) embedded in the edge of the detector collimator for
internal calibration of the detector in flight (cf. Covault 1991). Whereas
the original background image is noisy (Poisson statistics appropriate to its
count rate of ~ 10 cts/subpixel), the wavelet smoothed version is smooth
without loss of significant structure in the original image. Such a background
image should therefore be superior to the original for a flatfield image in
this band. (Note that the background image shapes, such as the contrast
in the background ring, are strongly energy dependent, and thus flatfields
are derived separately in each band.) We have verified this by using such

wavelet smoothed flatfields vs. either non-smoothed "local" backgrounds (i.e. for self-flatfielding) as well as vs. averaged "non-local" backgrounds (i.e. the original EXITE1 flatfielding technique) for both real and simulated EXITE1 data.

4. SIMULATIONS

We have carried out a full simulation of our EXITE1 detector and telescope system and thus far applied it to EXITE1 images. The simulation employs the EGS4 code (Nelson et al. 1985) to fully account for all electromagnetic interactions in the detector and intervening detector materials. In the energy range considered (10-300 keV), attenuation by the mask (50%) and collimator (15% on axis) are assumed to be total. Each photon is detected with full account for detector dead-layer effects (in EXITE1, surface roughness on the NaI crystal introduced small losses at low energies) as well as Compton scattering and measured spatial resolution. The actual source photon spectrum (power-law for the Crab), atmospheric transmission for actual grammage and elevation angles, and actual achieved aspect are included in detail so that a full simulation of a given observation (in this case, the Crab) can be carried out.

5. RESULTS

The simulations were run for three cases: "non-local" flatfielding (i.e. the original method of using an average background, in each band, derived from "blank-field" observations); self-flatfielding with no wavelet smoothing; and self-flatfielding with wavelet smoothing. The results of the simulations were compared with those of the actual Crab observation from the last EXITE1 flight (May 1989, from Australia; cf. Covault 1991).

Results for both the simulated and real Crab observations are given in Table 1. They clearly indicate that our simulated Crab images are consistent within the errors with those obtained (Covault 1991) for the 40 min observation when analyzed with non-local flatfields.

Having demonstrated by the previous results that the modelling of the EXITE 1 detector used in the simulations is accurate enough to reproduce a real observation, we have then simulated a stronger source with three times the Crab intensity so that differences in flatfielding techniques can be explored without statistical limitations. We have run three simulations: one with the Crab shifts (derived from the actual observation), one for a purely on-axis source (whithout shifts), and one with the "dither" shifts (the raster is a 3 × 3 grid with each element being offset by one mask pixel from the central on-axis raster point). The results of these simulations are summarized in Table 2.

6. CONCLUSIONS AND FUTURE WORK

From Table 2, four main conclusions can be drawn:

1. For any realistic aspect history self-flatfielding with or without wavelet smoothing gives signal to noise ratio (SNR) values lower than those achieved by non-local flatfielding, with a reduction in the most favorable case of only 20% ("dither shifts" in Table 2). The 20% loss of SNR in self-flatfielding (vs non-local flatfielding) is due to the negative PSF of the source in the dithered offset pointings, and may be recovered by a subsequent deconvolution with a bipolar-PSF.

2. As expected, the results of the self-flatfielding depend strongly on the aspect history of the observation. In particular, self-flatfielding with perfect pointing almost completely removes the signal.

3. The effect of wavelet smoothing is also strongly dependent on the aspect history of the observation. As anticipated, wavelet smoothing gives recovered SNR values larger than those achieved without smoothing. However, in the third case ("dither shifts") for which the dither strategy chosen makes the raw image (background + source) more uniform (flat), it does not significantly improve the SNR.

4. The aspect history of the Crab observation (azimuthal shifts only) was not optimized for self-flatfielding.

We are now extending the complete simulation package for the EXITE2 system. We shall use these results to devise the optimum dither strategy for the upcoming EXITE2 flight. These general results should be applicable to other coded aperture telescopes with fixed masks (e.g. the INTEGRAL spectrometer), which envision a dither strategy. Detailed results for both the EXITE1 and EXITE2 simulations, including various dithering strategies, and the expected additional improvement of deconvolving the PSF will be reported in a forthcoming paper.

Acknowledgements

We thank Corbin Covault for help with EXITE1 processing and Jean-Luc Starck for assistance with implementation of his wavelet smoothing software. This work was supported by NASA grant NAGW-624. D. Barret aknowledges the French Space Agency (CNES) for the support of a postdoctoral fellowship.

References

Althouse, W.E. et al 1985, Proc. 19th Intl. Cosmic Ray Conf., 3, 299.
Covault, C.E. 1991, *Ph.D. Thesis*, Harvard University.
Covault, C.E. et al 1991, IEEE Trans. Nucl. Sci., NS-38, 591.
Grindlay, J.E. et al., 1986, IEEE Trans. Nucl. Sci., NS-33, 750.

Laudet, Ph. and Roques, J.P. 1988, Nucl. Inst. Methods, A267, 212.

Lum K.S.K. et al (1994), IEEE Trans. Nucl. Sci., NS-41, 1354.

Manandhar R. et al (1993), Proc. SPIE, 2006, 200.

Manandhar R. (1995), PhD Thesis, Harvard University.

Nelson, W.R., Hirayama H. & Rogers, D.W.O., EGS4 user guide, CERN publications, 1985

Starck, J.L., Murtagh, F., and Bijaoui, A. 1994, preprint.

Fig. 1.   Comparison of original EXITE1 detector background image in band 5 (82-111 keV) (top) with wavelet smoothed version (bottom).

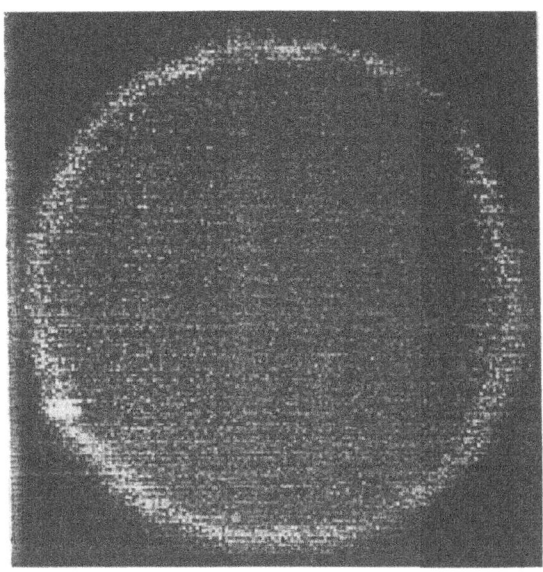

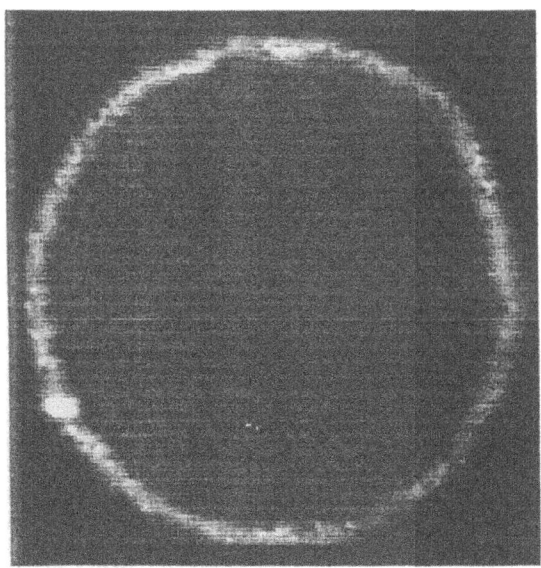

	Non-local FF	Self-FF	Self-FF + WS
Crab data (Alice Springs)	13.0	6.9	8.7
Monte Carlo (Actual aspect shifts)	11.5	4.7	8.4

TABLE I

Comparison of the SNR (number of sigma) derived from the simulations and the real data. FF stands for FlatFielding, and WS for Wavelet Smoothing. Results are presented for bands 1-5 or energy range 30-110 keV.

Simulation	Non-local FF	Self-FF	Self-FF + WS
Zero shifts	31.1	2.7	15.3
Crab shifts	32.1	13.4	20.8
Dither shifts	30.3	23.6	24.2

TABLE II

Simulations of a source with three times the Crab intensity. Comparison of the SNR between simulations made for three different aspect histories (perfect pointing, actual shifts observed during the Crab observation, and dither shifts; see text).

USEFUL CLASSES OF REDUNDANT ARRAYS FOR IMAGING APPLICATIONS

MARK H. FINGER

Compton Observatory Science Support Center, USRA

and

THOMAS A. PRINCE

California Institute of Technology

July 28, 1995

Abstract. We discuss several classes of redundant arrays. These arrays have applications for indirect imaging in a variety of fields including coded-aperture imaging, interferometric radio imaging, and optical imaging in the presence of atmospheric turbulence. The specific classes we will discuss are all based on Galois fields and include: antisymmetric redundant arrays (ARAs) which have as a subset the hexagonal uniformly redundant arrays (HURAs), non-redundant arrays (NRAs), and the general class of quadratic residue arrays (QRAs).

1. Introduction

Coded arrays were originally conceived for applications in X-ray imaging (Mertz and Young 1961 and Dicke 1968). A coded array is defined to be a pattern on a periodic two-dimensional lattice which associates with each lattice point a 0 or a 1 indicating whether the lattice point is "open" or "closed". In coded-aperture imaging, the open and closed lattice points become open and closed cells in an opaque mask which casts a shadow of the photon source on a position-sensitive detector. For a review see Caroli et al. (1987). A uniformly redundant array, or URA (Fenimore and Cannon 1978), is a particular form of coded array. For a URA, each possible vector displacement between pairs of inequivalent open lattice points occurs a uniform number of times. (Equivalent lattice points are separated by a period of the array.)

In this paper, we describe several classes of redundant arrays. All of these are based on Galois fields. Each of the classes was originally investigated because of a specific useful imaging property, e.g. antisymmetry, non-redundancy, or periodic pattern shape (such as square or hexagonal). The construction of these arrays and their properties are implicit in earlier mathematical treatments of cyclic difference sets (e.g. Baumert 1971), however the usefulness of these arrays for imaging was in most cases hidden.

221

L. Bassani and G. di Cocco (eds.), Imaging in High Energy Astronomy, 221–226.
© 1995 Kluwer Academic Publishers.

2. Antisymmetric Redundant Arrays (ARAs)

Arrays with rotational antisymmetry have attractive properties for applications such as coded aperture imaging (e.g. Cook et al. 1984). Finger and Prince (1985) showed that all antisymmetric URAs could be derived from skew-Hadamard cyclic difference sets. Since all two dimensional lattices have 180° symmetry, all skew-Hadamard cyclic difference sets can generate both hexagonal and rectangular URA's with 180° antisymmetry. A subclass of these with order v (the number of cells in the basic pattern) equal to a prime p with $p = 7$ mod 12 were shown to generate a hexagonal array (HURA) with an additional 60° rotational antisymmetry.

Modified uniformly redundant arrays (MURAs) having imaging properties similar to URAs were discussed by Gottesman and Fenimore (1989), and later Byard (1992) showed that a subset of these with $v = p$ with the prime $p = 5$ mod 8 or with $v = p^2$ with the prime $p = 3$ mod 4 had 90° rotational antisymmetry on a rectangular lattice. We will consider antisymmetric redundant arrays (ARAs) to include both the skew-Hadamard arrays and the anti-symmetric MURAs.

We are not aware of any other antisymmetric arrays. However, Gottesman and Fenimore (1989) have pointed out the existence of hexagonal arrays based on MURAs that have 60° and 180° rotational symmetry and rectangular arrays with 180° rotational symmetry. We will show later that additional sets of arrays called quadratic residue arrays (QRAs) can also have hexagonal and rectangular rotational symmetries. While antisymmetric arrays have useful imaging properties, it is not clear what advantages rotational symmetry might have for imaging applications.

3. Non-redundant Arrays (NRAs)

For several astronomical imaging applications, particularly those involving interferometry, we require the ability to generate large-order non-redundant arrays (NRAs). Applications include radio interferometry and removal of the effects of atmospheric turbulence from ground-based observations at visible and IR wavelengths. By an NRA we mean a set (array) of points on a regular lattice such that the vector difference between any two of the points in the set is unique. The advantage of such arrays for interferometry is clear: unique vector differences in the NRA correspond to unique interferometer baselines and thus unique spatial frequency (u-v plane) coverage.

A discussion of NRAs in the context of antenna arrays for radio interferometry is given by Golay (1971). Golay identifies several "non-redundant compact arrays" on both rectangular and hexagonal lattices. The construction involved a tree search and generated arrays with up to 12 elements. Other authors have discussed arrays with larger number of elements (e.g.

the 24 element array of Klemperer 1974). In contrast to these earlier techniques, the analytical techniques discussed here can be used to generate a much larger and richer set of arrays.

URAs are often based on cyclic difference sets (v, k, λ) (Gunson and Polychronopulos 1976, Fenimore and Cannon 1978, Finger and Prince 1985), where v is the number of cells in the lattice, k is the number of apertures, and λ is the redundancy of the array i.e. the number of times a given vector spacing repeats. In a similar fashion, cyclic difference sets can be used to generate a large class of non-redundant arrays (NRAs) for which $\lambda=1$. In this case there is an entire class of arrays based on Singer sets which have $v = q^2 + q + 1$ where $k = q + 1$ and $q = p^n$ with p prime. A list of Singer sets can be found in Baumert (1971) and the construction of aperture arrays is identical to the construction of a URA from a cyclic difference set (Gunson and Polychronopulos 1976, Finger and Prince, 1985). Only one cycle of the Singer set is used in order to retain the nonredundancy characteristic. Like other arrays discussed in this paper, the NRAs can be shown to be related to Galois fields $GF(p^n)$ (Baumert 1971).

The class of NRAs is very large. Choices for the number of apertures, k, in arrays of low-order include: $(k,v) = (3,7), (4,13), (5,21), (6,31), (8,57), (9,73),$ $(10,91), (12,133), (14,183), (17,273), (18,307), (20,381), (24,553), (26,651),$ $(28,757), (30,871), (32,993), (33,1057), \ldots$. We believe these arrays are near-optimal in the sense that they have close to the maximum number of non-redundant apertures for a given finite radius area on a lattice. Note that the number of apertures, k, approaches the square root of the number of points in the array, $\sqrt{v}$, as might be expected since the number of unique differences should go as k^2 and, for compact arrays, be of order v. For each choice of number of apertures, there is a large number of different non-redundant arrays available by permutation. For instance, for $k=24$, at least 78 different NRA patterns are available. In addition, each of the basic permutations can generate numerous other non-redundant arrays by translation of the periodic pattern before selecting a single cycle. We note that the set of non-redundant arrays generated by the techniques mentioned above include almost all the hexagonal point arrays with maximum compactness given in Golay (1971).

The large number of arrays generated by these analytical methods allows a wide selection for optimizing coverage in the u-v plane. As in Golay (1971), we may wish to choose a compact, quasi-complete sampling of u-v space out to a given spatial frequency. Alternatively, we could choose arrays with maximum coverage of high spatial frequencies. We can also trade off redundancy for completeness of baseline coverage in two ways: (1) Use arrays with $\lambda > 1$, or (2) use partial repetitions of $\lambda=1$ arrays to obtain more complete baseline coverage at the expense of some redundant coverage of u-v space.

An NRA of order $v=2863$ and $k=54$ and its autocorrelation (uv-coverage) are shown as an example in Figure 1.

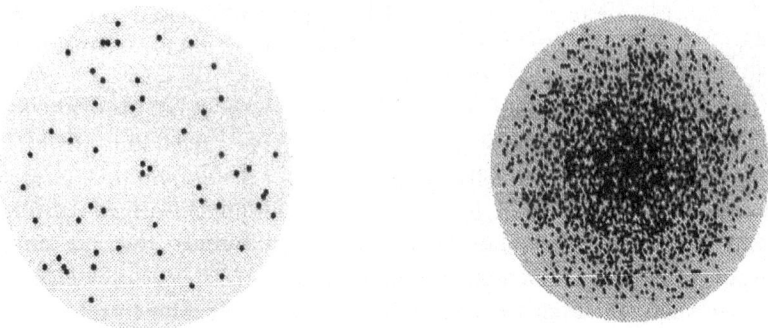

Fig. 1. NRA of order $v=2863$ and $k=54$ and its autocorrelation (uv-coverage).

4.   Quadratic Residue Arrays (QRAs)

We now discuss the general class of arrays that can be generated using the quadratic residues of Galois fields. We call this general class, QRAs. The QRAs are a superset that includes all the ARAs mentioned earlier as well the 1-D MURAs. The ARAs and 1-D MURAs are intrinsically one dimensional in that the cyclic patterns are based on one-dimensional Galois Fields mapped onto a two dimensional lattice. [There are two classes of MURAs (Gottesman and Fenimore 1979): MURAs based on tiling of the lattice with quadratic a residue set of order $v = p$ with prime $p = 1 \bmod 4$, and MURAs constructed from the exclusive or of two 1-d quadratic residue sets oriented orthogonally along the two axes of the lattice]. In addition, the QRAs contain a new subclass of intrinsically two dimensional patterns that result from the quadratic residues of two-dimensional Galois fields. We will call this subclass GFP^2As.

The imaging properties of the patterns we will describe result from the symmetry properties of the Galois fields that they are constructed from. Here we will describes this structures only to the degree needed for our construction. A more extensive treatment is given in the classic work by Dickson (1901).

A familiar example of a Galois field is the set $\{0, 1, ..., p - 1\}$ where p is a prime, under the operation of addition and multiplication modulo p. This set under addition modulo p is an Abelian group. In addition, because p is prime, the nonzero elements under multiplication modulo p also form an Abelian group. This algebraic structure is call $GF(p)$, the Galois field of p elements.

The more general structure $GF(p^n)$, the Galois field of p^n elements retains these two key properties: its elements form an Abelian group under addition, and its nonzero elements form an Abelian group under multiplication. An element $\mathbf{m}$, or 'mark', of $GF(p^n)$ may be represented as a polynomial of order $n - 1$:

$$\mathbf{m} = m_0 + m_1 x + \ldots + m_{n-1} x^{n-1} \qquad (1)$$

with coefficients m_i that are elements of $GF(p)$. Here x is a formal argument. Addition is accomplished vectorially, adding coefficients modulo p term by term. Multiplication is a more complicated process. Using arithmetic modulo p the polynomials are multiplied and the result is divided by an nth degree polynomial $f(x)$. Only the remainder from this division retained. The polynomial $f(x)$ must be irreducible, or unfactorable, using arithmetic modulo p. For the one and two dimensional fields concerning us this simply means that the equation $f(x) = 0$ must have no roots in $GF(p)$.

Given the complexity of the multiplication process, it is surprising that the multiplication group of a Galois field has the simple structure of a cyclic group. This means that we can find a primitive element $\mathbf{g}$ such that each nonzero mark can be represented by some power of $\mathbf{g}$. Multiplication of nonzero marks in $GF(p^n)$ is equivalent to addition of powers modulo $p^n - 1$.

If p is an odd prime, we can separate the nonzero marks into squares and not-squares. The squares can result from squaring and are even powers of $\mathbf{g}$, the not-squares cannot result from squaring another mark, and are odd powers of $\mathbf{g}$. The set of all squares is called the quadratic residue set.

We will construct a mask-antimask pair of coded aperture patterns from the squares and not squares of the the Galois field $GF(p^n)$. To do so we first map the marks in $GF(p^n)$ onto a two dimensional lattice in a way that associates addition in the field with translation in the lattice.

The mask minus antimask transparency function is associated with the following character function:

$$\chi(\mathbf{m}) = \begin{cases} 1 & \mathbf{m} \text{ a square} \\ -1 & \mathbf{m} \text{ a notsquare} \\ 0 & \mathbf{m} = \mathbf{0} \end{cases}$$

It can be shown (Williamson 1944) that

$$\sum_{\mathbf{m} \in GF(p^n)} \chi(\mathbf{m})\chi(\mathbf{m} + \mathbf{a}) = p^n \delta_{\mathbf{a},\mathbf{0}} - 1$$

which indicates that the arrays have the desired autocorrelation properties.

QRAs based on $GF(p^2)$, or GFP^{2}As, have inversion symmetry and some have rotational symmetry. One of the approximately $p(p - 1)/6$ hexagonal GFP^{2}As with $p = 5 \bmod 6$ has $60°$ rotational symmetry, while one of the approximately $p(p - 1)/4$ square GFP^{2}As with $p = 5 \bmod 4$ has $90°$ rotational symmetry. We do not know of any GFP^{2}As that have rotational antisymmetry. Examples of hexagonal and square GFP^{2}As are shown in Figure 2.

Fig. 2.   Hexagonal and square GFP^2As with $p{=}23$.

5. Conclusions

The Galois fields of a prime power, $GF(p^n)$, yield a rich set of redundant arrays for imaging applications. Quadratic residues on $GF(p)$ yield the skew-Hadamard URAs and the MURAs and therefore include all known ARAs. The $GF(p^2)$ yield a new set of square QRAs with a variety of symmetry properties. The 2-dimensional finite projective geometries of $GF(p^n)$ yield the Singer sets which give rise to NRAs. The redundancy and anti-symmetry properties of the Galois field arrays make them useful in a large class of astronomical imaging applications ranging from radio to gamma-ray wavelengths.

References

Baumert, L.D.: 1971, *Cyclic Difference Sets*, Springer-Verlag, Berlin

Byard, K.: 1992, *Exp. Astro.* **2**, 227

Caroli, E., Stephen, B., Di Coco, G., Natalucci, L., and Spizzichino, A.: 1987, *Sp. Sci. Rev.* **45**, 349

Cook, W.R., Finger, M., Prince, T.A., and Stone, E.C.: 1984, *IEEE Trans. Nucl. Sci.* **NS-31**, 771

Dicke, R.H.: 1968, *Ap.J.* **153**, L101

Dickson, L.E.: 1901, *Linear Groups and an Exposition of Galois Field Theory*, B.G. Teubner Publisher, Leipzig

Fenimore, E.E. and Cannon, T.M.: 1978, *Appl. Opt.* **17**, 337

Finger, M.H. and Prince, T.A.: 1985, *Proc. 19th Int. Cosmic Ray Conf.* **3**, 295

Golay, M.J.E.: 1971, *JOSA* **61**, 272

Gottesman, S.R. and Fenimore, E.E.: 1989, *Appl Opt.* **28**, 4344

Gunson, J. and Polychronopulos, B.: 1976, *MNRAS* **177**, 485

Klemperer: 1974, *A&A Suppl.* **15**, 449

Mertz, L. and Young, N.O.: 1961, *Proc. Int. Conf. Opt. Instr. and Tech.*, p. 305.

Williamson, J.: 1944, *Duke Math. J.* **11**, 65

SOLAR X-RAY IMAGING TELESCOPE

UPENDRA D. DESAI

NASA/Goddard Space Flight Center
Greenbelt, MD 20771 USA
(301)286-3032

and

CARL C. GAITHER, III

ANSER
1215 Jefferson Davis Highway
Suite 800
Arlington, VA 22202 USA
(703)416-3154

June 10, 1995

Abstract.
A design concept is presented for a new solar X-ray telescope for use on a small satellite. Imaging with high angular resolution will be achieved by using two Fresnel zone plates (FZPs), separated by a distance of up to a few meters, acting as a coder (Mertz, L., SPIE, 1159, 14, 1989). Such paired FZPs provide two-dimensional spatial coding in the form of parallel fringes whose frequency and orientation are dependent on the off-axis location of a point source. Extended sources have correspondingly more complex spatial coding patterns. One advantage of this scheme is the capability for all azimuth viewing over a range of source sizes. Also, the image-plane detector needs only moderate spatial resolution to achieve fine angular resolution. As compared with rotating modulation collimators, less stringent alignment is required and, given adequate flux, the time resolution is not restricted by the bi-grid rotation rate. With a 4-cm plate diameter and a separation of 2 meters, angular resolution of about 10 seconds of arc over a few degrees is possible. The proposed detector array will use avalanche photodiodes and/or silicon PIN diodes at the lowest energies, and CsI(Tl) viewed by APDs at higher energies. The instrument will be compact, low in power, light in weight, and ideally suited for solar studies on a small satellite.

Key words: Solar Physics – X-Ray Instrumentation – Telescope

1. Introduction

Mertz (1989) revived a dormant idea of using two Fresnel zone plates with spatial separation as a coder which he named an "FZP sandwich" in his book, *Transformation In Optics*. If a single zone plate is used as a coder and if its full angular resolution capabilities are to be exploited, one has to have an image plane detector system with a spatial resolution comparable to or better than the finest spacings of the outermost zones. For conventional high energy X-ray photon detectors (proportional counters, scintillators, etc.), the best spatial resolution is of the order of millimeters. If one uses two zone plates instead of one, the moire fringe pattern due to both zone plates

227

L. Bassani and G. di Cocco (eds.), Imaging in High Energy Astronomy, 227–233.
© *1995 Kluwer Academic Publishers.*

enables the use of low spatial resolution detectors in the image plane. The properties of the zone plates for non-coherent optical computing was worked out by Rogers (1977). He called this coder system of two FZPs a "Fourier-Transformer".

2.   The Telescope

2.1   THE CODER

The imaging portion of the telescope uses two identical zone plates separated by a few meters. Each zone plate consists of concentric, equal area circular zones, alternately transparent and opaque to X-rays. The circular zone edges are defined by

$$R_n = R_1 n^{1/2} \text{ where n = 1, 2, ... N.}$$

When two such zone plates are in contact, linear Moire fringes are observed. The fringe spacing depends upon the distance between the zone plate centers - the larger the distance the higher the spatial frequency. The fringes are orthogonal to the line joining the centers of the zone plates. Similarly when the FZPs are separated by a finite distance and illuminated by a point source at infinity, a linear fringe pattern is created in the detector plane. The fringe spacing, s, then depends upon the separation, l, between the FZPs and the angular offset, θ, of the source from the optical axis of the telescope.

$$S = R_1^2/l \ tan(\theta)$$

where R_1 is the radius of the first zone. Thus, the fringe spacing is inversely proportional to both the angular offset and the distance between the plates. The former dependence is quite unlike the situation encountered in all other coded aperture systems, where the smallest angular features in the object plane are necessarily resolved in the image plane with small sized detector elements. Instead, the angular resolution of a double FZP telescope is determined by how precisely the system records low (center of field) and high (edge of field) frequency spatial components. The angular resolution of the coding system is solely dependent upon the zone width of the outermost zone and the separation between the two zone plates. The width of the finest zone is given by

$$dR_n = R_1^2/D$$

where R_1 is the radius of the first zone and D is the diameter of the zone plate.

The field of view of the system is defined by the fringe pattern. It extends to an angle of $tan^{-1}(R_1/l)$. The fringe pattern contains satellite Fresnel zones along with parallel fringes at the outer periphery and thus puts a limitation on unambigous deconvolving. If instead of centro-symmetric zone plates one uses off-axis zone plates, one can avoid the formation of satellite Fresnel zone patterns and achieve a wider field of view. For narrow field of view telescopy, it is best to retain the complete Frensel zone pattern of the plates.

Two identical zone plates provide the symmetric Fourier cosine compenents of the object space. In order to get the sinusoidal components of the spatial frequencies one has to use phase-shifted zone plate pairs. Thus two pairs of coders are required for complete description of the object.

We have performed Monte Carlo simulations of the telescope design using a radius of 0.5 cm for the central zone and total of 400 zones. The finest zone width is 125 x 10^{-4} cm and separation between the plates is 1 meter. Fig. 1 shows the simulation done with identical zone plates. The pattern at the image plane reveals parallel fringes for a single source at infinity, while for two sources there are superposed fringe patterns. The deconvolved objects are shown in Fig. 2.

2.2 DETECTORS

2.2.1 Soft X-rays (1 - 20 keV)

Over the last couple of years, we have evaluated several types of detectors for both soft (1-20 keV) and hard (>20 keV) X-rays that can be used to form a matrix to enable position sensitive detection (Desai et al 1992). For soft X-rays, we have studied P-intrinsic-N (PIN) silicon diodes available at low cost commercially. They offer good spectral resolution (~1 keV FWHM) as well as long-term stability, even when operating at room termperature. The conventional Si(Li) diodes need cooling to achieve the same spectral resolution and they must be kept cool at all times, otherwise they will lose their detector capabilites. Fig. 3 shows the energy response of both a PIN and an APD detector. We have also looked into silicon diodes operating in the avalanche mode (APDs) for the direct detection of soft X-rays. With an internal gain of about 200, these detectors offer an energy threshold of less than 1 keV and an energy resolution of ~ 1 keV, comparable to those of PIN diodes. For the soft X-ray telescope, we intend to use a matrix of small PIN diodes or APDs. We will also investigate another type of X-ray detector consisting of a thin (2 mm thick) CsI(Tl) scintillator plate (10" diameter) coupled through a fiber-optic taper (10" > 1") to an intensified CCD detector.

2.2.2 Hard X-rays (> 20 keV)

We have coupled both PIN and APD silicon photodiodes to CsI(Tl) scintillators and studied their energy resolutions, stabilities, and low energy

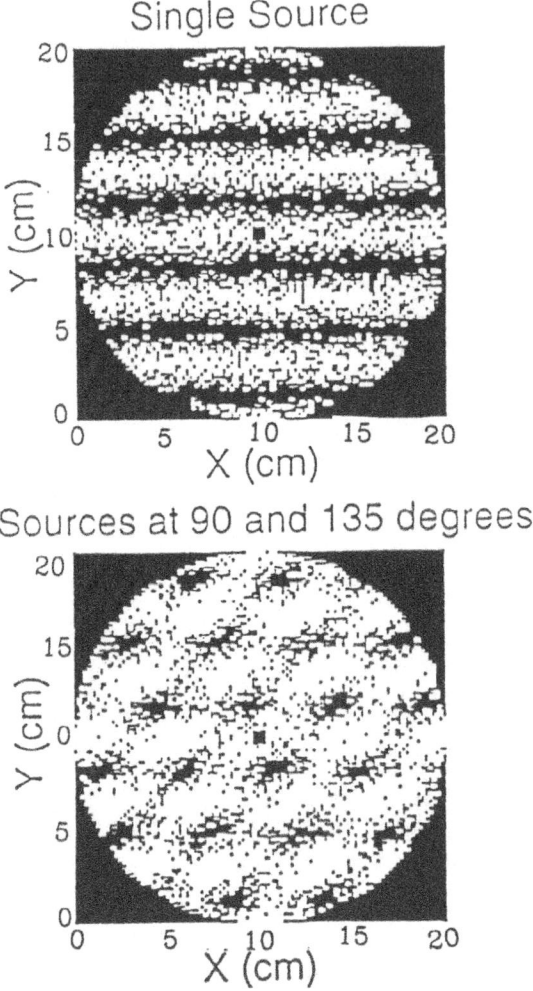

Fig. 1. Response of position-sensitive detector array.
Zone plate specifications: R_1 = the first zone radii, 0.5 cm; N = number of zones, 400; L = separation between zone plates, 100 cm; pixel size = 0.25 x 0.25 cm; 1 x 10^5 photons.

thresholds. With a CsI(Tl)/PIN combination, we have achieved a low ener-gy threshold of ~60 keV, while with a CsI(Tl)/APD combination we have achieved a low energy threshold of 18 to 20 keV and ~ 6 to 7 percent res-

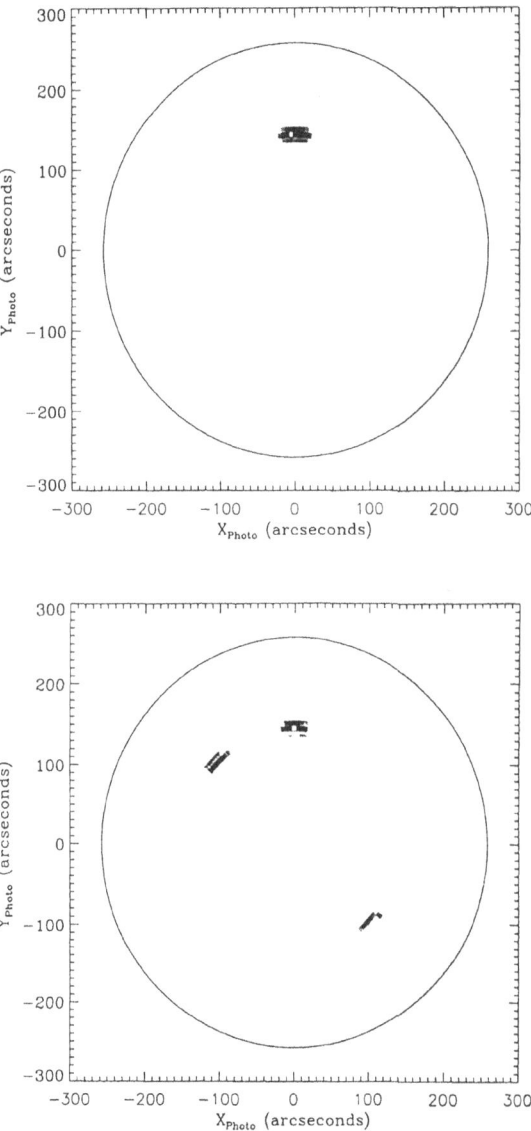

Fig. 2.   Deconvolved image for a single source and for sources at 90° and 135°.

olution at 662 keV. For a hard X-ray image plane detector, we can build a matrix of CsI(Tl)/APDs or CsI(Tl)/PIN detectors to obtain position sensitivity. We have also looked into using other solid-state detectors like HgI_2, CdTe, or CdZnTe to fabricate a position sensitive detector. These high-Z conduction counters are not yet commercially available at a low price, so

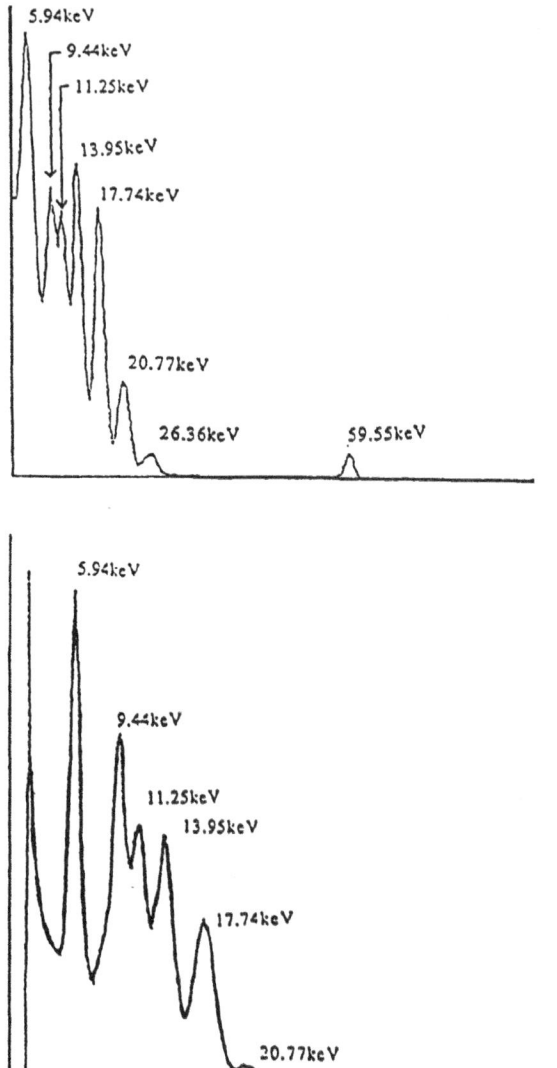

Fig. 3. X-ray spectrum at room temperature detected using
a PIN photodiode (without scintillator) connected to the Amptek A250
Charge-Sensitive Preamplifier and the X-ray spectrum at room temperature
detected using an Avalanche Photodiode (APD) connected to the Amptek A250
Charge-Sensitive Preamplifier.

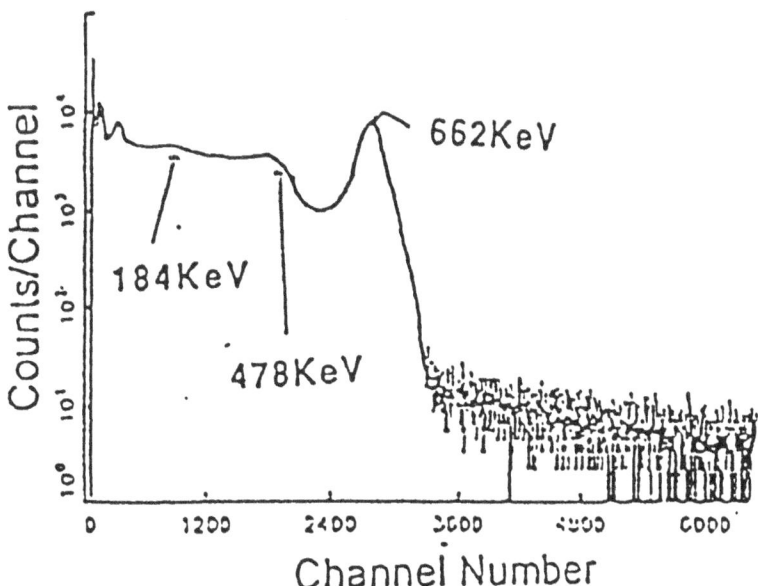

Fig. 4. Spectrum from an APD with a CsI(Tl) scintillator exposed to a Cs^{137} source.

they may not be cost effective. We have also considered using CsI(Tl) rods (1cm x 1cm x 10cm) looked at by either PIN diodes or APDs at both ends (1cm x 1cm). Putting such rods together, one can also make a hard X-ray detector array with adequate spatial resolution. Fig. 4 shows the response of such a CsI(Tl)/APD detector system with a threshold of ~ 20 keV.

The third concept we want to pursue is to use a large-area (~10" diameter) scintillator plate (1 cm thick) coupled through a fiber-optic taper to a proximity-type image-intensifier tube and CID camera readout system. Most of these ideas are being vigorously pursued by medical researchers. The solid-state detectors like HgI_2 or CdTe offer simplicity and good spectral resolution over a wide energy range (1-500 keV). Since these solid-state detectors are still not available commercially, the cost to make up large area detectors is very high.

Desai, U.D.: 1992, *ESA-356*, 285-287.

Mertz, L.: 1989, *SPIE* 1159, 14.

Rogers, G.: 1977, *Noncoherent Optical Processing*, Wiley: New York.

INCIDENT PIXEL RECONSTRUCTION FOR GAMMA RAY TELESCOPES

A. R. GREEN, F. LEI, A. J. BIRD, I. D. JUPP and A. J. DEAN

Astronomy Group, Physics Department, University of Southampton,
Southampton, SO17 1BJ ENGLAND

Abstract.
In this paper we show how the incident pixel reconstruction necessary in a scattering γ-ray telescope can affect the imaging performance and sensitivity of the telescope. We show that the incident pixel reconstruction algorithms are least efficient at around 600 keV due to the nature of Compton scattering at these energies.

Key words: Coded Mask Telescopes

1. Introduction

The sensitivity of a coded mask telescope relies not only on the ability to detect photons but also on the ability to position the photon correctly. The positional point spread function must be smaller than the size of the mask pixels or the shadowgram will appear unfocused and significant coding errors will result in a reduction in sensitivity. In this paper we present the results of simulations of a pixellated detector in which many of the events are actually scattered from one pixel to another. We examine the accuracy of the incident pixel reconstruction (IPR) and discuss the effects that the IPR has on the imaging capabilities of the telescope.

2. The Detector

The detector geometry used is a 3 layer design, with a semiconductor array top layer, and two lower layers which are scintillator/photodiode arrays. The semiconductor layer is made of CdTe 2mm thick and the scintillator arrays are 3cm of CsI(Tl). All the pixels are hexagonal with an across flats dimension of 11mm and a centre to centre spacing of 11.75mm. The detector is surrounded on all sides, and on the bottom by a 2cm thick BGO veto crystal.

3. The Simulations

This geometry has been fed into the Geant software which provides, for each event, a listing of the first 10 interactions in the detector, giving the pixel in which the interactions occurred and the energy of each interaction. If there are more than 10 interactions associated with any one incoming photon

235

L. Bassani and G. di Cocco (eds.), Imaging in High Energy Astronomy, 235–238.
© *1995 Kluwer Academic Publishers.*

then the energies of interactions 10+ are grouped into the 10th event and the positional information is lost.

For each interaction, the energy is broadened following a Gaussian distribution defined by the spectral resolution quoted in Table I. The relevant threshold is then applied, and finally, if any of the interactions occur in the veto the complete event is ignored.

Material	Lower Threshold (keV)	Upper Threshold (keV)	Spectral Resolution FWHM
CdTe	15	750	$0.026 + \frac{3.354}{E}$
CsI(Tl) Middle	80	4500	$72.84 \times E^{-1.0303}$
CsI(Tl) Bottom	80	4500	$72.84 \times E^{-1.0303}$
BGO (Veto)	100	–	$3.96 \times E^{-0.5}$

TABLE I

Spectral resolutions and thresholds used in this work.

4. Incident Pixel Reconstruction

In this paper we just study the simplest of IPR techniques, i.e methods where one of the interactions is chosen as occuring in the incident pixel, rather than centroiding techniques where fractional pixels can be used.

For a single layer detector one can choose the incident pixel by taking either the interaction with the maximum energy or with the minimum energy, however with a multilayer detector there are further choices where either the interaction with the least or greatest depth into the detector is selected. Here we use 6 different IPR methods :

E1 - Maximum Energy

E2 - Minimum Energy

D1E1 - Maximum Depth (i.e. bottom most layer) & Maximum Energy

D1E2 - Maximum Depth & Minimum Energy

D2E1 - Minimum Depth (i.e. top most layer) & Maximum Energy

D2E2 - Minimum Depth & Minimum Energy

Some of the IPR methods are redundant for some types of interactions.

The best reconstruction method depends upon the total energy of the event and the type of scattering which occurs in the detector. To determine which IPR method is best we split the events into different modes depending on how many interactions there are and where they occur. Then, for each mode, we determined the IPR efficiency of each method at a range of energies from 20 keV to 10 MeV, where the IPR efficiency is defined as :

$$\text{IPR efficiency } (\%) = \frac{\text{\# photons assigned to the correct pixel}}{\text{\# photons detected by the detector}} \times 100$$

Having determined which IPR method is best for each type of event the IPR efficiency vs energy has been calculated. The IPR efficiency as a function of energy is shown in Fig. 1(a). The main features of Fig. 1(a) are the gradual decline from 100 – 600 keV and then the sudden fall off in IPR efficiency above ∼ 4500 keV. The latter effect is due to the upper threshold of the detectors and is a result of many of the interactions not being detected. The 600 keV dip is the combination of a number of different factors. At

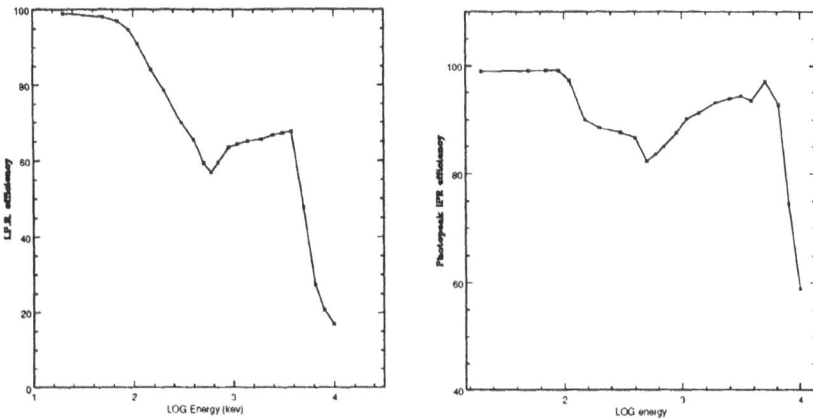

Fig. 1. (a) The IPR efficiency as a function of energy. (b) The photopeak IPR efficiency as a function of energy

lower energies most of the interactions are not scattered, and so the IPR efficiency is very high. However, above 300 keV many of the events are scattered events. For scattered events where the incident energy is a few hundred keV the scattering is mainly slightly forward meaning that the first interaction is less than half the total energy. For energies less than 600 keV and the thresholds and spectral resolutions shown in Table I, it is quite likely that the first interaction will not be detected, and so the incident pixel will not be accurately reconstructed. For incident photon energies greater than 600 keV all the interactions have a higher energy and are all detected.

A further effect which is evident in Fig. 1(a) is that of scattering out of passive material. The modelled detector contains about 25% dead area which is mostly the Aluminium support structure. Above a few hundred keV all of the interactions in Al are Compton scatters, and it is highly likely that the scattered photon will interact in the active pixels. The effect of scattered events can be removed by plotting the photopeak IPR efficiency which is shown in Fig. 1(b). One can clearly see in this figure where Compton scattering in the CsI becomes the dominant effect and that toward the higher energies the photopeak IPR efficiency nearly reaches 100% before the high energy threshold cuts in.

5.   The Effect of IPR on the Imaging Performance

If photons are not assigned to the correct pixel they are generally assigned to the adjacent pixel. This effect is shown graphically in Fig. 2 in which the positional point spread functions are shown at four different energies. The increase in IPR PSF as a function of energy leads to a decrease in sensitivity as the mask pixel/detector pixel sampling ratio is reduced due to the detector pixels being effectively slightly larger than their physical size.

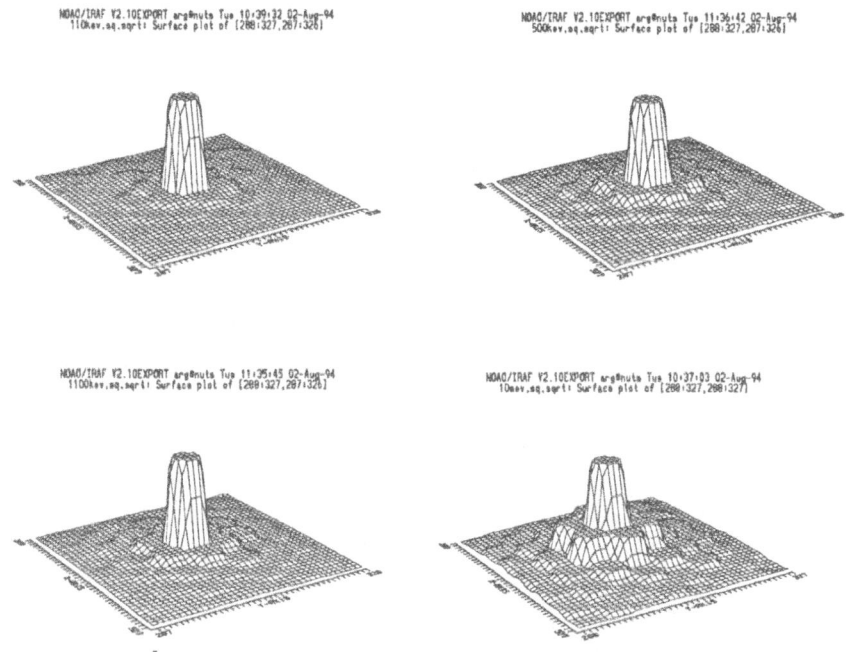

Fig. 2.   The positional point spread function for the incident pixel reconstruction at four different energies, 110 keV, 500 keV, 1100 keV and 10 MeV. The data has been rebinned into squares for plotting and is shown on a square root scale to emphasize the outlying pixels

6.   Summary

Even in a discrete pixel detector the effective pixel size might be slightly larger than the actual pixel size due to the reconstruction of scattered events. This can have a detrimental effect on the sensitivity of the instrument.

A Spaceborne Crystal Diffraction Telescope for the Energy Range of Nuclear Transitions

P. von BALLMOOS, J.E. NAYA, F. ALBERNHE, G. VEDRENNE

Centre d'Etude Spatiale des Rayonnements, 9, av. du Colonel-Roche, 31029 Toulouse, France

and

R. K. SMITHER, M. FAIZ, P. B. FERNANDEZ, T. GRABER

Advanced Photon Source, ANL, 9700 South Cass Ave., Argonne Ill, USA

Abstract. Recent experimental work of the Toulouse-Argonne collaboration has opened the perspective of a focusing gamma-ray telescope operating in the energy range of nuclear transitions, featuring unprecedented sensitivity, angular and energy resolution. The instrument consists of a tunable crystal diffraction lens situated on a stabilized spacecraft, focusing gamma-rays onto a small array of Germanium detectors perched on an extendible boom. While the weight of such an instrument is less than 500 kg, it features an angular resolution of 15", an energy resolution of 2 keV and a 3 σ narrow line sensitivity of a few times 10^{-7} photons s^{-1} cm^{-2} (10^6 sec observation). This instrumental concept permits observation of any identified source at any selected line-energy in a range of typically 200 keV to 1300 keV. The resulting "sequential" operation mode makes sites of explosive nucleosynthesis natural scientific objectives for such a telescope : the nuclear lines of extragalactic supernovae (^{56}Ni , ^{44}Ti, ^{60}Fe) and galactic novae (p$^-$p$^+$ line, ^{7}Be) are accessible to observation, one at a time, due to the erratic appearance and the sequence of half-lifes of these events. Other scientific objectives include the narrow 511 keV line from galactic broad class annihilators (such as 1E1740-29, nova musca) and possible redshifted annihilation lines from AGN's.

1. Introduction

Imaging combined with high resolution spectroscopy will be one of the major goals of the next generation of space borne gamma-ray telescopes. With the spectrometer on ESA's INTEGRAL mission, such an instrument will be available to the high energy community at the beginning of the next decade. High resolution spectroscopy will be performed by a bank of germanium detectors while the imaging is achieved by a coded aperture system [1]. The foremost objectives of this instrument will be the mapping of gamma-ray line sources emitting 10^{-4} photons s^{-1} cm^{-2} to a few times 10^{-6} photons s^{-1} cm^{-2}. Candidate sources of this intensity include the sites of recent nucleosynthesis, regions of e$^+$e$^-$ annihilation and clouds where nuclear de-excitation by energetic particles takes place. Many of these potential sources will be galactic. Some of them might appear as extended structures - either because of their truly diffuse origin, or because they are relatively closeby as the nucleosynthesis sites in the local spiral arm. A wide field of view and a mid-scale angular resolution make the INTEGRAL spectrometer adequate for such objectives.

In the future, experimental gamma-ray astronomy has to find ways to improve the observational performances. Yet, achieving sensitivities better than 10^{-6} photons·s^{-1}·cm^{-2} and resolutions better than fractions of a degree seems to be impossible with the presently practiced instrumental concepts: even larger collection areas are synonymous with larger detectors and thus again higher background noise.

L. Bassani and G. di Cocco (eds.), Imaging in High Energy Astronomy, 239–245.
© *1995 Kluwer Academic Publishers.*

A new type of gamma-ray telescope featuring a Laue-diffraction lens can overcome the impasse of present detectors where the collection area is identical to the detection area. As it was originally proposed, this focusing gamma-ray telescope [2],[3] has been designed to collect e^+e^- annihilation radiation on a large effective area (~ 150 cm^2) and focus the photons onto a Germanium detector matrix with a small equivalent volume for background noise (~ 14 cm^3) As a balloon-borne instrument it can provide high energy- and high angular resolution (2 keV, 15 arc sec, respectively) combined with an excellent sensitivity ($\sim 3 \cdot 10^{-5}$ photons s^{-1} cm^{-2} @ 511 keV). The performances of this Ge-lens/Ge-matrix system have been verified in June/July 1994 during laboratory measurements with a ground based prototype [4]. The instrument has first been proposed as a balloon-borne telescope with the lens tuned to diffract 511 keV photons only. Such a configuration makes possible the study of galactic "microquasars" and other broad class annihilators in the light of e^+e^- annihilation during a balloon flight.

Ultimately however, the concept should be put to use in space where longer exposures and steady pointing would result in outstanding sensitivities. Yet, as a satellite instrument, a monochromatic lens would clearly be a handicap since its scientific objectives are too exclusive - already e.g. the possible annihilation line of most extragalactic sources (AGN's, quasars) would be inaccessible because of cosmological redshift.

Here we present a space borne crystal diffraction telescope using a tunable gamma-ray lens for the energy range relevant for nuclear transitions 200 keV - 1300 keV. An "adaptative gamma-ray optic" permits observation of any identified source at any selected line-energy in a range of typically 200 keV to 1300 keV. The "sequential" operation mode resulting from such a concept makes the sites of explosive nucleosynthesis natural targets for a tunable crystal diffraction telescope.

2. The scientific case for a tunable crystal diffraction telescope

According to our present view of celestial gamma-ray sources in the energy range of nuclear transitions, narrow lines seem to be generally emitted from extended distributions while broad lines tend to be radiated by point sources. Besides of the supernovae 1987A [5] and 1991T [6] the evidence for point like sources of narrow gamma-ray line emission has been mostly implicit at this point. We therefore have to ask a) where the scientific potential of a tunable diffraction telescope is and b) how many source candidates can be expected for such an instrument.

a) the scientific potential of a tunable diffraction telescope : Sources of narrow line emission are though to have little angular extent if they are sufficiently distant or if the activity of their high energy processes is very recent. In either case the intensity of the emitted lines will be weak as the relatively rare nucleosynthesis events like SN or novae are more likely to occur at large distances. A crystal diffraction telescope with its narrow beam and excellent sensitivity is optimally suited for the detection of such sources. Besides of the sites of explosive nucleosynthesis (e.g. ^{7}Be, ^{13}N, ^{22}Na from novae, ^{56}Ni, ^{44}Ti, ^{60}Fe from supernovae), the scientific objectives include: narrow 511 keV lines from galactic broad class annihilators (such as 1E1740-29, nova Muscae) and from AGN's; nuclear de-excitation lines from energetic particle interaction with the ISM or dust grains (lines with energies above 1.3 MeV might become accessible to the lens if they are emitted by AGN's with high z); lines from the excited nuclei in solar flares.

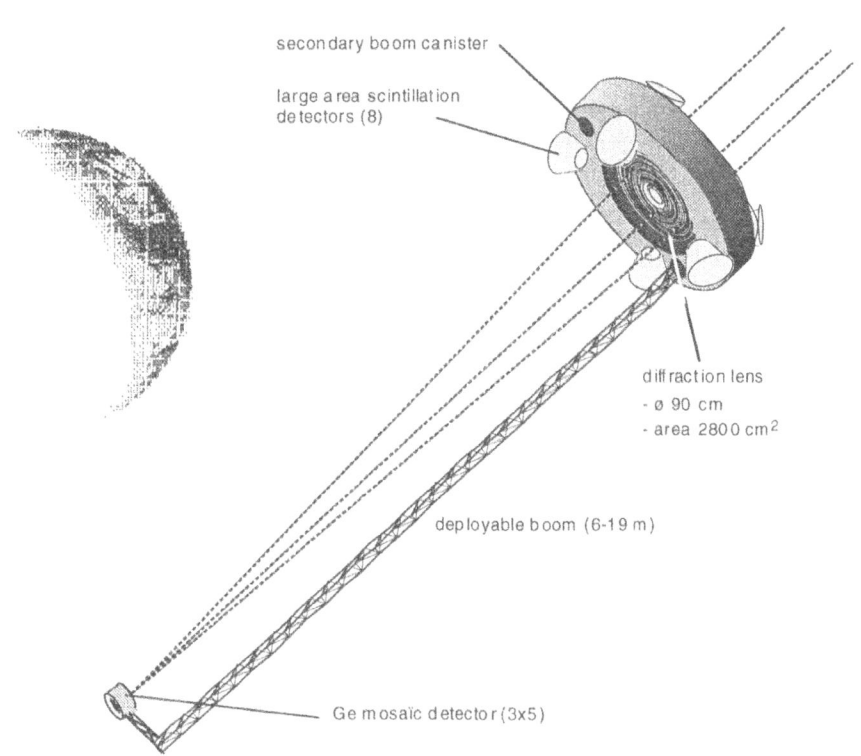

secondary boom canister

large area scintillation
detectors (8)

diffraction lens
- ø 90 cm
- area 2800 cm²

deployable boom (6-19 m)

Ge mosaïc detector (3x5)

Figure 1 : "artists view" of a space borne crystal diffraction telescope. As a counterpart to the extremely directional lens telescope, a full sky monitor could complement the payload. This possibility is indicated by the eight large area scintillators that would use the Earth or the Moon as "rotating modulation collimator" in order to pinpoint transient sources of interest.

b) How many source candidates for a crystal lens telescope ? Because of its narrow beam (~15") and energy band (typically 8-15 keV) the observed "astronomical area" $\Delta l \Delta b \Delta E$ (gal. longitude interval, gal. latitude interval, energy interval) of a crystal diffraction telescope is very small : this implies that typically only one source and one gamma-ray line can be observed at any one time. Yet, for sources such as the isolated events of explosive nucleosynthesis (novae, supernovae) the number of detectable sources is not related to the field of view, but to the sensitivity Δs which defines the "observable volume" $\Delta l \Delta b \Delta E \Delta s$. The sensitivity of a crystal diffration lens makes supernovae of the Virgo cluster detectable (about ten SN of type Ia per year with optical peak magnitude $m_V < 18$) while all galactic novae are accesible (the frequency of Nova explosions in our galaxy is about 40 per year).

The presented instrument also can help to clarify the long debated problem of the Galactic Center annihilation radiation. Since the intense bulge component (~10^{-3} ph cm^{-2}·s^{-1}, ~10° FWHM) of the galactic 511 keV emission [7] should not yield any line flux within the extremly narrow 15" field of view, the hypothesis of a possible contribution of isolated broad class annihilators can be tested.

3.   Characteristics and feasibility

A space borne telescope using an adaptative crystal diffraction lens will consist of three modules : the lens module, the detector module, and a boom. Optimally the lens module is located directly on the spacecraft, while the detector module is perched on the boom. The characteristics of a possible space borne gamma-ray lens telescope are summarized in Table 1 - an artists view of the concept is shown in Fig 1.

The *lens module* consists of a 90 cm diameter frame accommodating 700 germanium cubes. The single-crystal are organized in 11 rings, each ring uses a different set of crystalline planes to diffract the gamma rays. The crystals are oriented so that they all diffract the incident radiation of a certain energy to a same focal point. The 5 inner rings are composed of 1.5 cm thick crystals with an exposed area of 2 cm x 2 cm. Due to their thickness and position on the frame, these rings are optimized for the higher energies (a 1 MeV photon will still "see" ~$2 \cdot 10^5$ [220] planes - spaced at a distance 160 times larger than its wavelength - while the probability of its absorption is only 36%). The crystals in the outer 6 rings each have the same geometric area (4 cm^2) as the inner ones, yet they are only 0.5 cm thick. These rings are optimized for the lower energies. Above 600 keV they still can be used for diffraction with higher order planes - however with reduced efficiency.

Tuning the lens to an energy $E=E_{ref}+\Delta E$ requires that each of the single crystal is rocked by an angle $\Delta\Theta$ with respect to the position of a reference energy E_{ref} (ie. 511 keV) in order to satisfy the Bragg condition anew.

$$\Delta\Theta = \arcsin(hc/2d(E_{ref}+\Delta E)) - \arcsin(hc/2dE_{ref})$$

Tuning a crystal (e.g. the [220] planes) from 200 keV to 1300 keV implies a $\Delta\Theta$ of 0.75° corresponding to a displacement of 0.4 mm over a 3 cm lever of the crystal base plate. It is essential that precision/repeatability of this motion is of the order of <2 arc seconds - this is : better than the rocking curve for a single crystal.

These requirements comply with the performance of a device consisting of a piezo-driven actuator and an Eddy-current sensor that measures the displacement. The miniaturized closed loop system is being built and tested at CESR and ANL.

detector module : In order to take maximum advantage of the particular properties of a focused gamma ray beam, a germanium matrix will be used for the detector module. During the tests with our ground based telescope [4] a similar germanium matrix has been found to be ideally adapted to resolve the beam energetically and spatially. The matrix consists of 3x5 detector elements, each one with a geometric surface of 3x3 cm and a height of 7-8 cm. The 2.8 cm FWHM focal spot produced by the lens will optimally be pointed at one of the central detector elements. Using isotopically enriched ^{70}Ge as detector material will reduce the β$^-$ background component in our energy range while the enhanced β$^+$ production only effects the background above 1.5 MeV. Further reduction of the non-localized nβ components will be possible using the 15 matrix segments. The matrix also offers the possibility to monitor the remaining background simultaneously to the astrophysical observation. The low intensity of spacecraft induced background will allow us to use a detector shielded only by a very light anticoincidence shield.

In space, cooling of the detectors can be performed by a small sterling cryogenator, by a small tank liquid of liquid nitrogen, or passively by a radiator.

retractable boom : Since the focal length of the lens is increasing with energy, a retractable boom (ie. the coilable tube mast [8]) will be used to the vary the distance between spacecraft and detector along the optical axis of the lens. An energy of 200 keV and 1240 keV respectively corresponds to a change in focal length of 3 m to 20 m - for the 511 keV positron annihilation line the distance lens-detector is 8.3 m. Booms have been used in gamma-ray astronomy on Apollo 15 with a NaI detector and on Mars-Observer for the Ge detectors. In both cases, the extension was around 7 m. Deploying the detector on a boom instead of the diffraction lens has several striking advantages: The mechanical requirements on the mast rigidity are less severe since a Germanium detector array is small and lightweight and thus easy to handle on a boom; moreover, twists and bends of even up to a few cm's are tolerable, as the focal spot (ø 2.8 cm) can wander around on the detector array (total surface 9x15cm) without significant loss of sensitivity. On the other hand, the stringent requirements for the pointing of the lens (typically ~ 5") can be satisfied on board the pointed and stabilized spacecraft. Finally, moving the detector away from the spacecraft reduces the background by up to an order of magnitude (depending on the energy, see section 4). In order to have a mechanically redundant system, the spacecraft will feature two 'detector-boom systems'. If both detectors were to be operated at the same time, different energy-bands could be observed simultaneously, or, maximum sensitivity at one energy band can be achieved by combining the two collector-zones.

4. System performance

The imaging capabilities of a crystal diffraction telescope are defined by its beamwidth which is identical to the field of view of the lens : for compact sources discrete pointings of the object will be an appropriate observation mode, while extended structures as for example the jets of galactic microquasars will be scanned with the narrow beam. The field of view depends on the angular width of the mosaic structure of the crystals.

The calculations presented here assume a mosaic structure width of 10" resulting in a $\Omega \approx 16$" FWHM for the field of view.

Crystal lens		
diffracting medium	:	700 Germanium crystals (2cm x 2cm x 0.5/1.5cm)
diameter of lens frame	:	90 cm
tunable energy range	:	200 keV - 1300 keV
focal length	:	6.5 m - 19 m
diameter of focal spot	:	2.8 cm FWHM (at all energies)
energy bandwidth	:	6 keV FWHM @ 511, 10 keV FWHM @ 847 keV
diffraction efficiency	:	25% @ 511 keV, 11% @ 847 keV
effective collection area	:	2800 cm^2
Detector matrix		
detector type	:	15 high purity Germanium, coaxial, 3x3x8 cm
energy resolution	:	2 keV (@ 511 keV)
total detecting volume	:	1080 cm^3
total detector area	:	135 cm^2
efficiency at 511 keV	:	54% @ 511 keV, 44% @ 847 keV
Telescope system		
field of view / ang.res.	:	15" FWHM
system effective area	:	370 cm^2 @ 511 keV, 275 cm^2 @ 847 keV
effective volume for BG	:	72 cm^3

Table 2: Summary of instrument characteristics

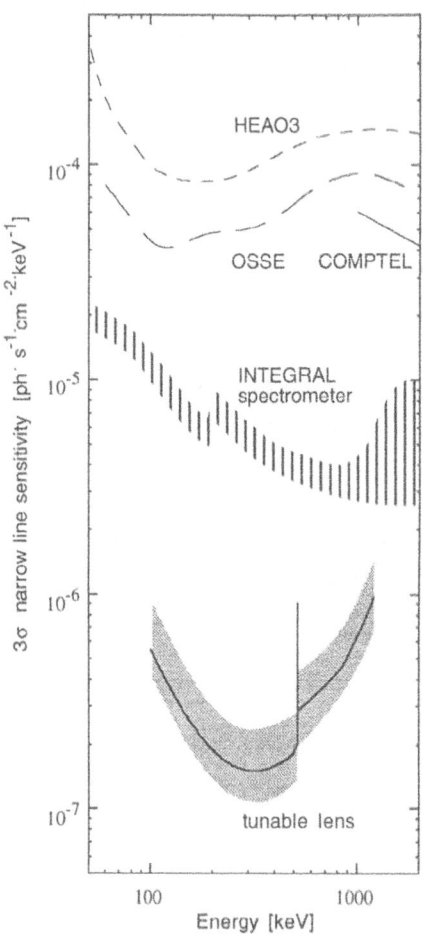

If a larger field of view is desirable for certain objectives, the beam can be widened by "detuning" the crystals with the individual closed loop servo systems.

Spectroscopy (Doppler shifts and broadening) of the lines is possible within the bandwidth of the crystals

$$\Delta E \;=\; E\,(\Omega/\Theta)$$

For the entire lens the bandwith is ~6 keV at 511 keV and ~15 keV at 1300 keV. The spectral resolution of present Ge detectors is typically 2 keV at 1 MeV. For a point source at infinite distance we estimate diffraction efficiencies of the order of 26% at 200 keV, 8% at 1000 keV. The full energy peak efficiency of the detector matrix has been calculated by GEANT (80% at 200 keV, 39% at 1000 keV).

We use a background based on the *measured* [70]Ge spectrum of the GRIS detectors during a balloon flight at Alice Springs in 1992 [20] - i.e. $2 \cdot 10^{-5}$ c·s[-1]·cm[-3]·kev at 200 keV, $2 \cdot 10^{-4}$ c·s[-1]·cm[-3]·kev at 511 keV. Yet, we assume that the [70]Ge background can be multiplied by a correction factor f<1 because of lower "shield leakage"- and "nβ"-contributions. The intensity of the background is decreasing when a detector is brought away from the "bright" spacecraft (or the earth). This solid angle effect has been demonstrated with a small scintillator that has been deployed on a boom on Apollo 15 [9]: Compared to the on board spectrum, at a distance of 7 m from the spacecraft, the background in the range 0.2-1.3 MeV was down by a factor of 4-8, at 511 keV even by a factor of 10. For our instrument, we have assumed that the spacecraft induced background will be strongly reduced for the above

Figure 2 : the estimated sensitivity of the presented instrument is shown together with the sensitivities of past, present and futur telescopes. For a crystal lens telescope, the point source sensitivity depends on the diffraction efficiency of the lens, the full energy peak efficiency and the background of the detector.

reason. Furthermore, the resulting low mass (light shield) of the detector module will again reduce the background (nβ[-] and nβ[+] components [10]) as less neutrons are produced compared to the present heavily shielded gamma-ray spectrometers.

Conclusion

The crystal diffraction telescope constitutes a breakthrough for the study of compact sources by combining an excellent sensitivity (a few 10^{-7} $ph\cdot s^{-1}\cdot cm^{-2}$) with high energy resolution ($E/\Delta E \approx 500$) and very good positioning (<15"). The performances of this concept will open new perspectives to gamma-ray astronomy.

Even though technically innovative, a tunable crystal diffraction telescope has become feasible today: 1) A monochromatic prototype lens suitable for an astronomical instrument exists and has been tested in the laboratory at energies up to 700 keV. 2) The energy-tuning of single crystals is possible using todays piezo-technology; an integrated closed loop system is being developped by our collaboration. 3) Germanium detector arrays are manufactured today and have demonstrated their advantages in conjunction with the prototype lens. 4) Various space experiments have alreay been carried out using extendable booms.

The project of a tunable crystal diffraction telescope is supported by the French space agency CNES since 1994 and has been selected by NASA for a mission concept study in 1995. This work is partially supported by the U.S DoE Contract No. W-31-109-Eng-38. The authors are grateful to Francis Cotin for his contributions to this project.

References

[1] INTEGRAL "redbook", ESA phase A study report, April 1993
[2] von Ballmoos P. and Smither R.K. 1994, ApJ Sup. S., 92, 663
[3] Smither R.K. et al., 1995, these proceedings
[4] Naya J.E. et al., 1995, these proceedings
[5] Matz S.M. et al., 1988, *Nature* **331**, 416
[6] Morris D. et al. 1995, proc. 17th Texas Symposium on Relativistic Astrophysics
[7] Purcell W.R et al., 1994, AIP Conf. Proc. **304**, p. 403, (ed. Fichtel C. et al., New York)
[8] Aguirre M., Bure R., del Campo F., Fuentes M., 1987, Proc 3rd Europ. Space   Mechanisms & Tribology Symp., Madrid, ESA SP-279, Dec 1987
[9] Trombka et al 1973, *ApJ* ,**181**, 737
[10] Naya J.E. et al. 1995, submitted to *Nucl. Inst. and Methods*.

A STUDY OF THE EFFECTS OF BACKGROUND
SUBTRACTION ON OCCULTATION IMAGING

W. S. PACIESAS
University of Alabama in Huntsville
Huntsville, AL 35899 USA

and

S. N. ZHANG,* B. C. RUBIN,* B. A. HARMON and G. J. FISHMAN
NASA/Marshall Space Flight Center
Huntsville, AL 35812 USA

Abstract. One of the crucial steps in the occultation transform imaging technique involves the removal of the time-varying instrumental background. Previous versions of this imaging technique applied to BATSE data have used simple high-pass filtering to eliminate background variations on timescales longer than the typical duration of an occultation step (of order 10 s). We have investigated an alternative technique in which the imaging algorithm is applied to the residuals generated from fitting the raw data with a semi-empirical model of the background. Comparison of the resulting maps shows that the latter does not significantly improve imaging performance.

Key words: X-Rays: Imaging – X-Rays: Background Studies

1. Introduction

Though observations of gamma-ray bursts are its primary objective, BATSE has considerable capabilities as a hard X-ray all-sky monitor. Pulsed sources may be detected by BATSE using Fourier and/or summed epoch analysis, and any type of source may be detected by Earth occultations, *i.e.*, measuring the difference in background rate between times when a source is visible and times when it is occulted by the Earth. The usefulness of BATSE for monitoring was increased substantially when we developed the method of occultation transform imaging [1, 2] which allows us to obtain more accurate positional information on transient sources and to discriminate better among sources in crowded regions. An overview of the imaging methodology is given elsewhere in these proceedings [3]. Here we describe the necessity for, and implementation of, background subtraction when producing images.

2. Background Effects in Occultation Imaging

Occultation imaging is possible only to the extent that the variations due to a source rising or setting may be distinguished from the time variability of the background due to other effects (*e.g.*, electron precipitation, cosmic ray

* also Universities Space Research Association

L. Bassani and G. di Cocco (eds.), Imaging in High Energy Astronomy, 247–250.
© *1995 Kluwer Academic Publishers.*

latitude dependence, solar flares, SAA activation). In the case of BATSE, these variations generally fall into one of two categories: 1) intermittent fluctuations on timescales shorter than, or comparable to, the duration of an occultation step, or 2) more regular, longer-timescale variations.

In practice, we remove the two different types of background fluctuations by different methods. The short timescale fluctuations are intermittent and relatively rare. These are quite effectively removed by visual scanning of the data: periods of activity on these timescales are flagged as part of daily BATSE operations and removed from subsequent occultation analysis. The longer timescale fluctuations are more difficult because they are ubiquitous, depending on the combined effects of the cosmic diffuse and atmospheric backgrounds, trapped cosmic rays, SAA activation, etc., whose relative contributions depend on energy as well as spacecraft position. During the initial development of the occultation imaging software, the available background models were not sufficiently mature to be useful, so a simple differentiation was applied to the data and the results modeled by the Radon transform [1, 2]. These early images clearly showed the usefulness of the technique and stimulated us to look for ways to improve the sensitivity.

One of the first significant improvements was to eliminate the simple differentiation scheme in favor of a high-pass filter to eliminate the lower frequency background components. Fig. 1 illustrates the effectiveness of this methodology. The uppermost curve in the figure shows slightly more than two orbits of raw data (1.024 s resolution) from a single BATSE detector. Gaps in the data are evident, typically resulting from either transmission errors, South Atlantic Anomaly passages, or filtering of short-timescale fluctuations. The next lower curve shows the same data with gaps filled in by a simple linear interpolation. The bottom two curves compare the data after differentiation (upper) with the data after application of a second order high-pass Butterworth filter (lower). Comparison of the lowest curve with the idealized point source signal (see Fig. 1 of ref. [3]) shows the obvious improvement in signal-to-noise for filtered vs. differentiated data.

In parallel with the development of occultation imaging, we have been developing semi-empirical background models for eliminating the longer timescale background variations [4, 5]. Fitting of the model described in Ref. [4] is now being performed daily as part of routine BATSE operations. The output files produced include not only the model terms but also the full set of data residuals.

3. Comparison of Images

It is natural to consider whether the use of a proper background model would further improve the occultation imaging relative to the simpler high-pass filter. Thus, we embarked on the present study to compare the two methods.

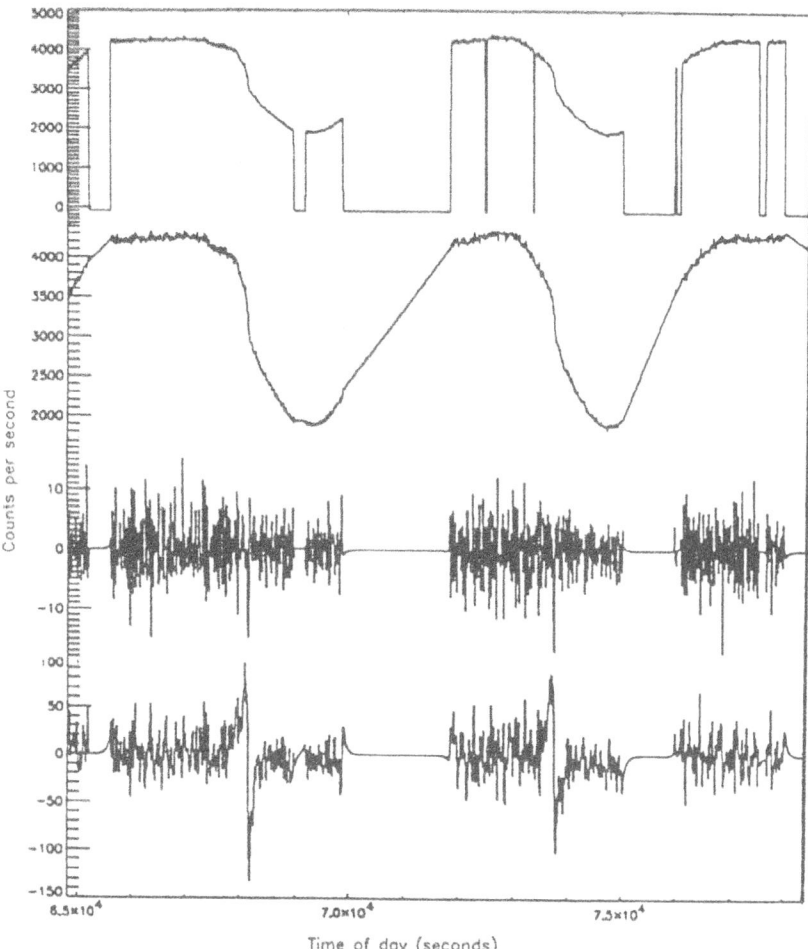

Fig. 1. Example of data treatment for occultation transform imaging. The uppermost pair of curves show raw BATSE data with gaps present and with gaps filled in, respectively. The lowermost pair of curves show the same data after differentiation and high-pass filtering, respectively.

We generated images of selected regions using both our standard imaging software which includes the Butterworth high-pass filter and a modified version which bypasses the filter and uses the background model residual files. Subtraction of the background model has the same effect on the point source signal as the high-pass filter; hence, the same mapping algorithm can be applied to both cases. Fig. 2 shows a typical result, in this case single-day images of a region in Scorpius which includes a strong source, the recent X-ray nova [6, 7]. The filtered image is slightly better in that it converges to a single source; however, the two images are effectively the same.

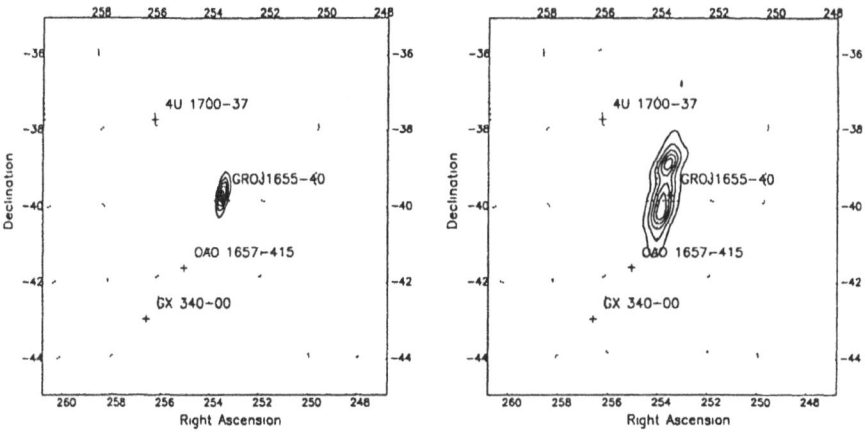

Fig. 2.   Images of Nova Scorpii 1994 on TJD 9575. The left hand image was produced with the high-pass filter; the right hand image was produced from the background model residuals.

We made additional comparisons for several different sky regions having a range of source intensities. Although some small differences were evident as in Fig. 2, we found no consistent reason to favor either method.

4.   Conclusions

For the current occultation imaging methodology, there is no advantage to using the semi-empirical background model. The images produced using the Butterworth filter, which is computationally less intensive, are at least as good as the images produced from the model fit residuals. Given this performance, we are implementing a similar high-pass filter in our other occultation software packages. We expect that this will significantly improve the BATSE sensitivity for detection and flux determination of hard X-ray and low-energy gamma-ray sources.

References

1.   Zhang, S.N. *et al.*: 1993, *Nature* **366**, 245
2.   Zhang, S.N. *et al.*: 1994, *IEEE Trans. Nucl. Sci.* **41**, 1313
3.   Zhang, S.N. *et al.*: 1995, these proceedings
4.   Rubin, B.C. *et al.*: 1993, in M. Friedlander, N. Gehrels & D.J. Macomb, eds., *Compton Gamma-Ray Observatory*, AIP: New York, 1127
5.   Skelton, R.T. *et al.*: 1993, in M. Friedlander, N. Gehrels & D.J. Macomb, eds., *Compton Gamma-Ray Observatory*, AIP: New York, 1189
6.   Zhang, S.N. *et al.*: 1994, IAU Circ. No. 6046
7.   Wilson, C.A. *et al.*: 1994, IAU Circ. No. 6056

HIGH SENSITIVITY IMAGING OF SOFT γ-RAYS
USING BRAGG CONCENTRATORS

NIELS LUND

Danish Space Research Institute
Gl. Lundtoftevej 7
DK 2800 Lyngby, Denmark

September 24, 1994

Abstract. An imaging telescope, capable of focussing soft γ-rays in the energy range between 300 keV and 2 MeV has been studied. This energy range encompasses many of the nuclear γ-ray lines of astrophysical interest, and, not the least, the electron-positron annihilation line and its wings. The study concludes, that a Bragg-concentrator system may provide order of magnitude increases both in sensitivity and in source localization compared to the INTEGRAL instrumentation, while at the same time allowing some rudimentary imaging over fields of about 10 arcminutes diameter. The feasibility of this concept still remains to be proven, however, since the telescope dimensions are very large, with focal lengths of the order 50 m.

Key words: Gamma-Ray Astronomy – Annihilation Radiation – Telescopes

1. Introduction

Soft γ-rays, in the energy range between 300 keV and 2 MeV are important tracers of high energy processes in the Universe. Here we find the signatures of recent nucleosynthesis in novae and supernovae, as well as annihilation photons from nuclear decay positrons and pair plasmas. The γ-photons have great penetration power, and they reach us. with little attenuation, from regions which are heavily obscured in the optical or the soft X-ray domains. The energy and spectral distribution of the photons brings information on gravitational redshifts and Doppler motions in the source regions. The intensity ratio between the annihilation two-photon line and the three-photon continuum constrains the temperature, the density and the ionization state of the medium, in which the annihilation takes place [3].

By its nature, Bragg diffraction is constrained to a small number of narrow energy bands. However. we have specifically directed our study towards a telescope capable of flux concentration over an extended band of energies. We expect a Bragg telescope to be a large and complex system, even if designed only to focus a narrow energy band. Thus we may just as well set ourselves ambitious goals. to be assured that the scientific capabilities will correspond to the effort needed to realize the project. During the study we have also realized, that the broadband approach brings an important added advantage, namely that the telescope acquires some surprising imaging capabilities.

251

L. Bassani and G. di Cocco (eds.), Imaging in High Energy Astronomy, 251–254.
© *1995 Kluwer Academic Publishers.*

2.   Telescope Concept

The concept of the Bragg telescope has been described previously by several authors [1,2,4]. The incoming γ-rays traverses a "lens" assembled from a large number of crystal facets. Each facet is individually adjusted so the Bragg diffracted beams are directed towards a common focus. The diffracted radiation from each crystal is a little parallel beamlet. Thus the size of the combined focus for such a lens is determined by the overlapping beamlets from all the facets. For energies of around 511 keV the typical diffraction angles are about 0.6°.

We intend to use only a few different types of crystals in our lens. All facets will be tuned to exploit those crystal planes which provides the highest Bragg diffraction efficiency (the 111-direction for crystals with face-centered cubic (fcc-) structure). The orientation of each facet is therefore uniquely determined from its position in the lens. Facets near the telescope axis are tuned for high energies (small Bragg angles) and facets farther away tuned for lower energies. We will only be interested in crystals with the highest possible integrated reflectivity. We will, therefore, not be using crystals with perfect crystal structure, but rather crystals with a finite and carefully controlled degree of mosaic structure. The optimum thickness of each facet will depend on the γ-ray energies for which it is used. The facets near the telescope axis should therefore be significantly thicker than those further out.

3.   Optimization Considerations

We have investigated how the properties of the crystals affects the telescope performance [2]. We have found that a high packing density of the atoms in the crystal increases the Bragg diffraction efficiency. The atomic density varies in a characteristic way with the atomic number, Z. So we have concentrated our studies to the elements around the peaks in this distribution, i.e. around $Z = 13, 29, 45$ and 76. We have also realized, that for γ-rays with energies above 400 keV crystals of heavy elements provides higher Bragg diffraction efficiencies than crystals of lighter elements.

Finally, we have consulted the experiences obtained in solid state physics laboratories concerning the growth of crystals from the various chemical elements. We have concluded, that copper seems to be the most promising material for the Bragg gamma ray lens for energies between 100 keV and 400 keV, and gold appears as the best choice for energies above 400 keV. Both elements form crystals with fcc-structure. and both are relatively easily grown as crystals of large dimensions.

The degree of mosaic structure required in our application will be of order 1 arcminute. This is a small value which may be difficult to achieve in crystals grown in crucibles, but can certainly be obtained by pulling the crystals from

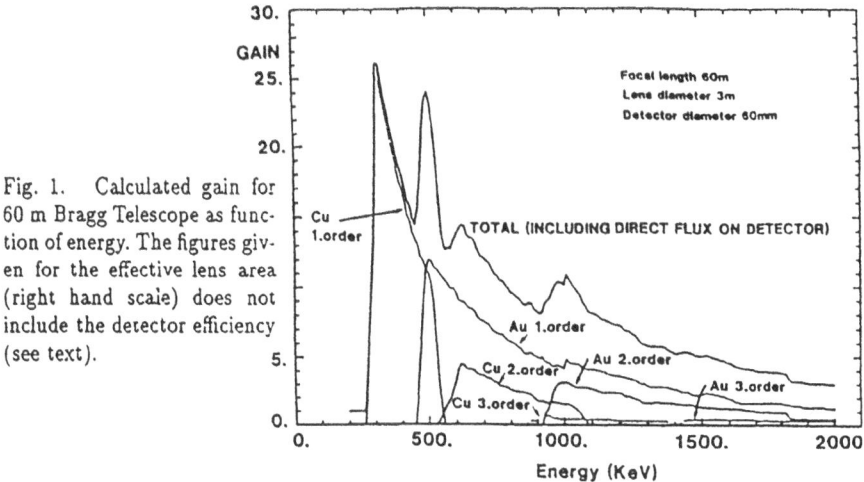

Fig. 1. Calculated gain for 60 m Bragg Telescope as function of energy. The figures given for the effective lens area (right hand scale) does not include the detector efficiency (see text).

the melt. The optimum value for the mosaic structure is determined by the ratio of the diameter of the detector to the focal length of the telescope.

4. Expected Sensitivity

When imaging is not required, we can use a a single, high-resolution germanium detector matching the size of the focal spot (in our simulations we have used a detector with 60 mm diameter). We define the gain factor, G, of a telescope as the factor by which the flux on the detector increases through the use of the lens. The sensitivity of a Bragg-telescope of gain, G, is equal to G times the sensitivity of the detector alone. To achieve the same sensitivity by operating many detectors in parallel one would need an array of G^2 similar detectors. We also note that any technique, which may conceivably be used to reduce the background in a cluster of detectors will be easier to apply to a single detector.

In figure 1 we show the calculated gain as function of energy for a Bragg lens with 60 m focal length. employing crystal facets made of copper (for energies below 530 keV) and gold (for energies above 490 keV). The Bragg angles for these two crystals are such that a certain overlap in the energy ranges covered is possible. we have chosen to use this overlap to increase the sensitivity around 511 keV.

When equipped with a modern low-background germanium detector, such a telescope will actually be signal limited for narrow line ($\approx$ 2 keV)-observations up to a duration of 10^6 s. The sensitive area will vary between $\approx$ 350 cm^2 at 300 keV and $\approx$ 25 cm^2 at 1300 keV. These values includes the expected efficiency for a germanium detector of 60 mm diameter and 70 mm length.

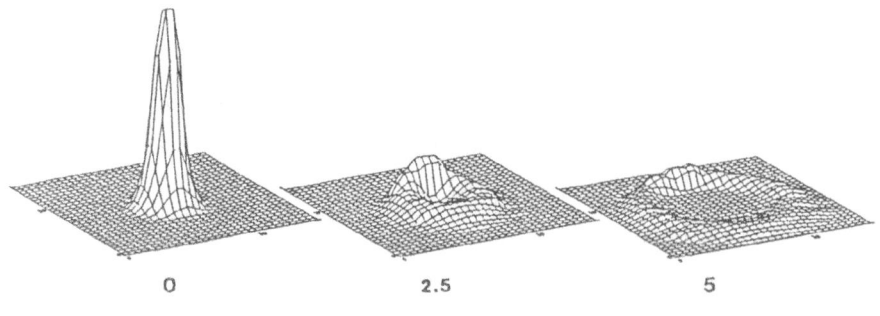

0   2.5   5

Arcminutes off axis

Fig. 2.   Point source images for Bragg-telescope as function of off-axis angle

5.   Imaging Properties

The response of a Bragg-telescope for an off axis source is reasonably constant as long as the image center is within half a detector radius from the telescope axis. For our 60 meter telescope with a 60 mm diameter detector this means within ±0.9 arcminutes. (This is, therefore, also the requirement for absolute pointing accuracy and stability for such a telescope). But beyond this value the response drops very rapidly, and is virtually zero when the image is more than twice the detector radius away from the axis.

However, when used with a large area, imaging detector at the focus, a Bragg-telescope may actually provide some rudimentary imaging over a 10′ field. The detector should have a useful diameter of about 250 mm and a position resolution of 20 mm to match the 60 m telescope.

Figure 2 shows some images computed for off axis sources. It will be seen how the strongly peaked image for the on-axis source is transformed into a low, circular ridge around the telescope axis for off-axis sources. The radius of the ridge corresponds to the off-axis angle, and the ridge is nonsymmetric with a high point where the image from the source would have fallen, had the laws of ordinary optics applied. It is remarkable, that the total signal is almost the same for all images, even when the off-axis angles are much larger than the mosaic spread for the individual facets. Despite the serious aberrations in the images of off-axis sources the sensitivity only drops by a factor 3 for sources 5′ off axis.

References

1. Lindquist, T. R. and Webber, W. R.: 1968, *Can. J. Phys.* **46**, 1103
2. Lund, N.: 1992, *Exp. Astr.* **2**, 259
3. Ramaty, R. and Lingenfelter, R. E.: 1981, *Phil. Trans. R. Soc.* **A 301**, 671
4. Smither, R. K.: 1982, *Rev. Sci. Instr.* **53-2**, 131

A NOVEL SYSTEM FOR THE LOCATION OF GRBS

F. LEI, M.J. PALMER, I.D. JUPP and D. RAMSDEN

Astronomy and Space Physics, Physics Dept, Southampton University
Southampton, SO9 5NH, ENGLAND

Abstract. A long standing problem in the identification of GRBs is the inability of current instruments to provide rapidly an accurate source location. An accuracy of a few arcminutes is required to allow follow-up observations at other wavelengths to be meaningful. An imaging system which employs one-dimensional coded masks in conjunction with silicon strip detectors can be used to locate bursts with an accuracy within a few arcminutes. The field of view of such an instrument could be more than 2 sr. This paper addresses the imaging principles and the design of a γ-ray burst telescope which uses this technology.

Key words: GRBs – Localization – Imaging

1.   Introduction

The identity of the source of Gamma Ray Bursts (GRBs) has remained a mystery ever since their first discovery more than 20 years ago. It has long been realized that the key to solving this problem is to identify a counterpart in other wavebands. Current GRB instruments are not capable of providing the accurate burst location rapidly, with the accuracy of a few arcminutes that would enable one to make follow-up observations at other wavelengths. For example, the BATSE instrument on board CGRO can only locate the site of an GRB to within a few degrees[1]. Alternative methods such as the triangulation technique which makes use of the timing data provided by an interplanetary network of burst detectors, has provided much better localisation to a few arcminutes and has permitted follow-up observations to be made in the radio, optical and soft X-ray wavebands. However, the fact that data were collected from several different instruments means there is an unavoidable delay in obtaining the position information. The best time delay achieved is of the order of a couple of days. The long delay between the onset of the gamma-ray burst and the follow-up observations using VLA radio and optical measurements may explain the failure to find a counterpart at these wavelengths.

It is now clear that in order to have better success in detecting the counterpart of a GRB, one needs to have both a GRB monitor which can provide the burst localization with a precision better than a few arcminutes shortly after the onset of the burst and a rapid response telescope operating at other wavebands. The Rapidly Moving Telescope (RMT) which has 1 second response time, a 9×12 arcminute field of view and 1 arcsecond angular resolution, is more than adequate to meet this requirement[2]. In this paper, we propose a GRB telescope which is capable of locating the source to better

255

L. Bassani and G. di Cocco (eds.), Imaging in High Energy Astronomy, 255–258.
© 1995 *Kluwer Academic Publishers.*

than $\sim$ 2 arcminutes whilst retaining a large field of view (> 2 sr). A GRB can be instaneously located using a computer on-board the satellite, thus the only delay in performing the follow-up observation will be in the communication link between the satellite and the other ground or space based telescopes in radio or optical wavelengths.

2. A Low Energy Gamma-Ray Burst Telescope

Although most GRBs have been discovered at energies above 20 keV, their emission spectra are believed to span from keVs to MeVs or even wider. In the energy range from a few keV to $\sim$ 20 keV, silicon detectors have already been widely used in astronomy (eg. BBXRT, ASCA). Silicon-strip detectors may be used to provide a relatively large area, highly efficient detector with excellent 1-D spatial resolution[3]. In the design of a compact, wide field imager such characteristics enable one to avoid parallax affects. It is now possible to have large area ($\sim$ 100 cm^2) detectors with $\sim$ 1000 strips. Such a detector, if used in conjunction with a one-dimension coded mask, could provide 1-D images of a large area of the sky with arcminute angular resolution. With 2 modules of this type of 1-D imager mounted orthogonally on a satellite, any GRBs within the FOV can be located rapidly.

The leakage current noise generated within each silicon strip detector at 0 ° C within an integration time of 1.5 μs is less than 100e (1σ). This could be reduced significantly by cooling the detector further using a passive radiator. The noise contribution from the amplifier having a peaking time of 1.5 μs would be about 180e (1σ). Such performance is consistent with that achieved with the 64 channel Viking chip which has a power consumption of 3mW/channel. Such a combination of cooled detector and amplifier should be capable of detecting a 3 KeV photon at the 4σ level. This suggests that a 2-D imaging system constructed using two such 100 cm^2 detector modules, and a dedicated data processing computer, could be designed to consume less than 12 watts. The coordinates of the source of the GRB could therefore be available for transmission and distribution within a few seconds of its detection.

3. The Predicted Performance

A summary of the main characteristics of a single detector is provided in Table I. The mask strips are designed to be slightly wider than the detector ones in order to maintain a uniform response at all angles within the FOV. The background in the detector is expected to be dominated by the aperture flux of cosmic diffuse X-rays. According to experience of BBXRT and GINGA, the total background in the energy range (3 - 20 KeV) will not be more than 1.8 times the aperture flux. The $F_{3-20keV} = 0.9\ ph.cm^{-2}s^{-1}$ sen-

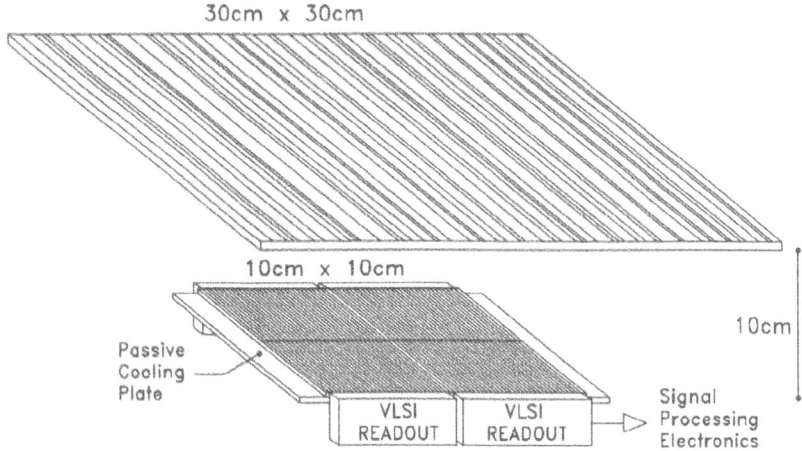

Fig. 1.   A schematic diagram of one detector module of the proposed GRB telescope

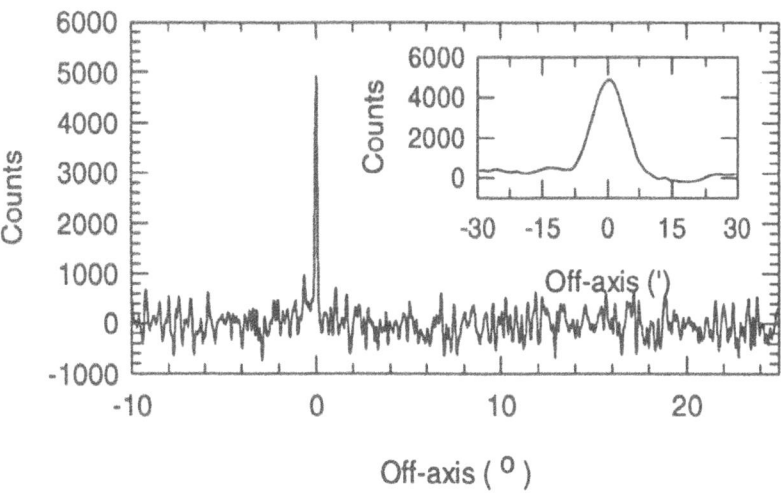

Fig. 2.   A simulated 1-D image of GB870303, a well known GRB detected by GINGA. The source was assumed to be on-axis and can be easily identified within a large FOV and be located with an arcminute precision (see the in-set figure).

sitivity means that more than 50 GRBs per year will be detected according to the $logN$ - $logP$ curve of BATSE[4]. If one assumes that the GRBs have the same 'classical' spectral shape[5] and a typical duration of 20 seconds[4]. With a $7'$ intrinsic angular resolution and detection threshold set at the 5σ

TABLE I

A summary of the telescope design and performance:

Detector Area (×2):	10×10 cm²	Mask Size:	30×30 cm²
Det Strip width:	200 μm	Mask Strip Width:	200 μm
Number of Strips:	512	Basic Mask Pattern:	511 PRS
Detector Thickness:	300 μm	Mask Detec. Sepa.:	10 cm
Energy range:	3 -20 keV	Detection efficiency:	50%(10 keV)
Field of View:	2.3 sr	Angular resolution:	$\sim 7'$
Background:	$\sim$ 2600 counts/sec	Sensitivity (5σ, 20 sec.):	
		$F_{3-20keV} =$	$0.9\ ph.cm^{-2}s^{-1}$
		or 1.4×10^{-8}	$ergs.s^{-1}cm^{-2}$

level it will be possible to locate a source to better than $2'$.

Figure 2 shows the simulated 1-D image of GB870303 detected by GIN-GA. The simulation demonstrates that a GRB of this magnitude can be detected at 18σ level and its site located to better than $1'$.

4. Discussion and Conclusion

A low energy GRB telescope based on the use of silicon strip detectors and 1-D coded masks is proposed. It has the ability to locate a GRB rapidly with accuracy better than $2'$. It therefore should permit rapid follow-up observations to be made at other wavelengths and may enhance the probability of making an identification with counterparts in the optical waveband. The system can be extended to higher energies by the use of a second, thicker layer of silicon strip detectors. The telescope can also be used as an all sky X-ray monitor, by utilising the satellite orbit precession and Radon transforms [6].

References.

[1] Fishman, G.J., et al.: 1989, Proc. of the GRO Science Workshop ed. W.N. Johnson (Greenbelt: NASA/GSFC),2-39.

[2] Barthelmy, S.D., et al.: 1992, Proc. of the 1st Compton Symposium, AIP 280, p1137.

[3] MICRON SEMICONDUCTIR catalogue.

[4] Fishman, G.J., et al.: 1994, ApJS, 92, p229-283.

[5] Schaefer, B.E.: 1993, Proc. of the Huntsville gamma-ray burst Workshop (Oct. 1993).

[6] Zhang, S.N., et al.: 1995, This Procceding.

NEW-CONCEPTION-MULTILAYER HARD X-RAY
TELESCOPE FOR X-RAY ASTRONOMY

O. CITTERIO

Osservatorio Astronomico di Brera, Merate, Italy

F. FRONTERA

Dipartimento di Fisica, Università di Ferrara and Istituto TESRE, CNR, Bologna, Italy

K. JOENSEN, P. GORENSTEIN

Harvard-Smithsonian Center for Astrophysics, Cambridge, MA 02138, USA

and

G. PARESCHI

Dipartimento di Fisica, Università di Ferrara, Italy

Abstract. Several reflection techniques of hard X-rays (>10 keV) are currently under study. Among these, the reflection from two material multilayer coatings with graded d-spacing appears very promising to focus hard X-rays of energy up to about 80 keV. Taking account of the present technology to fabricate multilayers, Kirkpatrick-Baez configurations appears already feasible. We propose here new techniques to fabricate multilayers that can permit to build a Wolter I hard X-ray concentrator with great advantages in imaging performance and sensitivity. The new techniques are an extension of the replication technology by electroforming that are successfully employed for the SAX and JETX missions. A concentrator configuration based on these techniques is proposed.

Key words: Multilayers – X-ray Telescopes – X-ray Astronomy

1. Introduction

The flux sensitivity limitations of the current generation of direct-viewing (imaging and non imaging) telescopes can be overcome by focusing hard X-rays collected over a large passive area onto a small area detector. Indeed in this case the telescope sensitivity is proportional to S^{-1}, where S is the photon collecting area, rather than to $S^{-0.5}$, as in the case of direct-viewing detectors. Hard X-ray concentrators not only can achieve much higher flux sensitivity but also higher imaging performance with respect to the mask imaging systems.

Techniques for hard X-ray reflection include, among others, the reflection from multilayer structures with graded d-spacing. The present status of multilayer coating development allows to fabricate high quality flat mirrors, for which a large body of experience exists at the OSMIc Inc., Troy, MI, USA (Joensen et al., 1994). With flat mirrors these authors have demonstrated the feasibility of Kirkpatrick-Baez multilayer hard X-ray telescopes.

It is well known that Wolter I telescopes provide larger effective area with smaller focal lengths and give better imaging capability and sensitivity. How-

L. Bassani and G. di Cocco (eds.), Imaging in High Energy Astronomy, 259–262.
© 1995 *Kluwer Academic Publishers.*

ever, the attempts to fabricate multilayers with cylinder-like geometries that are needed to build a Wolter I concentrator have been thus far unsuccessful(Joensen et al., 1994). The attempts performed thus far (Marshall, 1984) to coat the inside of a cylinder by sputtering the multilayer material through its open end rose uniformity problems of the bi-layers across the surface, that have not been solved yet. Also the feasibility of a conical shape by bending a flat multilayer coating has not been demonstrated yet.

In order to overcome the mentioned problems in the fabrication of an (approximated or not) Wolter I multilayer mirror, we intend to investigate new techniques. They are an extension of the replication technology by electroforming successfully employed for the SAX and JETX missions (Citterio et al., 1989)

2. Techniques to fabricate Wolter I multilayer mirrors

We are primarily considering the following method to deposit multilayer coatings on mirror supports with Wolter I geometry:

- Deposition of the multilayer structure on a mandrel, electroforming of Nickel on the multilayer coating, separation of the mandrel via colding of the assembly.

This method presents the advantage of a very simple deposition of the multilayer structure. However we have to verify the possibility to separate the multilayer coating plus support structure from the mandrel.

For mirror radii greater than 10 cm, the following method will be also investigated:

- Deposition via sputtering of the multilayer coating directly on the mirror support previously replicated for electroforming.

This method presents the difficulty of a uniform deposition of the multilayer structure, but it does not require to separate the mandrel from the mirror.

3. Telescope configuration proposed

By assuming to be capable to build a multilayer mirror with Wolter I (approximated or not) geometry, we have investigated a hard X-ray telescope configuration with bandwidth from 2 to 60-70 keV. We assumed a multilayer coating with the features given in Table 1.

We assumed a four-focus telescope made of 4 concentrator modules with Wolter I approximated geometry with the module features given in Table 2. With this assumption the expected total effective area at 60 keV is about 100 cm^2, while that at 2 keV is about 1600 cm^2.

The effective area at high energies can be increased by increasing the focal length, that now is assumed to be 6.5 meters.

Table 1. Multilayer parameters assumed

Multilayer material	W/Si
Number of Bi-layers	600
Range of d-spacings ($\mathring{A}$)	20–140
W-thickness/Bilayer thickness	0.35
Roughness ($\mathring{A}$)	4.3

Table 2. Concentrator module parameters

Geometry	conical approx. Wolter I
Number of mirrors	72
Total mirror height h (cm)	60
Diameter of the outermost mirror (cm)	34
Diameter of the innermost mirror (cm)	7.6
Focal length (cm)	650
Multilayer material	W/Si
Mirror mechanical support material	nickel
Mirror support thickness (mm)	0.7
Effective area @ 2 keV (cm^2)	400
Effective area @ 60 keV (cm^2)	25

4.   Prospects and conclusion

We intend to implement the new fabrication techniques of conical-like multilayer shells, by exploiting both the know-how already acquired to build flat multilayer coatings and that acquired for building the SAX and JETX mirrors.

If the test results will give satisfactory results, the fabrication of a (approximated or not) Wolter I multilayer hard X-ray telescope becomes actually feasible.

References

Citterio, O. and Jensen, P.:1989, 'High throughput replica X-ray optics', *SPIE Proc.* **1140**, 337.

Joensen, K. D., Gorenstein, P.,Wood, J., Christensen, F.E., and Hoghoj, P.: 1994, 'Preliminary results of feasibility study for a hard X-ray Kirkpatrick-Baez telescope' *SPIE Proc.* **2279**, in press.

Marshall, G. F.: 1984, 'A unified geometrical insight for the design of toroidal reflectors with multilayered optical coatings', *Proc. SPIE* **563**, 114.

REPLICATED REFLECTORS FOR WIDE FIELD X-RAY IMAGING TELESCOPES

RENÉ HUDEC

Astronomical Institute, Czech Academy of Sciences, CZ-251 65 Ondřejov, Czech Rep.

ADOLF INNEMAN

KOMA Composite Materials, K lesu 965, CZ-142 00 Praha 4, Czech Republic

LADISLAV PÍNA

Faculty of Nuclear Engineering, Czech Technical University, Břehová 7, CZ-115 19 Prague, Czech Republic

and

P. GORENSTEIN

Harvard–Smithsonian Center for Astrophysics, 60 Garden Street, Cambridge, MA 02138, USA

September 16, 1994

Abstract. At energies between 0.1 and 10 keV, wide-field imaging can be achieved by lobster-eye type reflecting X-ray optics. We summarize possible approaches and suggest an innovative technology for production of X-ray reflecting flats and cells necessary to develop one- or two-dimensional wide-field X-ray optics. The technology is based on double–sided replicated reflecting foils produced by electroforming and CF/composite technologies. **Key words:** X-ray optics–X-ray imaging–replica technology

1. Introduction

In 1970, we have started to design and develop Wolter type replica X-ray optics at the Astronomical institute in Ondřejov (Hudec and Valníček 1984, 1986, Hudec et al., 1988). More recently, this technique has been applied also to other geometries such as conical systems and bent foil mirrors. The recent application is the development of lobster-eye type X-ray optics.

2. The replicated grazing incidence X-ray mirrors

The idea of replica technology is to create a perfect copy of a negative shaped master. The replication process starts with the production of high quality glass or glass ceramic masters. The final mirrors are then produced by using electrodeposition (galvanoplastics) of a nickel layer on to a master which is then removed. In some cases, the electroformed nickel deposit will be reinforced by composite material (external epoxy-carbon fibre layer). The material of reflecting surfaces is electroformed nickel with good reflectivity, an application of supplementary reflective film (e.g. gold coating) is also possible. For flat mirrors, float glass can be used instead of polished

263

L. Bassani and G. di Cocco (eds.), Imaging in High Energy Astronomy, 263–266.
© 1995 Kluwer Academic Publishers.

masters. Recently, different specimens of very high-quality float glasses are considered, with surface microroughness below 1 nm.

3. The foil mirrors and reflecting flats

The recent stage of electroforming allows to produce both cylindrical as well as flat and bent mirrors and also numerous applications. While in the first decade mainly the classical Wolter objectives were used, later numerous modifications have been proposed.

The example is the XSPECT SODART X-ray telescope with double conical approximation to the Wolter 1 telescope based on bent thin flats (Byrnak et al., 1987). Numerous materials have been studied for this type of optics including electroformed reflecting foils (Hudec et al., 1989). It appeared that the production of perfect electroformed flats is a difficult task and so it was necessary to initiate a development of modified technology. It was necessary to solve the problem of thickness nonuniformity and long scale errors. During the developing period, we were able to decrease the average thickness nonuniformity from more than 10 % to below 2 %. The laser scattering tests confirmed the expected higher quality of the surface smoothness when compared with lacquer and gold-evaporated Al-foils (Hudec et al., 1989). The X-ray reflectivity of replicated foils is high, roughly 0.8 for incidence angles below 0.4 deg at FeK_α. This development has lead to preparation of technology for production of X-ray reflecting flats with thicknesses between 1 mm and 100 microns.

4. The X-ray optics of lobster-eye geometry

The lobster–eye geometry X-ray optics offer an excellent opportunity to achieve very wide fields of view while the classical Wolter grazing incidence mirrors are limited by about 1 deg FOV.

4.1. ONE DIMENSIONAL LOBSTER SYSTEMS

One dimensional lobster-eye geometry was originally suggested by Schmidt (Schmidt, 1975). The device consists of a set of flat reflecting surfaces. The plane reflectors are arranged in an uniform radial pattern around the perimeter of a cylinder of radius R. X-rays from a given direction are focussed to a line on the surface of a cylinder of radius R/2. The azimuthal angle is determined directly from the centroid of the focused image. At glancing angle of X-rays of wavelength 1 nm and longer, this device can be used for the focusing of a sizable portion of an intercepted beam of X-ray incident in parallel. Focussing is not perfect and the image size is finite. But a one dimensional focusing device offers a wide field of view, up to maximum of 2π with the coded aperture. It appears practically possible to achieve an angular resolution of the order of one tenth of a degree or better. Two such systems is sequence, so as to form a double–focusing device, should offer a field of view of up to 1000 square degrees at moderate angular resolution.

Innovative very wide field X-ray telescopes have been suggested based on these optical elements but have not been flown in space so far. One of the proposals is the All Sky Supernova and Transient Explorer (ASTRE) (Gorenstein, 1987). This proposal also includes a cylindrical coded aperture outside of the reflectors which provide angular resolution along the cylinder axis. The coded aperture contains circumferential open slits 1 mm wide in a pseudo–random pattern. The line image is modulated along its length by the coded aperture. The image is cross–correlated with the coded aperture to determine the polar angle of one or more sources. The field of view of this system can be, in principle, up to 360 deg in the azimuthal direction and nearly 90 % of the solid angle in the polar direction. There is potential for extending the wide field imaging system to higher energy by the user of multilayer coatings in analogy to those described by Joensen et al, 1994, for flat reflectors in the Kirkpatrick-Baez geometry. These coatings exert a great deal of stress upon the substrate. The system must meet severe weight limitations and so the new development of double–sided flats reinforced by composite material to keep the minimal weight still at good mechanical stability must be initiated.

This is the goal of the new development in which innovative technologies for double–sided flats are tested. The basis of the sandwich-type construction of the X-ray flats is an electroformed nickel layer which is deposited on plates of float glass. The nickel-coated plates of float glass are connected by means of carbon/fibre composite material and after hardening the set of connected plates will be cut/ground off. Subsequently, the plates of glass and the produced composite sandwiches with double-sided nickel mirror foils and inner composite reinforcement will be separated. These foils are lacquered on both sides so that the surface microroughness can reach values under 1 nm. The foils will be covered by a thin gold layer at the final stage.

4.2. Two dimensional LOBSTER systems

The idea of two dimensional lobster–eye type wide–field X-ray optics was first mentioned by Angel (Angel, 1979).

The lobster–eye optical grazing incidence X-ray objective consists of numerous tiny square cells located on the sphere and is similar to the reflective eyes of macruran crustaceans such as lobsters. The field of view can be made as large as desired, and it is practical, to achieve good efficiency for photon energies up to 10 keV. Spatial resolution of a few seconds of arc over the full field is possible, in principle, if very small reflecting cells can be fabricated.

This idea was however never been further developed because of difficulties with production of numerous polished square cells of very small size (about

1x1 mm or smaller at lengths of order of tens of mm). This demand can be also solved by electroformed replication and first test cells have been already successfully developed this way. The recent approach is based on the electroforming and composite material technology to produce identical triangular segments with square cells while these segments will be aligned in quadrants onto a sphere.

5. Discussion

The use of very wide field X–ray imaging system could be without doubts very valuable for many areas of X–ray and gamma–ray astrophysics. The production of corresponding optical elements can be reasonably achieved by methods of electroforming and composite replication as an alternative to other methods. The results obtained with the development of technology for production of large area and high quality one–sided X-ray foils are very promising and together with composite material technologies represent an important input for the development of double–sided flats needed for lobster eye geometries of X-ray optics.

Acknowledgements

The development of double–sided reflecting X-ray foils is supported by a grant within the US-Czech Science and Technology program, No. 930 37.

References

H. Wolter: 1952, *Ann. der Phys.* **10**, 94.

R. Hudec and B. Valníček: 1984, 'Development of X-ray mirrors for high-energy astrophysics in Czechoslovakia', *Adv. Space Res.* **3**, No. 10-12, 545.

R. Hudec and B. Valníček: 1986, 'Czechoslovak Replica X-ray mirrors for astronomical applications', *SPIE Proc.* **Vol. 597**, 111.

R. Hudec, B. Valníček, B. Aschenbach, H. Braeuninger and W. Burkert: 1988, 'Grazing incidence replica optics for astronomical and laboratory applications', *Appl. Optics* **27**, 1453.

B. Byrnak et al.: 1987, 'XSPECT-an X-ray spectroscopy and timing mission concept', submitted to Interkosmos by DSRI Lyngby.

R. Hudec, B. Valníček, L. Svátek and V. Landa: 1989, 'New developments in replica X-ray grazing incidence optics', *Proc. of the Conference X-ray instrumentation in medicine and biology, plasma physics, astrophysics and synchrotron radiation*, Paris, 24-28 April 1989.

W. K. H. Schmidt: 1975, 'A proposed X-ray focusing device with wide field of view for use in X-ray astronomy', *Nucl. Instr. and Methods* **127**, 285.

P. Gorenstein: 1987, 'All sky supernova and transient explorer (ASTRE)', in *Variability of Galactic and Extragalactic X-ray Sources*, A. Treves Ed. Associazione per L'Avanzamento dell'Astronomia, Milano-Bologna.

K.D. Joensen, P. Gorenstein, J. Wood, F.E., Christensen, and P. Høghøj: 1994, *SPIE* **Vol. 2279**, in press.

J. R. P. Angel: 1979, 'Lobster eyes as X-ray telescopes', *Astroph. J.* **233**, 364 (1979).

DETECTION OF GAMMA RAY POLARIZATION
WITH INTEGRAL

G. L. HILLS, A. J. DEAN and F. LEI

Astronomy and Space Physics, Physics Dept, Southampton University
Southampton, SO9 5NH, ENGLAND

and

B. M. SWINYARD

DRAL, Didcot, Oxfordshire, OX11 0QX, ENGLAND

Abstract. A method of determining the degree of polarization and the polarization electric vector angle from polarized cosmic sources in the energy range 100 keV to 10 MeV, has been developed for a pixelated CsI(Tl) photodiode detector plane by the determination of the Compton polarimetric modulation factor. The feasibility of the application of this method to the INTEGRAL mission to obtain a significant modulation factor and the resulting sensitivity of a detector to the above energy range is discussed. It is concluded that the INTEGRAL Imager will be sensitive at the three sigma level to a 100% polarized source of 55 mCrab in the 200-600 keV band, for a detector energy threshold of 120 keV, observed in 10^6 seconds.

Key words: γ-rays – Polarization – Polarimeter

1. Introduction.

The International Gamma Ray Astrophysics Laboratory (INTEGRAL) has recently been accepted as the next mission in the European Space Agency's (ESA) Horizon 2000 medium mission plan. It is designed to work in the 10 keV to 10 MeV energy band. This work concerns the gamma ray Imager, one of the four instruments that make up the INTEGRAL payload.

The Imager consists of three layers, where the bottom two layers each consist of 2880 thallium doped caesium iodide (CsI(Tl)) scintillation crystals. Each of these 3 cm deep crystals is viewed by a photodiode. The CsI(Tl) crystals are hexagonal in cross section and have a cross sectional area of 1cm^2, forming a pixelated array of independent detectors.

Between 300 keV and 3 MeV, the primary mechanism for gamma ray interaction with the CsI(Tl) detector elements is via Compton scattering. It can be shown that the scattering cross section is related to the electric vector of the incident gamma ray [1], such that upon interacting within the crystal, the photon will be preferentially scattered orthogonally its electric vector. By sampling the distribution of Compton scattered gamma rays over the INTEGRAL Imager detection plane, the nature of the polarization of the emission from a cosmic gamma ray source can be determined. In this paper the results from the simulation of the polarization response of the Imager detector are combined with simulations of its photo peak detection

L. Bassani and G. di Cocco (eds.), Imaging in High Energy Astronomy, 267–270.
© *1995 Kluwer Academic Publishers.*

efficiency and background to elucidate the possibility of making polarization sensitive measurements in the 300 keV to 3 MeV energy band.

2.　The INTEGRAL Imager as a Compton Polarimeter.

To evaluate the sensitivity of the Imager for use as a polarimeter, a Monte Carlo Compton scattering routine [2] has been incorporated into the GEANT-Detector Description and Simulation Tool package developed at the CERN laboratories [3]. To calibrate the routine, a simulation of the experiment performed by Ohya et al. [4] has been performed and compared to the experimental data. The simulation successfully reproduced the experiment showing the validity of the routine.

The response of a gamma ray polarimeter is described by the polarimetric modulation factor, Q:

$$Q = \frac{N_\perp - N_\|}{N_\perp + N_\|}$$

where $N_\perp$ and $N_\|$ are the detected counts that fall into the area produced by 90° beams placed perpendicular and parallel to the plane defined by the incident photon direction and its electric vector. The INTEGRAL Imager can be used as a polarimeter by selecting those events that produce multiple site interactions and by subsequently sampling the distribution over the detection plane. Figure 1 shows the variation of the modulation factor with the incident photon energy for no detector energy threshold and for a threshold of 120 keV, with the error bars produced from the analysis of twenty bootstrap samples [5].

The polarization sensitivity of a polarimeter in Crab [6], ie the minimum source strength necessary to detect the degree of polarization at n sigma, is given by:

$$\Delta P_{n\sigma} = \frac{n}{Q\varepsilon C}\sqrt{\frac{\varepsilon C + B}{At}}Crab$$

where C is the source flux, ε the detection efficiency at the energy of observation, B the background count rate per unit area of the detector, A the area of the detector and t the length of the observation. The detection efficiency and the background count rates have been determined by further simulations. The source flux was calculated from a phase-averaged Crab Pulsar spectrum of the form:

$$F = 8.6E^{-2.2}photons\, cm^{-2}\, s^{-1}\, keV^{-1}$$

Figure 2 shows the 3 sigma sensitivity to a 100% polarized source for a 10^6 second observation for no detector energy threshold and for a threshold

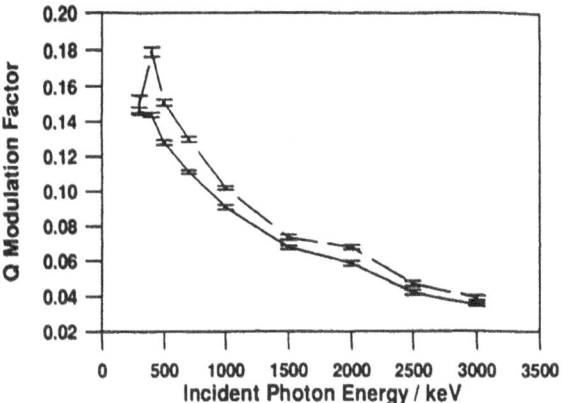

Fig. 1. The modulation factor vs. incident photon energy of the INTEGRAL Imager. The solid line shows the modulation factor for no threshold, while the dashed line shows the modulation factor for a 120 keV threshold.

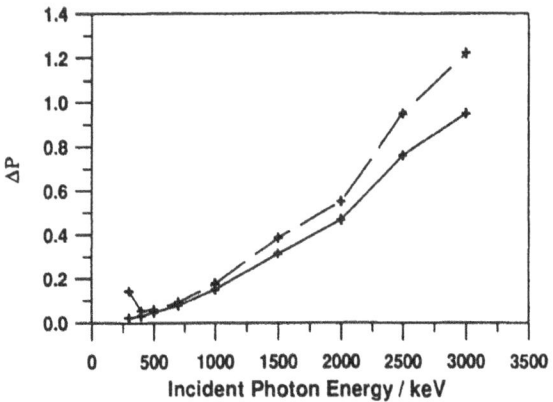

Fig. 2. The minimum detectable degree of polarization in Crab vs. incident photon energy for the INTEGRAL Imager detector observing an un-pulsed source with the time averaged spectrum of the Crab pulsar. The solid line shows the sensitivity for no threshold, while the dashed line shows the sensitivity for a 120 keV threshold.

of 120 keV. The sensitivity was calculated for energies, E, between 300 and 3 MeV, with energy bands of $\Delta E = E$.

3. Discussion and Conclusion.

The increase in Q upon the application of a threshold can be clearly seen in figure 1. The angle scattered through on interaction increases with the energy deposited. Hence applying a threshold increases the average scattering angle. As the modulation factor is greatest for scattering angles approaching $90°$, the threshold will lead to a larger modulation factor. However, the threshold decreases the efficiency of detection faster than the modulation factor is increased leading to a reduced sensitivity.

The polarimetric modulation factor and the polarization sensitivity are both highly dependent upon the energy detection threshold of the CsI(Tl) detectors. For energies below 500 keV where the effect of polarization is most apparent, it is critical for the energy threshold to be as low as possible.

Figure 2 shows that for a 120 keV threshold, a 3 MeV photon source would have to have a source strength greater than 1 Crab for the polarization to be detected. To overcome this problem and identify the polarization of the source, an observation time greater than 10^6 s would be required. Figure 2 also shows that the INTEGRAL Imager will be sensitive at the 3σ level to a 100% polarized source of 55 mCrab in the 200-600 keV band, observed in 10^6 s.

The fact that the sensitivities calculated here are significantly worse than results published before [6], appears to be due mainly to the previous under-estimation of the detector background. The sensitivity has also suffered as a result of the inclusion of passive material in this calculation.

Both of the figures show worse than expected behavior at the 300 keV level when a 120 keV detector energy threshold is applied. This appears to be due to the requirement that in each of the two interactions almost half of the incident energy needs to be deposited, leading to a large restriction on the nature of the possible interactions.

The data represented here should be viewed as a conservative estimate of the sensitivity of the INTEGRAL Imager. Methods of preferential event selection, with the aim of increasing the polarimetric modulation factor are currently under investigation. It is believed that with these analysis methods the sensitivity can be improved by at least a factor of two.

References.

[1.] Berestetskii V.B. et al.: 1971, 'Relativistic Quantum Theory Part 1', (Pergammon Press: Oxford)

[2.] Swinyard B.M. et al.: 1991, 'Production and Analysis of Polarized X-rays', ed. D.P. Siddons, *Proc. SPIE* **1548** 94

[3.] Brun R. et al.: 1994, 'GEANT-Detector Description and Simulation Tool, CERN Program Library Long Writeup', **W1503**

[4.] Ohya S. et al.: 1989, 'A Compton Polarimeter Constructed with a Large Si(Li) Scatterer and Two Ge Analyzers', *Nucl. Instr. Meths. Phys. Res.* **A276** 223-227

[5.] Simpson G. and Mayer-Hasselwander H.: 1986, 'Bootstrap Sampling: Applications in Gamma-Ray Astronomy', *Astron. Astrophys.* **162** 340-348

[6.] Swinyard B.M. et al.: 1994, 'The INTEGRAL Imager Detector as a Gamma Ray Polarimeter', accepted by, *Nucl. Instr. Meths. Phys. Res.*

SENSITIVITY AND EFFICIENCY OF THE INTEGRAL IMAGER

F. SANCHEZ

Instituto de Física Corpuscular, Universidad de Valencia-C.S.I.C.
E-46100 Burjassot, Valencia, Spain.

F.J. BALLESTEROS and V. REGLERO

Departamento de Matemática Aplicada y Astronomía, Universidad de Valencia
E-46100 Burjassot, Valencia, Spain.

and

G. MALAGUTI and G. DI COCCO

Istituto Studio e Tecnologie delle Radiazioni Extraterrestri/CNR
Via P. Gobetti 101, 40129 Bologna, Italy.

February 15, 1995

Abstract. A detailed simulation program of the INTEGRAL Imager has been written and implemented using the GEANT-3 Monte Carlo code. The expected detection efficiency and continuum sensitivity have been evaluated. The results obtained for the CsI configuration of the Imager are compared with those obtained with the new configuration which foresees a top plane made of CdTe solid state detector elements.

Key words: artificial satellites – instrumentation.γ-ray detectors – Monte Carlo simulation.

1. Introduction

INTEGRAL [1] is a satellite mission for γ-ray astronomy recently selected by ESA within the Horizon 2000 program. The payload consists of two main instruments, one optimised for spectroscopy (Spectrometer) and the other for fine imaging (Imager). Both detectors are coupled with a coded mask. The INTEGRAL payload is completed by two monitors, one operating in the X-ray band (XRM) and the other in the optical window (OTC).

The Imager detector consists of three planes and each plane is made out of 24 triangular modules. Each module contains 120 CsI hexagonal shaped detection unit each viewed by a silicon photodiode. The top plane is 1 cm thick, while the other two are 3 cm thick. At the end of the Phase-A Study this configuration has been changed, and the top CsI plane has been substituted with a 2 mm layer of CdTe elements in order to achieve a lower energy threshold and improve the energy resolution at low energy.

In this work, the detection efficiency and sensitivity of the Imager obtained by Monte Carlo simulation for both configurations are presented.

L Bassani and G. di Cocco (eds.), Imaging in High Energy Astronomy, 271–275

TABLE I

Operational modes in the Imager detector.

Mode	number of interaction(s)/layer			Mode	number of interaction(s)/layer		
	1st layer	2nd layer	3rd layer		1st layer	2nd layer	3rd layer
1a	1	-	-	4a	1	1	-
1b	2	-	-	4b	1	>1	-
1c	>2	-	-	4c	>1	≥1	-
2a	-	1	-	5a	1	-	1
2b	-	2	-	5b	1	-	>1
2c	-	>2	-	5c	>1	-	≥1
3a	-	-	1	6a	-	1	1
3b	-	-	2	6b	-	1	>1
3c	-	-	>2	6c	-	>1	≥1
7a	1	1	1	7b	≥1	≥1	≥1

2. The INTEGRAL Imager

The Imager detector operates between 50 keV and 10 MeV, while with the CdTe option the lower energy threshold goes down to 20 keV. The total active area in both configurations is $\simeq 3000$ cm^2. However, only the inner 2500 cm^2 are used, due to background reasons [2]. The cross-sectional dimension of the detector pixel is $\simeq 1$ cm^2. In the case of CdTe the detection units of the top layer are further subdivided into six triangular sections so that the intrinsic spatial resolution is improved. The main detector is then shielded on six sides and the bottom by a 2 cm thick BGO crystal shield. The multi-layer discrete elements geometrical assembly provides the possibility for the Imager to work in different interaction modes (Table I), which are defined by the number and location of the interaction(s) caused by the primary incident γ-ray.

3. Simulation and results

The Monte Carlo simulation of the Imager was performed using a program based on the GEANT-3 software package [3]. In order to reproduce as closely as possible the real operational conditions, the whole detector, including all the passive materials, was considered in the simulation.

The energy thresholds for individual signals used in the simulation correspond to those under consideration for the Imager: 50 and 3000 keV (lower and upper) for the top layer in the CsI option, being such values 15 and 1000 keV in the CdTe option. For the second and third layer the values

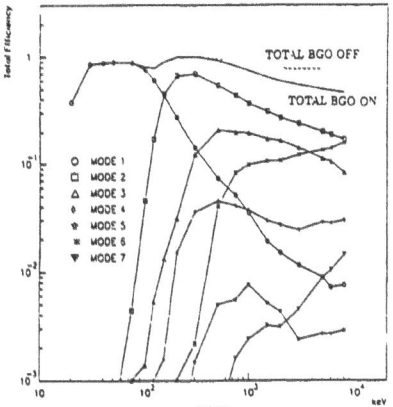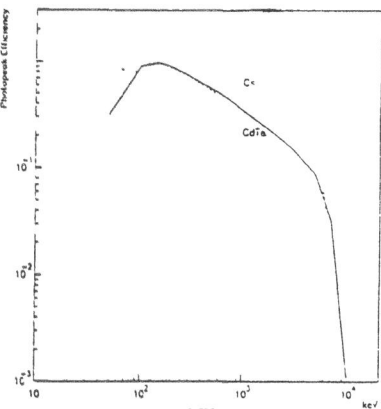

Fig. 1.   Left: Total efficiency of the Imager for the CdTe option. The effect of the BGO veto shield (BGO "on", BGO "off") is displayed.Right:Comparison between photopeak efficiencies of the Imager for the CdTe option (dashed line) and CsI option (full line).

considered are 120 and 5450 keV. The low energy threshold for the BGO veto is 100 keV.

3.1 DETECTION EFFICIENCY

The calculated total and full-peak efficiency profiles are shown in Figure 1. The effect of the BGO veto shield (BGO A/C "ON") is negligible below 500 keV, increasing its importance as energy increases. On the other hand we have found that single events (only one detection unit triggered) dominate at incident energies below 1 MeV whereas the multiple events (more than one unit triggered) dominate above that energy.

In Fig. 1 (right panel) the photopeak efficiency obtained with the CdTe configuration is compared with those obtained with the CsI option (CsI for the three layers of the Imager). As it can be expected there is a significant improvement in the photopeak efficiency for energies below 60 keV when the CdTe option is considered, being similar for both options the efficiencies in the rest of the energy interval considered.

3.2 SENSITIVITY

The continuum sensitivity is one of the key scientific parameters of the Imager. The background spectrum has been taken from Monte-Carlo results obtained using the same code [4]. The model for background cal-

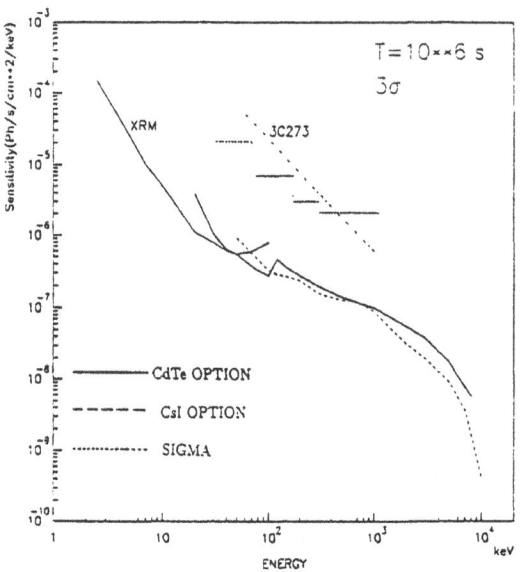

Fig. 2.   Imager continuum sensitivity for a significance level of 3σ for the two options considered in this work. The X-Ray Monitor (XRM) and SIGMA sensitivities are also shown (see text).

culations takes into account the contributions from cosmic diffuse background, gamma-rays produced in the material spacecraft, cosmic ray induced radioactive spallation products within the material of the gamma-ray detector itself and events derived from protons trapped within the earth's radiation belts.

In Fig. 2 we compare the expected continuum on-axis sensitivity for the Imager CdTe option with that obtained for the CsI option. A statistical significance of 3σ (σ=standard deviation) was considered in these calculations. Ojo intentar ver lo que hay de Pino y Cocco sobre esto. In both cases there is a significant improvement with respect to previous γ-ray missions (SIGMA sensitivity is showed for comparison together with the 3C273 spectrum). As it can be seen in Fig. 2, in the CdTe option the sensitivity shows an improvement of a factor $\sim$1.5–2 at energies less than 100 keV with respect to the CsI configuration. Furthermore, at low energy the overlap between the sensitivity curve of the Imager with the X-Ray Monitor one ensures a coverage of the critical energy band comprised between 10 and 80–100 keV. Nevertheless, further studies are under development in order to increase the sensitivity of the Imager at low energies, mainly related to the suppression and/or minimization of passive materials that interfere in the field of view of the Imager.

This work has been supported by the Spanish Comisión Interministerial de Ciencia y Tecnologia under grant ES-92-0855-E.

References

[1] Bergeson-Willis, S. et al.: 1993, 'INTEGRAL Phase-A Report', *ESA SCI(93)I* ,

[2] Dean, A.J.: 1994, 'European Imaging Detector for Observing γ-ray Sources. Scientific and Technical Plan.', *Proposal submitted to the ESA for the INTEGRAL M2 Mission* ,

[3] Brun, R. et al.: 1987, *Data Handling Division. C.E.R.N. DD/EE/84-1* ,

[4] Imager background team: 1994, *Private Communication* ,

INTEGRAL X-RAY MONITOR:
A PROPOSAL FOR THE HARD X-RAY IMAGER
ON-BOARD INTEGRAL

P. Ubertini[1], L. Bassani[2], A. Bazzano[1], R. Cole[3], J. Lapington[4], M. Mas[5], L. Natalucci[1], B. Ramsey[6]
E. Soggiu[1], R. Staubert[7], M. Turner[3], L. Waldron[1], M. Weisskopf[6]

1. Istituto di Astrofisica Spaziale, Frascati, Italy
2. Istituto TESRE, Bologna, Italy
3. Leicester University, Leicester, U.K.
4. MSSL, Dorking, U.K.
5. LAEFF, Madrid, Spain
6. MSFC, NASA, Huntsville, USA
7. Astronomisches Institut, Universitaet Tuebingen, Germany

Abstract

The inclusion of a broad band hard X-Ray imager is important to the success of the International Gamma-Ray Astrophysics Laboratory (INTEGRAL) mission, the next ESA blue box M2 mission, due to be launched during April, 2001. The observatory payload comprises two main instruments, the IMAGER and the SPECTROMETER, and two monitors, the X-RAY MONITOR (XRM) and the OPTICAL TRANSIENT CAMERA (OTC).

The four instruments are co-aligned and will simultaneously observe the same sky region. The X-Ray Monitor will provide images with arcminute angular resolution in the 3−80 keV band, with minimum detectable flux of one milliCrab at 80 keV for a typical 10^5 second observation.

The baseline detector system is a high pressure imaging proportional counter placed behind a coded mask placed at 4 meters from the detector window. The instrument view angle is limited by a square passive collimator with a FOV of 6 degrees FWHM.

This instrument, as a monitor, will provide fine imaging and simultaneous X-Ray flux measurements. The spectral resolution and sensitivity will be optimised to make it a competitive spectroscopic instrument for medium to high energy X-Rays. The scientific rational and the experiment configuration are discussed.

1. Introduction

The INTErnational Gamma-Ray Astrophysical Laboratory (INTEGRAL), will address beyond the year 2000 high energy Astrophysics phenomena providing high sensitivity sky images in the 15 keV to 10 MeV range with high resolving power ($E/\triangle E > 1000$) and few arcminute spatial resolution. The payload comprises a germanium based Spectrometer, a large area coded mask Imager, an X-Ray Monitor (XRM) and an optical - CCD transient camera (1).

L. Bassani and G. di Cocco (eds.), Imaging in High Energy Astronomy, 277–281.
© 1995 *Kluwer Academic Publishers.*

In accordance with the scientific objectives of this mission (2), the X-RAY MONITOR (XRM) will extend the spectroscopic, positioning and timing capabilities of the observatory by providing images and spectra of fields with arcminute angular resolution over the 3−80 keV range, simultaneously with the primary gamma-Ray instruments. In fact, the XRM is required to provide images with an angular resolution better than those of the Imager and the Spectrometer, in order to resolve source confusion and possibly provide optical identification of the gamma ray source (3). In addition it must be sensitive up to 80 keV, in order to provide sufficient overlap with both the two other instruments. It must have sufficiently good energy resolution to resolve lines, particularly cyclotron lines between 10− 80 keV. The mission also requires significant resolution sensitivity at 6 keV, so that the iron emission of X-ray sources can be monitored.

2. Historical Highlights

During the 30-year old history of balloon, rocket and satellite observations, proportional Counters have emerged as the favored detection instruments of X-Ray (1-100 keV and beyond) astronomy (4). In the early 1960's, a pioneering group led by Riccardo Giacconi obtained the first evidence of X-Ray beyond the solar system with their rocket-borne instruments. Within two years of the launch of the UHURU satellite in the 1970's, a catalogue containing more than 400 sources in the range 3-6 keV band became available to astronomers. More recently, the ROSAT survey has revealed in excess of 60.000 sources. each with a positional accuracy of a few arcseconds.

The proposed photon detection system of the XRM experiment is a high-pressure bi-dimensional imaging Counter behind a coded mask aperture system, and will be used with fluorescence (or escape) gating technique, thereby permitting a substantial improvement in sensitivity, if compared with former experiments above 35 keV.

Already, these potential capabilities of the XRM have been partially demonstrated by both satellite (ART-P onboard GRANAT, MART-LIME onboard SPECTRUM X-Γ) and balloon experiments such as the MIXE Marshal Space Flight Center imager (5). Moreover, these programs have enabled the clarification of several problem areas and key technological issues: spectral resolution above the xenon K-edge. background rejection efficiency, photon detection efficiency in the "fluorescence-gated" mode, microphonics susceptibility, high strength uniformly-transparent windows for low energy operation, capable of withstanding high internal pressures without degradation of imaging capability of the detection cell, detector tightness with high pressure filling mixture, detector lifetime. and finally, aging effects on the quenching gas in the space environment.

The angular resolution of the XRM (3′ FWHM) will be superior to both that of the IMAGER and SPECTROMETER. Moreover, the limiting sensitivity, in a typical 10^6 second observation, will be better than one milliCrab. That is, the sensitivity of XRM will enable the detection of most of the sources that are detectable with the other instruments, assuming a spectral shape within the range exhibited by so far known sources. In this context, XRM will enable the location of gamma-ray sources positioned with the IMAGER to be improved to an accuracy in better than 1 arcmin, thereby resolving any potential source confusion and facilitating the optical identification of high energy sources. The sensitivity of the XRM over the 3−80 keV band will also provide the needed overlap with the IMAGER (CdTe) and SPECTROMETER to avoid the very well known problem of different instrument calibrations.

In addition to its supporting role on INTEGRAL, the spectral, positional and timing capability of XRM will make it a valuable stand alone instrument.

3. The Experimental Configuration

The XRM experiment comprises three major subsystems: the detector assembly, the coded mask and the electronics.

3.1 The Detector

The detector is based on a high pressure chamber filled with a xenon based mixture at 5 bar, with an active area of 1000 cm^2 , and sensitive over the 3-80 keV energy range. This ensures both adequate quantum efficiency in the vicinity of the xenon K-edge at $\sim$ 35 keV, and sufficient overlap with the SPECTROMETER and the IMAGER. It will be operated in a fluorescent-gated mode wherein those events for which a xenon K fluorescence photon is produced and detected within the sensitive volume of the chamber (drift or multiplication regions) are tagged and analyzed separately to give low-background and high spatial and energy resolutions.

3.2 The Coded Mask

Since it is impossible to focus hard X-Ray photons, the most effective way for extracting directional informations from the incoming radiation is to cast a shadowgram onto the detector, a well known technique for producing images of X-Ray sources (6). The imaging capability of the instrument will be provided by a self-supporting coded mask aperture system. This mask system will comprise a Tungsten foil, etched with a pattern of square holes (with a basic pixel of $\sim$ 4 mm) such that there is a $\sim$ 50% overall transparency to cosmic X-Rays. The location of this Uniformly Random Array (URA) mask will be 4 m infront of the detector window, in order to match the field of view with that of the gamma-Ray instruments, and to enable sufficient angular resolution for the imaging requirement of the mission. The result will be a fully coded FOV of 4.0 square degrees. Once obtained, images of the sky will be reconstructed by decoding the shadowgram with the pre-determined aperture function of the mask (7).

3.3 Electronics

The digital electronics of XRM together with the interfaces to the Spacecraft, will be contained within a box that is isolated from the detector. The tasks to be performed by this multi-functional micro computer controlled system include the event selection and validation, background rejection, collection and formatting of scientific data and the data compression required by the low telemetry rate.

It will also provide the interface to the spacecraft, command and control interface, detector control, the collection of housekeeping data, power conversion and conditioning, overall experiment health supervision and temporary data storage in the built-in mass memory.

3.4 The Instrumental Background and Rejection

Because of the low source flux levels and high induced background, intrinsic to X-Ray astronomical measurements, the minimum flux that can be measured is ultimately limited by the level of the diffuse cosmic ray flux at low energies ($E < 30$ keV) and of instrumental background at higher energies. Therefore, in order to maximise the sensitivity of the instrument, it is essential to minimize the background noise associated with the aperture flux, shield leakage and X-Ray emission from passive materials within the spacecraft environment. The various contributions to this background noise include the cosmic diffuse X-Ray flux, the neutron induced component, the spallation component, mask secondary production, cosmic rays (predominantly electrons at low energy and high energy protons) and natural

radioactivity of the detector materials and filling gas mixture. In making accurate estimates of both the
background noise level and XRM sensitivity, use has been made of knowledge gained from numerous
balloon-borne and satellite experiments. On the basis of this knowledge and a detailed understanding of the physical processes leading to the production of the background noise, the detection system
design (architecture, fabrication materials and associated electronics) will be optimised for effective
background suppression. To this end, several techniques will be employed, including passive shielding,
multi-wire veto fluorescence "gating", background minimization via data reduction software analysis
procedures etc.

4. The Predicted in Flight Performances

In order to correctly evaluate the in-flight instrumental capability to detect weak cosmic X-Ray
sources the XRM sensitivity to continuum and line emission was computed estimating the expected
background in the very eccentric 4-day orbit provided by the baseline PROTON Russian launcher
injecting the payload in a 40,000 km perigee and 300,000 km apogee orbit. The ultimate continuum
sensitivity of the XRM experiment, for a 10^6 s observation period, at 3σ level of statistical significance,
is shown in Figure 1.

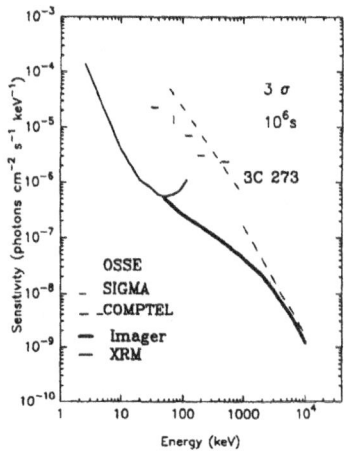
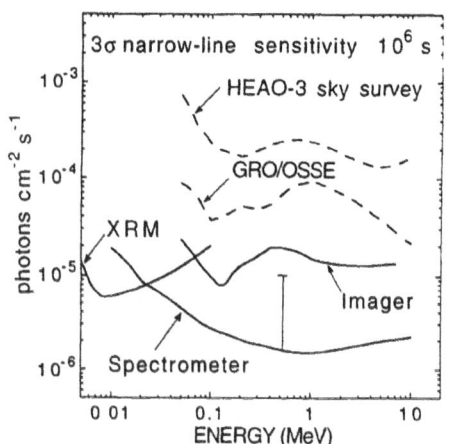

Figure 1: *XRM continuum sensitivity*
(3σ). ΔE = E has been assumed. Ob-
servation time: $10^6 s$.

Figure 2: *XRM narrow-line sensitivity.*
Observation time: $10^6 s$.

This sensitivity was evaluated with the well known formula

$$F_{cont} = n\sigma/e(\sqrt{4\,B/(A\,T\,E)})\ \ [photon\,cm^{-1}\,s^{-1}\,keV^{-1}] \tag{1}$$

for the ideal case of a source located at the center of an image element. In this case, $n\sigma$ is the statistical significance of the observation, in standard deviations, A is the sensitive area of the detector, e is the quantum detection efficiency, E is the instrumental energy range considered (assumed equal to E in our computation), B is the instrumental intrinsic background in [$counts\,cm^{-1}\,s^{-1}\,keV^{-1}$], and T is the period of observation.

Defining "narrow" as a width comparable to the instrumental energy resolution, the narrow-line emission sensitivity, for the same period and level of statistical significance, is shown in Figure 2. This was evaluated using the expression

$$F_{line} = 2Fn\sigma/e\sqrt{(EB/AT)} \quad [photon\,cm^{-1}s^{-1}] \tag{2}$$

where F is a parameter of near unity which depends weakly on the signal-to noise ratio.

To summarize, the XRM continuum sensitivity during a typical 4-day orbit will be $1\times10^{-6}\,ph\,cm^{-2}s^{-1}\,keV^{-1}$, while the line sensitivity at 50 keV, for a narrow line emission (line-width $\triangle E = 2$ keV), will be $\sim 2\times10^{-5}\,ph\,cm^{-1}s^{-1}$.

References

1. Bergerson-Willis B., et al., "INTEGRAL - International Gamma Ray Astrophysics Laboratory", ESA - SCI(93)1, 1993.

2. Skinner G., et al., SPIE, Vol 1945, 112-123, 1993.

3. Ubertini P., et al., Astron. Astrophys., Suppl., 97, 389, 1993.

4. Ubertini P., Space Science Review, 46, 1, 1987.

5. Ramsey B.D., et al., SPIE Vol. 2006, 97-101, 1994.

6. E.E. Fenimore and T.M. Cannon, Appl. Opt., 337, 1978.

7. Soggiu M.E., Bazzano, A., Ubertini, P., and Waldron, L., SPIE, Vol. 1948, 109-118, 1993.

COMPTEL AS A GAMMA-RAY POLARIMETER

F. LEI, A. J. DEAN and G. L. HILLS

Astronomy and Space Physics, Physics Dept. Southampton University
Southampton, SO9 5NH, ENGLAND

and

B. M. SWINYARD

DRAL, Didcot, Oxfordshire. ENGLAND

Abstract. COMPTEL. the imaging telescope on board CGRO, uses Compton scatter kinematics to reconstruct the direction of the incident photon. On the basis of the gamma-ray detection principle, it can also be used as a gamma-ray polarimeter. A series of Monte Carlo (M-C) simulations have been carried out using a simplified geometrical model of COMPTEL. The aim is to investigate COMPTEL's response to polarized gamma-rays. including the dependence on incident photon energy and incident angle. M-C simulations show that COMPTEL can detect the polarization of Crab pulsar (if it is 100% polarized) at a marginal level and determine the polarization angle to within a few degrees.

Key words: γ-rays – Polarization – Polarimeter

1. Introduction

COMPTEL, the imaging γ-ray telescope on board CGRO. reconstructs the direction of incident photon using the Compton kinematics [1]. To a known source like the Crab pulsar, it however can be used as a Compton polarimeter. Polarization features of the incoming photons can be revealed by the non-symmetrical distribution of the scattered photons in the scatter plane. In this paper the characteristics of COMPTEL as a γ-ray polarimeter are studied by M-C simulations. These include the efficiency in detecting linear polarization as a function of incident photon energy and angle, as well as the angle of the electric vector. The sensitivity of COMPTEL to polarized γ rays is evaluated based on the measured background level in space. And finally, a simulated polarization observation of the Crab pulsar is presented.

2. COMPTEL As a Polarimeter

Soft γ-ray photons are largely detected via Compton scattering interactions. In this process the photon is more likely to be scattered at right angles to its electric vector, hence the non-symmetry of the photon distribution in the scatter plane reveals the polarization information of the incoming photons. For a Compton polarimeter its response is defined as: $Q = \frac{N_\perp - N_\parallel}{N_\perp + N_\parallel}$, where $N_\perp$ and $N_\parallel$ are the counts in two analyzer detectors placed perpendicular and parallel to the electric vector direction of a 100% polarized beam of γ rays scattered from a central detector.

283

L. Bassani and G. di Cocco (eds.), Imaging in High Energy Astronomy, 283–286.

For a more general case when the two orthogonal analyzers are placed at an angle (Φ) and the incident electric vector is at an angle (η) to the local coordinates, the response of such a Compton polarimeter is given by: $M(\Phi, \eta) = Q \cos 2(\Phi - \eta)$. In the case of COMPTEL which records both the scattering position and position where the scattered photon is stopped, it is possible to project the scattering onto a scatter plane and by applying the Moving Mask Technique (MMT) the Q and η values can be obtained [2].

The $M(\Phi, \eta)$ curve derived directly from COMPTEL data contains a large contribution from systematic modulation. This systematic effect can be removed by using the polarimeter's response to unpolarized photons. In Figure 1 are the $M(\Phi, \eta)$ curves of COMPTEL to both polarized and unpolarized on-axis 1 MeV photons, as well as the true modulation curve which follows the $Q \cos 2(\Phi - \eta)$ form.

To investigate COMPTELs polarization response, an simplified COMPTEL model consists of the 7 top detectors and the 14 bottom detectors in GEANT3 has been developed. The GEANT3 code itself has been modified so that it is sensitive to polarized photons. M-C simulations of both polarized and unpolarized photons with different incident energy and angles have been carried out. The nominal COMPTEL energy thresholds(E1 > 70 KeV and E2 > 650 keV) were applied [1].

i) Photons incident at an angle (θ) to the telescope axis. The Q value is expected to vary as a function of θ. In figure 2.a are the M-C simulation results for 1 MeV photons incident at different angles. It seems with θ between 10° to 15° yields the best Q values. In the simulations the electric vector of the polarized photons are at 0° to the x-axis. The size of the data sets are typically 10^5 unpolarized and 3×10^4 polarized event. The errors were derived by bootstrapping with 10 samplings.

ii) Q as a function of incident photon energy. The polarization effect in Compton scattering process decreases with increasing γ-ray energy. In figure 2.b is the Q vs. E curve for the normal incident photons. It is clear that COMPTEL, as a polarimeter, is most sensitive at energies below 5 MeV.

iii) Uniformity of Q over the electric vector angle η. Due to the discrete nature of the COMPTEL detector planes its polarization response varies with different electric vector angles. Figure 2.c is the case for 1 MeV normal incident photons, it shows the derived Q values can change by as much as 20%, and $\eta = 0°$ gives the best result.

iv) Analysis using MMT also identifies the position of the electric vector of the incoming polarized photons. In all cases discussed above η have all been located to within a few degrees of its input values.

The Crab pulsar is the most likely source to emit polarized γ-rays which might be detectable to COMPTEL. Its γ-ray emission is best described as (phase averaged) $8.6 E^{-2.2}$ $ph.s^{-1}cm^{-2}kev^{-1}$, which gives a source flux of $1.24 \times 10^{-3} ph.s^{-1}cm^{-2}$ between 0.75 and 5 MeV.

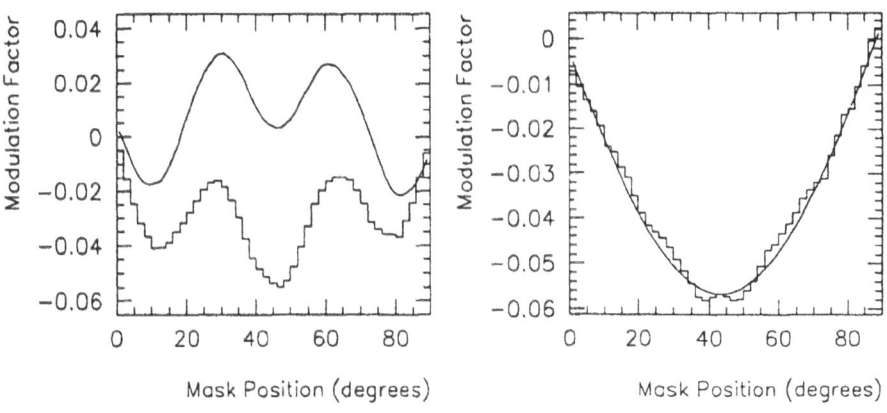

Fig. 1.   *left) Modulation curves of unpolarized (smooth) and polarized photons (histo); right) System free modulation curve of polarized photons.*

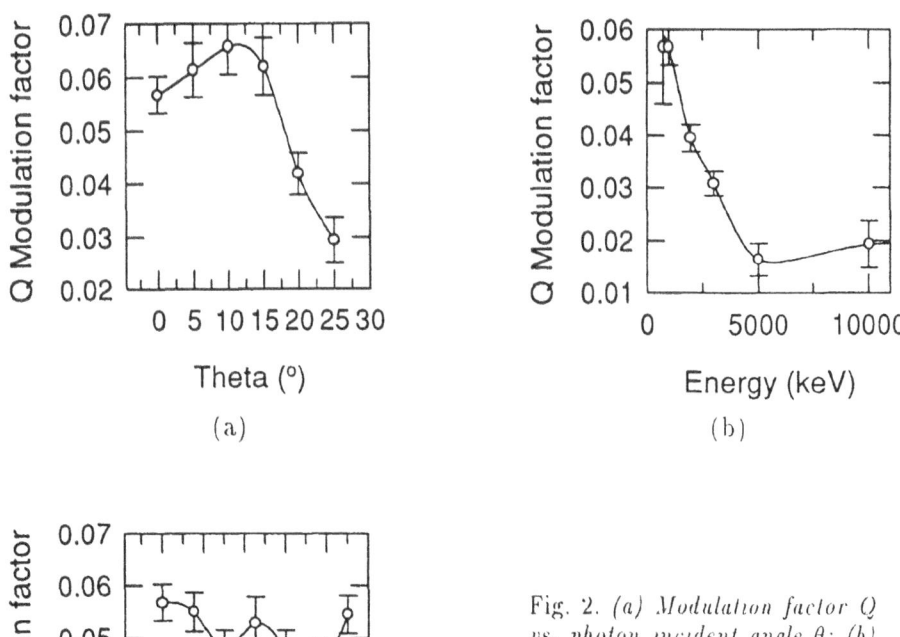

Fig. 2. *(a) Modulation factor Q vs. photon incident angle θ: (b) Modulation factor Q vs. incident photon energy E: (c) Q as a function of electric vector angle η.*

A simulated observation of the Crab pulsar was performed in the energy band between 0.75 and 5 MeV. assuming the γ-rays are 100% linearly polarized with the electric vector η at 0° to the x-axis of COMPTEL. Also it was being observed 15° off axis with the azimuth angle as 0° to the x-axis. The data sets obtained contain 10^5 unpolarized and 10^5 polarized events.

A standard MMT analysis yields the Q and η values as $Q = 0.0475 \pm 0.0020$, $\eta = -0.27° \pm 1.71°$. The Q value of a polarimeter like the COMPTEL can be improved in several ways. Studies have shown that Q can be nearly doubled by simply reducing the mask size used in MMT. Also by selecting only events scattered at large angles increases the Q value significantly as photons scattered as small angles carry little information of the polarization. Using the following additional criteria in the event selection: i) $20° < \phi_{geo} < 40°$; ii) $|\bar{\phi} - \varphi_{geo}| \leq \sigma_\phi$, and 10° mask size in MMT a improved set of Q and η values were obtained: $Q = 0.110 \pm 0.007$, $\eta = 0.99° \pm 3.30°$. This improvement, however, is at the cost of a reduced detection efficiency as the data sets have been reduced to ~ 20000 events each.

COMPTEL's in-flight performance shows that, with the above event selection criteria, for a two week observation the background rate is about 51000 counts in the energy band 0.75-5 MeV, and for a Crab like source the effective detection area and time is $A_{eff}T_{eff} = 3 \times 10^6 cm^2 s$ [1]. The positive detection of polarization is given by: $SNR_\sigma = \frac{QN_{source}}{\sqrt{N_{BGD} + N_{source}}}$. This leads to a marginal detection, at 1.75σ level, of polarization in Crab pulsar if it is 100% polarized. This detection can be improved to a 2.4σ level by selecting 'on-pulse' events only as the pulsed emission is only ~ 0.5 the phase.

3.   Discussion and Conclusion.

COMPTEL as a γ-ray polarimeter has been investigated by M-C simulations and its polarization response been obtained. It is most sensitive to low energy γ-rays, < 5 MeV. While it may only have a modest Q value, $\sim \leq 0.10$, the Crab pulsar can be detected in a two week observation if it is 100% polarized. The fact that there are more than two months data available on the Crab source means there is a very good chance now to tell. at least, whether the γ-rays are polarized or not.

References.

[1] Schonfelder V. et al.: 1993. Ap.J Suppl., 86, p657-692.
[2] Swinyard B.M. et al.: 1991. Proc. SPIE, 1548, 94.

GLAST – A BROADBAND HIGH ENERGY γ-RAY TELESCOPE USING SILICON STRIP DETECTORS

K.S. WOOD

Naval Research Laboratory, Washington, DC

Abstract. The Gamma Ray Large Area Space Telescope (GLAST) has been under development since 1992. GLAST is an imaging γ-ray telescope sensitive to photon energies from 10 MeV to 300 MeV . Scientifically, it follows up the highly successful EGRET instrument on CGRO. Like EGRET, it is a pair conversion telescope consisting of a tracker and calorimeter that give respectively the direction and energy of the incident photon. The area, angular acceptance, and energy range of GLAST are each an order of magnitude beyond corresponding EGRET values, and it will provide much better angular resolution (20 arcmin for single photons at 1 GeV). Key innovations are the use of silicon strip technology in a solid state tracker design, a segmented CsI calorimeter, and an onboard computation scheme to handle the large volume of events. GLAST is light enough to be launched by a Delta 2 rocket. Its capabilities can be applied to studies of AGN, gamma ray bursts, pulsars, and γ-ray sources in the Galactic Plane. GLAST will continue the dramatic development of extragalactic high energy γ-ray astronomy and will be particularly powerful in determining the origin of the high-latitude diffuse background at GeV energies.

1. Introduction

The Gamma-ray Large-Area Space Telescope (GLAST) concept originated in 1992, when a proposal to NASA established the original GLAST collaboration of Stanford University, Stanford Linear Accelerator Center (SLAC) and the Naval Research Laboratory (NRL), with Principal Investigator P. Michelson (Stanford). * This group initiated feasibility studies and optimizations that have since developed GLAST into a candidate mission concept. Areas that have been explored include (i) analysis and characterization of the expected scientific returns, (ii) design optimization through extensive Monte Carlo simulations, (iii) calculation of performance parameters for the optimized design, and (iv) preliminary scaling of mission requirements.

The case for GLAST is straightforward. EGRET is a highly successful experiment that demands a follow-on mission. GLAST is such a mission. It will yield positions, spectra, and temporal variability for $> 10^3$ sites where particle acceleration extends to extreme high energies. GLAST exploits new technical approaches to enhance the pair conversion telescope

---

* The collaboration of institutions studying the GLAST mission concept includes Stanford University, the Stanford Linear Accelerator Center, Naval Research Laboratory, Los Alamos National Laboratory, University of California at Santa Cruz, University of Chicago, University of Maryland, University of Washington, Lockheed Research laboratory, Sonoma State University, Max Planck Institut für Extraterrestrische Physik, Istituto Nazionale di Fisica Nucleare, University of Tokyo, and Kanagawa University.

L. Bassani and G. di Cocco (eds.), Imaging in High Energy Astronomy, 287–291.
© *1995 Kluwer Academic Publishers.*

design. GLAST's tracker is a solid state design based on silicon strip technology. The time of flight (TOF) system used in EGRET is eliminated. Event discrimination functions of the EGRET TOF are handled through onboard computation. GLAST's increases over EGRET in effective area and field of view (FOV) are each a factor of ~ 10. The time for which a source is monitored is in practice proportional to the size of the FOV. The total photon count collected from a source scales as the product of the FOV and effective area integrated over the energy range (which for GLAST is a broader range than for EGRET). The large GLAST FOV (1.8π sr) does not compromise angular resolution. Rather, GLAST's silicon strip tracker improves angular resolution substantially over EGRET. Its segmented calorimeter and veto design provide response to photons from 10 MeV to 1 TeV. (Effective area as a function of energy is constant above 10 GeV; hence the maximum detectable energy for any source is set by the point where count rate becomes unobservably low. The Crab can be seen to 1 TeV in $\sim 10^7$ s.) It is practical with GLAST to position point sources using photons from ~ 1 GeV energies, where the finest angular resolution is available. Results anticipated from these improvements are many more source detections, better monitoring, better positions and better mapping. The source count increase over EGRET will be comparable to that experienced in X-ray astronomy in going from Uhuru to the Einstein Observatory.

2.   Scientific Program

Kanbach [1] reviews EGRET's impressive scientific achievements. EGRET has found new high energy γ-ray source classes and given a wealth of information on particle acceleration sites from solar flares to active galactic nuclei (AGN). GLAST's performance improvements – summarized above and justified in section 3 – will serve for estimating expected GLAST results by scaling from EGRET results. Source classes GLAST will study include AGN, diffuse high latitude emission, normal galaxies, gamma ray bursts, the Galactic Center, diffuse emission from the Galactic plane, molecular clouds, supernova remnants, spin-powered pulsars, and solar flares. Scaling from EGRET restricts discussion to phenomena already discovered, leading to conservative but still impressive estimates of GLAST results, as representative examples will show.

Establishment of the blazar class of AGN as a GeV source class is one of the great accomplishments of EGRET. The discovery that these sources are also highly variable, most likely with beamed emission, points the way to future studies, which must further lucidate the physics responsible for the γ-ray emission. The extended energy range of GLAST supports searches for spectral breaks and features that might discriminate among models for emission mechanisms. EGRET increased the number of detected high-energy γ-ray AGN to > 40 [2]. Most are blazars, that is, OVV quasars or BL Lacertae objects. Conservative extrapolation of the number-flux (N(S)) relation from EGRET combined with GLAST's hundredfold increase in photon collection yields an estimated GLAST total AGN count well in excess of 1000 objects. For typical AGN spectra (power laws with photon spectral indices of 1.6-2.6) GLAST will collect about as many photons above 1 GeV as EGRET collects above 100 MeV, hence GLAST positions can be based on GeV photons. Single-photon projected positional accuracy at 1 GeV is 20 arcminutes. Positional accuracy of a source improves roughly as the square root of the number of photons collected. Counterpart identification is crucial to determinations of distance and intrinsic parameters.

The existence of a diffuse background at high galactic latitudes was established by SAS-2 [3]; it is known to have a comparatively steep spectrum, but its origin is not understood. It could be a truly diffuse emission or could result from unresolved discrete sources. GLAST may establish the nature of the extragalactic diffuse component by resolving it, if it is composed of point sources. GLAST can reach flux levels where discrete sources could explain as much as 50% of the diffuse background in a manner consistent with extrapolation of the EGRET N(S) curve, according to unpublished simulations done by T. Willis at Stanford University.

Spin-powered pulsars are a galactic source class that has grown with EGRET discoveries. Before EGRET there were two known and now there are six [4]. EGRET light curves are used to refine and test models of the pulse formation including the outer gap model and the polar cap model[5,6]. GLAST is likely to find more pulsars from our Galaxy (the projected yield being highly model-dependent) and it can see a pulsar with the characteristics of the Crab at the distance of the LMC. The fact that one pulsar seen by EGRET – Geminga – is primarily a GeV source and is not seen in radio means there may be an important class of GeV spin-powered pulsars for which the only option for observational monitoring of spin period evolution is that provided by GLAST.

EGRET found emission from gamma ray bursts at energies > 100 MeV and in one case up to 18 GeV, with burst persistence in this band to later times than at lower energies. The 1.8π sr field of view of GLAST will increase the number of high energy burst detections. The number expected is sensitive to details of the extrapolation, but a hundred bursts per year is a reasonable estimate.

3. Instrument Concept, Event Discrimination, and Simulations

The primary components of the GLAST pair conversion telescope are the solid state (silicon strip) tracker and the segmented CsI calorimeter. The tracker has 12 conversion layers (0.5 radiation lengths) plus a front veto layer, and the calorimeter is 10 radiation lengths. There is a further major subsystem to handle event discrimination.

Key advantages of the silicon strips are that they measure tracks of converted electrons and positrons with ten times better position resolution than the EGRET spark chambers, do not use gas or consumables, and are essentially 100% efficient. There is no large pressure vessel such as that required on EGRET. Using the top silicon layer as a segmented shield for veto avoids the need for a monolithic anticoincidence dome. The segmented CsI calorimeter permits event showers to be imaged through shower maximum, while measuring their energy. This aids discrimination based on distinctive signatures of photons and cosmic rays .

A challenge for any pair conversion telescope is overcoming the scarcity – by a factor of $\sim 10^4$ – of gamma-rays relative to cosmic ray events. In GLAST's event discrimination by analysis cuts, each cut reduces the surviving fraction of particle events by a factor of 3 - 5, while rejecting very few valid gamma ray events. The cumulative result of this method of analysis cuts is near-total elimination of particles (rejection factor 3×10^{-6}) with only 20% reduction of effective area below the maximum established by the total cross section of the tracker. The residual rate is equivalent to a diffuse flux of 10^{-6} cm^{-2} s^{-1} sr^{-1} at E > 100 MeV, or $\sim 1/20$ of the diffuse background at high latitudes. Analysis cuts recognize and test for patterns in the event distribution in the tracker and calorimeter. One test requires valid events to materialize inside the tracker; three others test on the quality of the track produced, and so forth for a total of eight cuts altogether. Some cuts are implemented in onboard computation to reduce downlinked data. Space qualified computers with 32-bit words and 20 Mips throughput now being prepared for the USA Experiment on ARGOS [7] are sufficient for the onboard processing.

Design optimization has been done by Monte Carlo simulation with the GISMO package [8] which uses QED interactions adapted from the EGS4 code and hadronic interations adapted from the Gheisha code. Using GISMO, the design geometry is varied and performance parameters are calculated. Effective area as a function of energy is determined from the ratio of photons correctly characterized to photons incident on the array. Self-veto effects cause EGRET's effective area to decline above 1 GeV. In contrast GLAST's effective area vs energy curve continues almost flat to the highest energies (4000 cm^2 at 50 MeV; 8000 cm^2 above 1 GeV). Angular response is determined from the distribution of the difference between estimated and actual values for the direction of origin. For normal incidence, single photon angular resolution is 6.6 deg at 25 MeV, improving to 2 arcmin above 20 GeV. Sensitivity estimates result from simulating known backgrounds, then applying analysis cuts and computing signal to noise for photon fluxes.The limiting point source sensitivity is 1.5x 10^{-9} cm^{-2} s^{-1} at E > 100 MeV.

Photon events need not be confined to a single tower to be correctly characterized. Removing the TOF creates the potential for a large FOV, but realizing that possibility fully calls for preserving accurate characterization of photons incident at large zenith angles, where they affect several towers. The simulations show performance declines very gradually going off axis. There is a plausible case for sometimes being able to point the array for prolonged times at confused regions such as the galactic plane, hence for housing GLAST in a 3-axis stabilized spacecraft. However for many prime scientific objectives it is sufficient to have an instrument that is always pointed to zenith, scanning at the orbital rate.

4. Mission Concept

GLAST meets the requirements for an intermediate class NASA mission. A baseline GLAST has been scoped, assuming a 7x7 array of towers. Each tower has a frontal area 24 cm x 24 cm and consists of a 13-layer silicon strip tracker and segmented calorimeter. The weight, dominated by the calorimeter, is 3000 kg. The power requirement is kept below 1 kW, a number dominated by power requirements of preamps for the $\sim 10^6$ tracker channels. The envelope is 170cm x 170cm x 80 cm. The FOV is 1.8π sr. The attitude determination requirement is 20 arcsec, in order to support the best source position determinations. A GLAST instrument with these characteristics can be launched by a Delta 2 vehicle. The ability to realize a GLAST that goes beyond EGRET by a substantial performance margin using a low-cost launch vehicle means that it should not be necessary to wait many years to realize a worthwhile follow-on to EGRET. A mission concept study team recently has been formed to examine and develop the GLAST concept to the next stage of design and planning.

References

1. Kanbach, H., 1994, these proceedings.
2. von Montigny et al., 1995, ApJ 440,525
3. Fichtel, C., Simpson, G.A., and Thompson, D.A, 1978, ApJ 222, 833.
4. Fierro, J. et al., 1993, ApJ 413, L27.
5. Cheng, K.S., Ho,C., and Ruderman, M.A., 1986, ApJ 300, 500.
6. Daugherty, J.K., and Harding, A.K., 1982, ApJ 252, 337.
7. Wood, K.S. et al., 1993, Proceedings of S.P.I.E. 2280,19.
8. Atwood, W.B., et al., Int.J. Mod Phys. C, 3, 459.

AGATE: EXPANDING ON THE SUCCESS OF EGRET

B. L. DINGUS, J. A. ESPOSITO and R. MUKHERJEE

Universities Space Research Association, NASA / Goddard Space Flight Center, Code 662, Greenbelt, MD 20771

and

D. L. BERTSCH, R. CUDDAPAH, C. E. FICHTEL, R. C. HARTMAN, S. D. HUNTER and D. J. THOMPSON

NASA / Goddard Space Flight Center, Code 662, Greenbelt, MD 20771

Abstract.
The Energetic Gamma-Ray Experiment Telescope (EGRET) on the Compton Observatory has increased our understanding of active galactic nuclei, pulsars, diffuse emission, solar flares and gamma ray bursts. In order to continue these advances in high energy gamma-ray astronomy, development of a follow-on telescope to EGRET is in progress. The new experiment, named the Advanced Gamma-Ray Astronomy Telescope Experiment (AGATE), has the same basic components which have made EGRET a successful, low background telescope for 30 MeV to 30 GeV gamma rays; however, the sensitivity has been increased by an order of magnitude, the energy range extended to higher energies, and the angular resolution improved.

The large area, 2m x 2m, of AGATE has lead to the selection of drift chambers for the tracking detector, rather than the spark chambers used in EGRET. Drift chambers have fewer wires and much less deadtime per event. The power per wire is low enough to use many layers in order to reduce the multiple scattering of the electron and positron before the gamma-ray direction can be measured.

A 16 layer prototype of 1/2m x 1/2m drift chambers has been built. Muon tracks are used to study position resolution, efficiency, and noise. Preliminary work has also begun on the spacecraft requirements.

Key words: gamma-ray detectors – pair production – drift chambers

1. Scientific Justification

Prior to the launch of the Compton Gamma-Ray Observatory only a handful of astrophysical sources were known to emit photons of energy > 100 MeV. However, the catalog [Fichtel *et al.* 1994] of the first year and a half of data from the Energetic Gamma-Ray Experiment Telescope (EGRET) lists 124 sources detected above 100 MeV. These objects range from distant active galactic nuclei to pulsars in our galaxy to our own sun to gamma-ray bursts of unknown distances. Also, a large fraction of the sources in this catalog are not identified with objects detected at lower energies. While EGRET has provided new information about these astrophysical sources, many new questions have been raised. In order to answer these questions a new instrument is required. The capabilities needed for this new instrument are summarized in Table 1.

L. Bassani and G. di Cocco (eds.), Imaging in High Energy Astronomy, 293–296.
© 1995 *Kluwer Academic Publishers.*

2.   Basic High Energy Gamma-Ray Detector Design

The Advanced Gamma-ray Astronomy Telescope Experiment (AGATE) follows the tradition of the three successful gamma-ray satellites SAS-2, Cos-B, and EGRET, by retaining the four principal components of their designs: active anticoincidence, directional coincidence, energy measurement, and tracking systems as seen in Figure 3. However, in order to fulfill the scientific requirements listed in Table 1, a much larger instrument of ~ 2 m x 2 m area with improved angular resolution and smart trigger system is necessary. While a larger gamma-ray telescope requires some modification to the anticoincidence, directional coincidence, and energy measurement system (Cuddapah et al., 1993), the major changes are needed in the tracking detector.

The tracking detector is composed of many position measurement layers interleaved with high Z material which allows the gamma ray to pair produce. The amount of high Z material between measurements should be small in order to reduce multiple scattering of the electron and positron before their direction can be measured. Yet the total amount of material has to be of the order of half a radiation length in order to get a sufficient probability of the photon converting. EGRET contains spark chambers between 0.02 radiation lengths of tantalum foil. However, spark chambers do not have sufficient position resolution (0.8 mm) and have too much deadtime (100 msec) for a 4 square meter area telescope. Drift chambers are a good alternative since at accelerators even larger area detectors have been constructed which have low power requirements, position resolution of order 100 microns, and deadtimes of less than a few microseconds. The power requirements are so low that over 100 measurement planes are possible with a total spacecraft power of 2 kW.

3.   Drift Chamber Prototypes

A stack of 16 planes of 1/2 m x 1/2 m drift chamber frames has been built. Each frame has 3 planes of wires – 2 planes of 100 micron diameter Cu/Be on a 2 mm pitch seperated by 1 cm which create the uniform electric field and 1 plane half way between with alternating anode (20 micron diameter Au plated W wire) and cathode (100 micron diameter Cu/Be) with 4 cm pitch. Various voltages and gas mixtures have been used, but typical gains are a few thousand and drift velocities of few cm/microsec. A typical muon track is shown in Figure 1, and the position resolution for two gas mixtures is shown in Figure 2. In order to determine the position resolution, no high Z conversion foils were in the stack. Further results are given by Mukherjee et al., 1994. Eight planes of 1/2m x 2 m drift chambers have also been constructed.

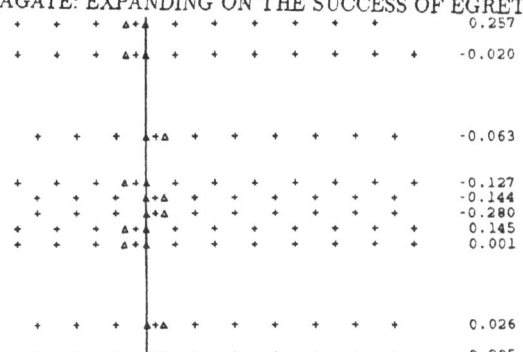

Fig. 1. Typical muon event recorded with the 1/2 m x 1/2 m prototype drift chambers with 90% Xe and 10 % methane at 1 atm. The crosses indicate the position of the anodes which are separated by 4 cm. Each horizontal row of anodes constitutes a frame which is 1 cm high, and the total stack height is 50 cm. Half of the frames are offset by half a drift cell to eliminate the left/right ambiguity. The triangles are the hit positions corresponding to left or right drift towards the anode. The numbers on the right hand side are residuals in units of mm for each point used in the reconstruction.

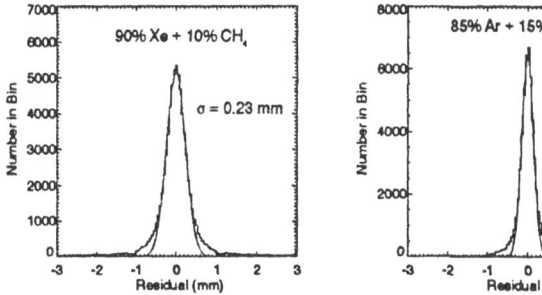

Fig. 2. Distribution of the residuals to a straight line fit of muon tracks for two different gas mixtures. The data are fit by a Gaussian function. The standard deviation is a measure of the spatial resolution attainable with these drift chambers.

4. Future Plans

The 1/2 m x 1/2 m prototype will be taken to an accelerator during the spring of 1995 to measure the angular resolution for gamma rays. Improvements on the drift chamber design include decreasing the thickness of the individual drift cells thereby increasing the field of view. The feasiblity and scientific benefits of several configurations of a gamma-ray telescope (Figure 3) are being considered.

References

Cuddapah, R. et al.: 1993, 'Development of a High Energy Gamma-Ray Telescope Using Drift Chambers', *1992 IEEE Nuclear Science Symposium* **Vol. 1**, 643-46

Fichtel, C. E. et al.: 1994, 'The First Energetic Gamma-Ray Experiment Telescope Catalog', *Ap J Supp* **94**, 551-581

Mukherjee, R. et al.: 1994, 'Development of Large Area Drift Chambers for High Energy Gamma-Ray Astrophysics' in Aprile, E., ed(s)., *SPIE Proceedings*, Vol. 2305, 2-12

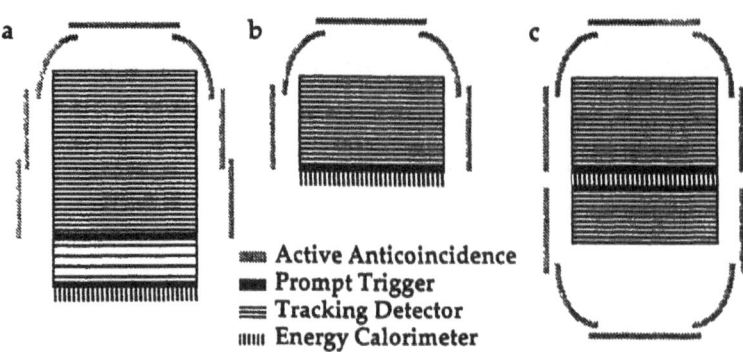

Fig. 3. High energy, gamma-ray telescope options using drift chambers as the tracking detector. (a) Best angular resolution is achieved by having the pair conversion material spread over a large number of measurement planes; however, the aspect reduces the field of view and low energy sensitivity. (b) Increasing the field of view is achieved by shortening the stack height and measuring the directionality of the particle using pattern recognition and dE/dx in the drift chambers which may not be as effective as a time of flight system in rejecting background. (c) Because of the low power of drift chambers, both sides of the calorimeter can be instrumented with a tracking detector. The livetime of a pointed telescope is doubled since one side can observe while the other is occulted by the earth. Also, background rejection is improved.

TABLE I

Science driven characteristics of the successor to EGRET.

REQUIREMENT	WHY	HOW
Increased Sensitivity	•Weaker Sources •Faster Time Variability •Source Localization •Transient Sources	•Larger Area •Low Energy Sensitivity •Better Angular Resolution •Larger Field of View
Improved Angular Resolution	•Multiple Source Confusion •Source Localization	•Reduce Multiple Scattering (Low Energies) •Improved Position Resolution (High Energies)
Broad Energy Range	•Low and High E Spectral Breaks or Continuations •Broad Pion Bump at 70 MeV	•Reduce Radiation Lengths (Low Energies) •Increase Area and Prevent Self Veto (High Energies)
Low Background	•10,000 to 100,000 Particles per Gamma Ray •Particles Masquerade as Gamma Rays •Albedo Rejection	•Active Anticoincidence Shield •Directional Veto (e.g. time of flight)

THE ALL-SKY HIGH ENERGY GAMMA-RAY MONITOR
(CYGAM PROJECT)

V.V.AKIMOV and N.G.LEIKOV

Space Research Institute, Moscow, Russia

Abstract. Being launched in a high-apogee orbit the proposed omnidirectional gamma-ray telescope will provide contemporary observations of a broad sky region, which may include the whole Galaxy, with sensitivity in GeV range by an order of magnitude higher and angular resolution by 2-3 times better than in the current EGRET/CGRO mission. Discovery of thousands of new sources is expected but their spectra can be measured only approximately.

1. Introduction

The impressive astrophysical results, provided by the ongoing EGRET/CGRO mission [1] inspire optimism about the perspectives of the high energy gamma-ray astronomy (50 - 10000 MeV). At the same time, to get a real progress we must ensure that a gamma-ray telescope for the next mission would be by an order of magnitude more sensitive than EGRET just as the latter in comparison with its predecessor COS-B [2]. If we assume that the more or less complete coverage of the sky will be again one of the main tasks then typical point source flux in the catalog of a future experiment must be of an order of 10^{-8}, cm^{-2} s^{-1} (E>100 MeV).

The required level of sensitivity makes even more critical the painful question about the angular resolution of gamma-telescopes. If we take into account the statistical relation $\log(N)/\log(S)=-1$ for galactic sources and transparency of the Galaxy for gamma-rays then the gain in minimal detectable flux by two orders (compared to COS-B) would result in detection of thousands of new sources. It means that angular resolution of the future telescope must be not worse than, say, 0.2° - 0.3°. Otherwise a bulk of potentially detectable sources could not be resolved and the gamma-ray astronomy would become essentially an extragalactic science. The pair production kinematics imposes the principal lower limit on the angular resolution which, however, can be reached only asymptotically at zero converter thickness. The limit of 0.2° corresponds to the photon energy of about 1 GeV. Therefore in the design of a future telescope special emphasis should be laid on the detection of gamma-rays with higher energies. The problem is in a small number of gamma-quanta of such energies. Being devoted to the traditional design of a gamma-ray telescope we would come to an instrument with many cubic meters of the track detector volume and with total weight of the order of 10 tons. Technical problems and extremely high cost of such experiment, especially of its launch, are evident. We show that there is a possibility to meet the requirements to the sensitivity and angular resolution without enormous enlargement of the detectors.

L. Bassani and G. di Cocco (eds.), Imaging in High Energy Astronomy, 297–300.

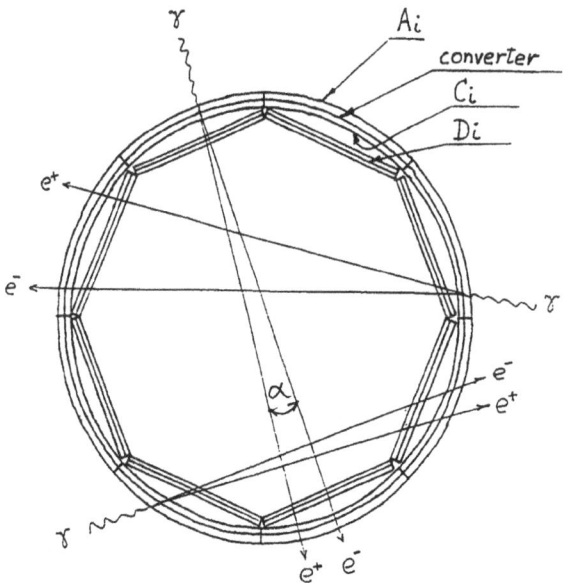

Fig.1. Schematic diagram of the proposed Cylindrical Gamma-Monitor (CYGAM). A_i : anticoincidence counters; C_i : coincidence counters; D_i :

2. Omnidirectional telescope

Normally the total observational time is shared between many targets which can not be combined in the telescope field of view. Therefore, expansion of the telescope aperture, in practice, would result in the increase of total exposure, i.e. the experiment sensitivity. An ideal instrument, one could imagine, should be an all-directional telescope observing all the sources in the sky simultaneously.

A possible variant of the instrument, which we would like to call Cylindrical Gamma-Monitor (CYGAM) is schematically shown in Fig.1. Eight elementary track detector modules (2 planes, X- and Y-coordinate sensitive each) are mounted on the surface of the imaginary octagonal prism of the outer diameter about 2 m. All the material where conversion of the photon into the electron-positron pair may take place is concentrated in a surrounding cylindrical layer. From inside of the converter we place a segmented cylindrical layer of coincidence counters (C_i, i=1,..8) and from outside - similar segmented layer of anticoincidence counters (A_i). In this axially symmetric design photons incident from any azimuthal direction can be detected. The electron and positron tracks initiated in the converter are measured each in the two points near the vertex and two points at the opposite side of the prism. This provides a well defined gamma-ray signature and high accuracy of angular measurements. In practice the useful high energy gamma-ray event appears as a single (merged) track in the input module and two close but distinctly separated tracks in the output one. Directions and coordinates of these three elementary tracks must be in agreement with a V-type image within the measuring errors. Admissible is the image with an apparent opening angle α less than some predifined value α_{max}. The time-of-flight logic coupling the coincidence segments (TOF(C_i C_j)) prevents self vetoing of events in an anticoincidence segment A_j at the pair exit. Any two coincidence segments except of adjacent ones can initiate the trigger signal in the absence of veto from the anticoincidence segment close to the vertex:

$$M_{i,j} = TOF(C_i C_j)A_j \bar{A_i}$$

Note that A_j opposite to A_i operates in coincidence. Additional selection criterion demands the signal amplitude from A_j to correspond to the two minimum ionizing particles.

Easy to see that there is no place for a calorimeter in the proposed omnidirectional telescope design. It is sacrificed to the maximal sensitive area. However, we will show below that some information about the spectrum slope can be derived from the distribution of angles α. Characteristics of the described cylindrical telescope were obtained by Monte-Carlo calculations with the following parameters: track detector prism side length - 80 cm, height - 160 cm, converter thickness - 0.2 r.l., spatial resolution of two tracks - 0.2 mm. Results are presented in Table 1 in comparison with the EGRET telescope [3].

Table 1.

	Angular resolution (67%-con). deg		Effective area at 1/2 apert., cm² (10 GeV)	Field of view (FWHM), ster	Effective geom. factor (10 GeV), ster*cm²
	3 GeV	10 GeV			
EGRET	0.7	0.5	600	0.38	228
CYGAM	0.2	0.06	1840	4.6	8464

The gain of CYGAM in angular resolution by 3-8 times at GeV energies is an evident consequence of the favorable geometry for angular measurements. Mainly due to the considerable increase of the field of view, we have the gain by 37 times in the effective geometrical factor, i.e. in the number of registered gamma-quanta. Of course, to avoid screening by the Earth CYGAM must be launched into a high-apogee orbit.

Characteristics of the optimized CYGAM version are given in [4] in detail. Here we only quote that for a point source at the mean galacic background the miniumum detectable flux above 1 GeV at the confidence level of 5σ is $8*10$, cm^{-2} s^{-1} for a year of observation. The accuracy of localization of this source is about 4 arcmin.

The weak point of CYGAM is its poor energy resolution. But for extragalactic and strong enough galactic sources there is a possibility to derive spectral index from the distribution of angles α

Fig.2 shows these distributions for two sources with equal fluxes above 1 GeV but different spectral indices: -2 and -3, background is not included. It is important that spectral measuremens with CYGAM can be always made relative since the large field of view ensures an appropriate "standard candle" (Crab for example) any time.

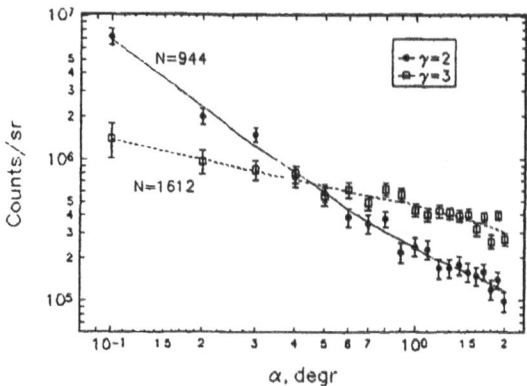

Fig. 2. Distribution of events versus the apparent pair opening angle for two spectral slopes for the source with flux F(> 1 GeV)=5*10⁻⁹ cm⁻² c⁻¹ ; T=1 year.

3. Instrumental background questions

The cosmic ray intensity above 500 MeV outside the Earth's magnetomosphere is usually reffered as $1 \, cm^{-1} \cdot sr^{-1} \, s^{-1}$. If we assume a quite realistic value of 10^{-4} as inefficiency of the anticoincidence shielding and a value of 0.1 as inefficiency of amplitude restriction in A then the total background counts rate from CR (yet with no respect to the track detector information) will be of the same order as from the diffuse galactic gamma-radiation. These CR-induced events as well as low energy photon events, with one pair component absorbed near the vertex, can imitate the high energy photon pictures in the track detectors via energetic secondaries (δ-electrons and also cascade particles for electrons) born along the path inside the telescope. Preliminary estimations show very low probability of such imitations (~10⁻⁴) but the final answer will be obtained with a laboratory prototype of the telescope. We plan to use there the large-area drift chambers (4 planes in each module) with spatial resolution not worse than 0.2 mm. Such accuracy must be sufficient for effective rejection of the parazite tracks not matching the true gamma-ray image.

References

1. G.Kanbach et al.: 1988, Space Sci. Rev. 49, 69

2. G.F. Bignami et al.: 1975, Space Sci. Instr. 1, 245

3. D.J. Thompson et al.: 1993, Astrophys.J.(Suppl.), 86, 629

4 G.Bisnovatyi-Kogan and N.Leikov: 1993, Astroph. Space Sci. 204, 181

IMAGING PERFORMANCE OF FIGARO-IV, A LARGE AREA γ-RAY TELESCOPE ABOVE 100 MeV.

B. SACCO, O. CATALANO, M.C. MACCARONE
Istituto Fisica Cosmica e Applicazioni Informatica, CNR,
Piazza G. Verdi 6, Palermo, Italy

G. GERARDI
Istituto Fisica, Università di Palermo, Palermo, Italy

E. COSTA
Istituto Astrofisica Spaziale, CNR, Frascati, Italy

E. MASSARO
Istituto Fisica, Università "La Sapienza", Roma, Italy

and

E. MORELLI
Istituto Tecnologie e Studio Radiazioni Extraterrestri, CNR, Bologna, Italy

Abstract. We are developing a new telescope, named FIGARO-IV, for γ-ray astronomy above 100 MeV, in which the electron-positron pairs, produced by photons in lead converters, are tracked in several independent planes of Limited Streamer Tubes (LST). Because of its large sensitive area and good angular resolution, this telescope is well suitable, and competitive with respect to satellite-based detectors as EGRET, to localise discrete γ-ray sources in a relatively short observation time, to detect high-energy γ-ray bursts and to investigate both periodic and random time variability on γ-ray sources.
In this contribution we present the results on the imaging performance, evaluated by means of many numerical simulations based on the GEANT code, for a $16m^2$ telescope, as FIGARO-IV that we project to fly by means of a stratospheric balloon.

Key words: γ-ray astronomy - imaging

1. Telescope Characteristics

FIGARO-IV telescope [1] will use standard Limited Streamer Tubes LST [2] modified to ensure a good performance in the environmental conditions of the balloon flights such as pressure, temperature, mechanical stresses. LST planes will be used as tracking detectors to identify the arrival of γ-ray photons after the pair conversion in thin layers of lead.

The present version of FIGARO-IV telescope (see Fig.1) is basically composed by nine planes of LST, ~4mx4m each, and of two 1mm thick lead γ-ray converter placed at a distance of 1m each other. Every LST plane consists of several units, each of them made by 8 square tubes of $1cm^2$ in section, ~4m long. The LST planes are supported by a honeycomb styrofoam structure, ensuring their parallelism. Two guard rings are moreover placed around the planes close to the lead converters. Three LST planes above the upper lead converter and the relative guard ring constitute the anticoincidence system to discriminate charged particles from photons. Photons converted in the upper lead layer are

L. Bassani and G. di Cocco (eds.), Imaging in High Energy Astronomy, 301–304.
© 1995 *Kluwer Academic Publishers.*

triggered by the coincidence between the signals detected in the first LST plane below the lead and in one of the two next LST planes. Photons which are converted in the lower lead layer are triggered in a similar way with the further condition that all the upper planes must be in anticoincidence.

The read-out of each LST plane is performed by two independent sets of strips with 1cm pitch: one parallel to the tubes (X-direction), and the other orthogonal to them (Y-direction), for a total of 6400 read-out channels in the present version of the telescope. A chain of read-out electronics boards placed on both X and Y sides of each plane transfers the information about the track position into the data acquisition system. The directions of the incoming photons can therefore be derived from the bisecting lines of the pair tracks, weighted with the scattering amplitudes. The pair opening angle and the electron scattering produced by the lower lead converter give a rough estimate of the γ-ray energy, sufficient to permit flux evaluation at least in two energy intervals.

The main telescope characteristics are summarised in Table I.

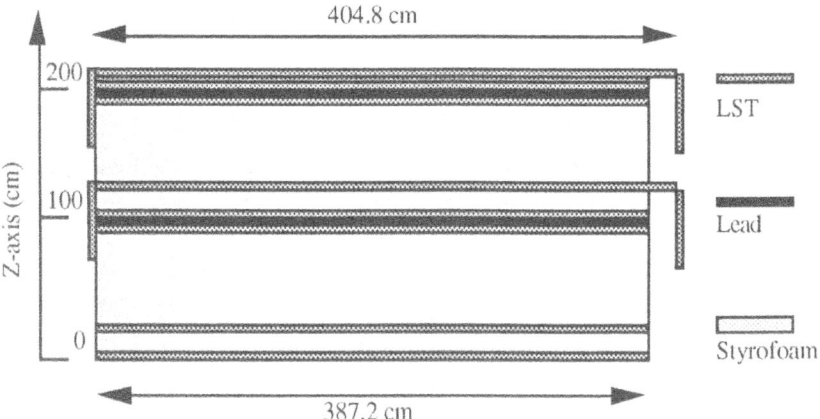

Figure 1: FIGARO-IV telescope cross-section. The asymmetry on the right side is due to the read-out electronics allocation.

TABLE I
FIGARO IV characteristics

Detector type	LST (1 cm² cell)
Number of LST planes	9
Gamma-ray converters	Lead (1 mm thick)
Number of lead converters	2
Geometric area	~ 16 m²
Pick-up	Strips (1 cm pitch)
Read-out channels	6400
Spatial resolution	0.5 cm
Telescope height	~ 2 m
Telescope weight	~ 2.4 tons
Field of view	60° radius
Effective area (at 300 MeV)	~ 32000 cm²
Angular resolution (at 300 MeV)	~ 2°

2.   Telescope Imaging Performance

Figure 2 shows the field of view at 300 MeV of the FIGARO-IV telescope in comparison with that of the EGRET [3]. To evaluate the global efficiency, we consider the quantity Aeff.*$\Delta\Omega$ which is 90,000 cm^{2}sr in the case of FIGARO-IV and 380 in the case of EGRET. One can see that FIGARO-IV is well competitive with EGRET for a class of scientific items in which the time variability plays a relevant role: searching for periodic signals in the galactic γ-ray sources, studying the hour-day variability of the γ-ray emission by AGN, looking for high-energy γ-ray bursts.

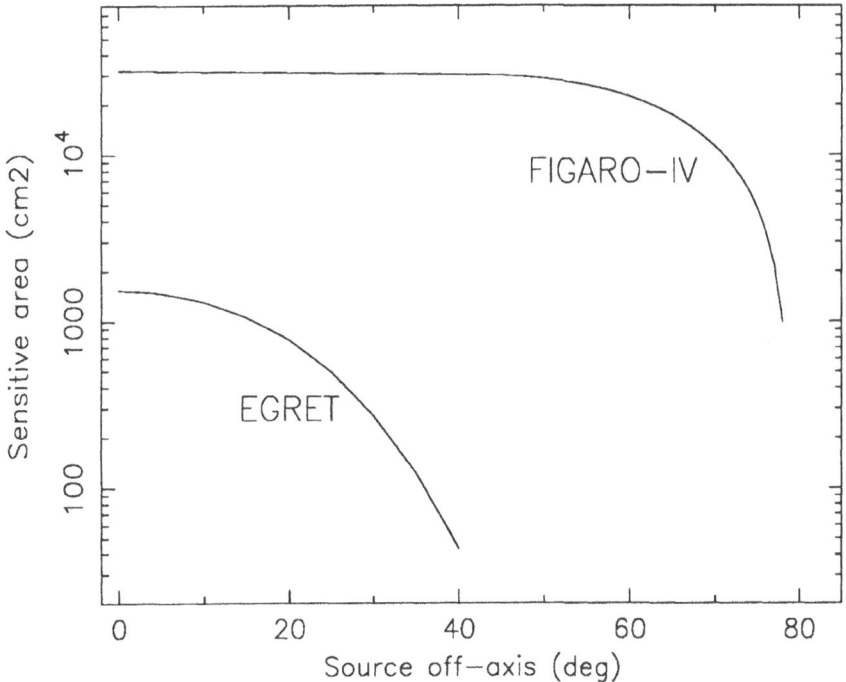

Figure 2: Comparison between EGRET and FIGARO-IV sensitive area at 300 MeV. EGRET data are relative to the primary operating mode with TASC in coincidence.

To evaluate the imaging capability of FIGARO-IV telescope, we simulated an observation of the Cygnus region by using GEANT/3 code of CERN which takes into account the physical processes of the γ-rays and of the e$^+$e$^-$ pairs produced by them. In the Cygnus region the sources J2019+40, J2021+37, and J2032+40 [4] are located within a circle of two degrees of radius. The simulation has been performed in the hypothesis of a balloon flight at a ceiling altitude of 4mbar at a geomagnetic cut-off of 8.1 GV (as in the Mediterranean area); during the observation (8 hours long) the telescope axis is pointed towards the Zenith position and the sources transit across the field of view.

The result of the simulation, when the atmospheric and galactic background is neglected, shows (see Fig.3a) the intrinsic imaging capability of FIGARO-IV telescope. As a second step we introduce the expected background which is of the order of 6*10^{-3} ph cm^{-2}s^{-1}sr^{-1}, as derived from previous balloon experiment [5],[6] and normalised for the altitude and the geomagnetic cut-off. At this background level, the three Cygnus γ-ray sources are detected at a significance of 9, 10, and 8 σ, respectively. The image is now computed by means of a fitting procedure which uses the telescope Point Spread Function (energy averaged) and assuming uniform background in a small sky region, few degrees around the three sources. The fit results are shown in Fig.3b: the source intensities are well reconstructed and the source positions are in agreement with the nominal ones within 0.2 degrees.

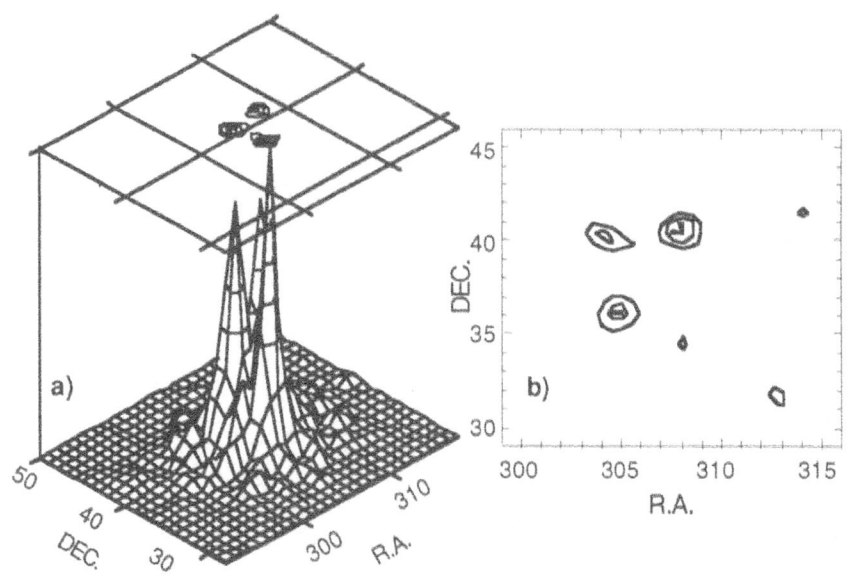

Figure 3: Simulation of the Cygnus region (see text for explanation).

3.   Conclusions

FIGARO-IV is a γ-ray telescope designed to fly by means of a stratospheric balloon; its large sensitive area, the large field of view and the good imaging performance, make it competitive with EGRET/CGRO for what concerns the search for pulsars in the γ-ray galactic sources, the study of random variability of galactic and extragalactic γ-ray sources in hour/day time scale with particular attention to the AGN, and the detection of high energy γ-ray bursts. Its use in long duration balloon flights, as transatlantic flights, may give answers to important questions open by the EGRET/CGRO mission.

References

[1] SACCO, B. et al., 1993, *Nuovo Cimento*, **16 C-6**, 715.
[2] IAROCCI, F. et al., 1983, *Nucl. Instrum. Methods* , **217**, 30.
[3] THOMPSON, D.J. et al., 1993, *Astroph. Jou. Suppl.* , **86**, 629.
[4] FICHTEL, C.E. et al., 1994, *The First Energetic Gamma-Ray Experiment (EGRET) Source Catalog* , pre-print.
[5] BEUERMANN, K.P., 1971, *J. Geophys. Res.* , **76** , 4291.
[6] SCHONFELDER, V. and LICHTI, G. , 1975, *J. Geophys. Res.* , **80** , 3681.

HARD X-RAY IMAGING USING MICROCHANNEL PLATE OPTICS

J.E. Lees, G.W. Fraser, A.N. Brunton and R. Willingale

X-ray Astronomy Group, Department of Physics and Astronomy,
University of Leicester, University Road, Leicester LE1 7RH, England.

Abstract Recently there has been much interest in the X-ray focusing properties of microchannel plates (MCPs). MCPs with both round and square pores have been shown to focus soft X-rays (Fraser *et al.* Nucl.Inst.Meth.A 324 (1993) 404 and Nucl.Inst.Meth.A 334 (1993) 579). In particular, images have been obtained from point sources positioned at the foci of round-pore MCP "X-ray lenses" for 0.28-17.4 keV X-rays. A Monte Carlo ray trace model has been developed which gives good agreement with experimental images.

In this paper we describe recent experimental progress in the development of MCPs for focusing hard X-rays (10-100 keV).

Keywords Microchannel plates – X-rays – Focusing

1 Introduction

Microchannel plate (MCP) optics [1] offer the possibility of low mass and relatively low cost focusing elements for use over a very wide X-ray energy range (0.1–100 keV).

The main technical challenges in fabricating high quality MCP optics for hard X-ray astronomy can be summarised as:
(a) the production of MCPs with channels of square cross-section
(b) the development of techniques to slump MCPs to suitable spherical figures
(c) the reduction of channel surface roughness to minimise scatter
(d) the etching of MCPs with long channel lengths (aspect ratios 500:1) to ensure good focusing efficiency at high X-ray energies.

In this paper we report significant progress towards these goals and describe a novel MCP geometry which might form the basis for a sensitive, multifocus 2-60 keV hard X-ray telescope.

2 MCP optics

The requirement for MCPs with square cross-sections is a consequence of their higher X-ray focusing efficiency when compared to MCPs with conventional circular pores [1]. Focusing measurements with planar square pore MCPs have been reported by Fraser *et.al* [2] who quote a best angular resolution of ~5 arcminutes FWHM from Si-K (1.74 keV) X-rays. These authors note that channel misalignment, especially angular alignment of the channel long axis, is the main factor presently limiting the angular resolution for soft X-rays.

Improvements in channel misalignment without sacrificing channel open area are now being pursued by a number of MCP manufacturers.

L. Bassani and G. di Cocco (eds.), Imaging in High Energy Astronomy, 305–308.
© 1995 *Kluwer Academic Publishers.*

To focus a parallel beam to a point, as in X-ray astronomy, the MCP must be spherically "slumped" to a radius of curvature, R, dependent on the focusing application.

Such a slumped channel plate optic obeys the X-ray lens formula: [3, 2]:

$$1/L_s - 1/L_f = 2/R$$

where L_s is the source distance and L_f is the focal distance.

For an X-ray source positioned at $L_s=R/2$, on the concave side of the optic, the output from a slumped MCP optic will be a quasi-parallel beam [4]. Such an experimental geometry was used to investigate the response of a 4mm thick (length-to-channel diameter ratio, L:D, 320:1), round pore Philips Photonics MCP slumped to a radius of 1.4m [5]. The top two images of figure 1 were measured with Cu-K (8.05 keV) and Mo-K (17.5 keV) X-rays. Both images are in good agreement with predictions of our Monte Carlo ray trace model, also shown in figure 1. X-ray source sizes were 5.6mm diameter for Cu-K and 0.8×2.2mm for Mo-K. The ray trace model assumes rms surface roughness 25Å, a correlation length of 30μm with rms long axis misalignment of 0.5 mrad.

We note that the slump radius of 1.4m is about the right magnitude for a soft X-ray wide-field monitor but a factor of about 5 **more** severe than would be required for a high energy X-ray telescope.

Willingale and Fraser [6] have calculated the critical surface roughness σ_c, which gives a total integrated scatter (TIS) of 1, for a typical lead glass used in MCPs (Corning 8161 ($Si_6O_{17}Pb_2K$; density 4.0 gcm^{-3}) over the range 1-100 keV. In order to reduce the scattering to acceptable levels the channel surface roughness must be much less than σ_c. For the hard X-ray range (10-100 keV) σ_c is approximately 26Å and Willingale and Fraser suggest an acceptable rms roughness of 10Å[6].

Measurements of channel roughness by Atomic Force Microscopy for two Philips Photonics MCPs, "etched-only" and "etched and weak-acid-polished", found rms surface roughness of 50Å and 22Å respectively [5]. There is also some evidence [5] that the hydrogen reduction process used in MCP manufacture also reduces the surface irregularities. It therefore does not seem unreasonable that manufacturers will be able to produce MCPs with surface roughnesses of ∼ 10Å as required for efficient hard X-ray focusing.

Focusing X-rays at high energies by total external reflection requires, finally, the photon incident angle to be small. The critical angle for reflection for 40 keV X-rays for MCP lead glass is ∼ 3′. For MCP optics to work efficiently this implies "long" channels satisfying , for the minimum of a single reflection, the equation $L = D/\theta_c$, where θ_c is the energy dependent critical angle for external reflection. We believe that current etching processes are capable of etching L:D ratios of order 600:1.

3   Discussion

All the individual components for a high energy MCP optic are available or will be with modest improvements in manufacturing techniques. What remains to be done is their combination into a square pore, slumped, high L:D, low surface roughness element. To this end, ESA have recently approved a program in their

Technology Research Programme scheme for the development of MCP optics for X-ray focusing.

Presently under consideration is a novel "two-stage" MCP focusing element [6]. The new optic design comprises of two square pore MCPs, with channels radially packed (figure 2 left), having different radii of curvature (R and R/3; figure 2 right). Such an optic closely approximates a Wolter Type 1 system and provides a factor ~4 higher geometric area than the regular square-grid square channel geometry. Furthermore the two-stage optic will focus a parallel beam to a point as opposed to the cruxiform focus of a conventional square pore MCP.

It is this optic which is envisaged as the focusing element in a high energy X-ray telescope, the concept of such a telescope which will be discussed in a companion paper at this workshop [7].

4   Acknowledgements

The work of the Leicester X-ray Astronomy Group is supported by the UK PPARC. The authors also wish to acknowledge the contributions of David Emberson (Philips Photonics) and Bruce Feller (Nova Scientific Inc.).

References

[1] H. N. Chapman, K. A. Nugent, and S. W. Wilkins. *Rev.Sci.Instrum.*, **62** (1991) 1542.

[2] G. W. Fraser. A. N. Brunton, J. E. Lees, J. F. Pearson, and W. B. Feller. *Nucl.Instr.Meth.*, **A324** (1993) 404.

[3] H. N. Chapman. K. A. Nugent, S. W. Wilkins, and T. J. Davies. *J.X-ray Sci. Tech.*, **2** (1990) 117.

[4] G. W. Fraser. A. N. Brunton, J. E. Lees, and D. L. Emberson. *Nucl.Instr.Meth.*, **A334** (1993) 579–588.

[5] G. W. Fraser. A. N. Brunton, J. E. Lees, J. F. Pearson, R. Willingale, D. L. Emberson. W. B. Feller, M. Stedman, and J. Haycocks. Development of microchannel plate (MCP) X-ray optics. In *Multilayer and Grazing Incidence X-Ray/EUV Optics III*. R.B.Hoover and A.B.C.Walker Jr, editors. Proc.SPIE. 2011. 1993.

[6] R. Willingale and G. W. Fraser. Broad band X-ray imaging with microchannel structures. In preparation.

[7] A. D. Holland. G. W. Fraser, R. Willingale, and M. J. L. Turner. The Hard X-ray Telescope: instrument concept and detectors. These proceedings.

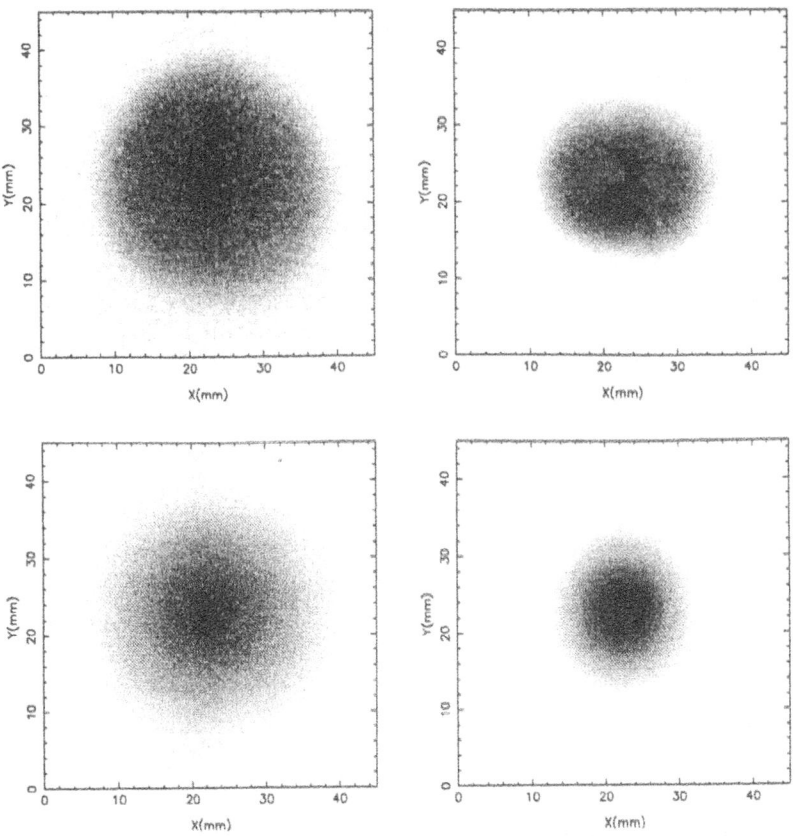

Figure 1: Measured (top) and predicted (bottom) X-ray intensity distributions from a slumped, 320:1 MCP in beam expander mode at Cu-K (8.05 keV), left, and Mo-K (17.5 keV), right.

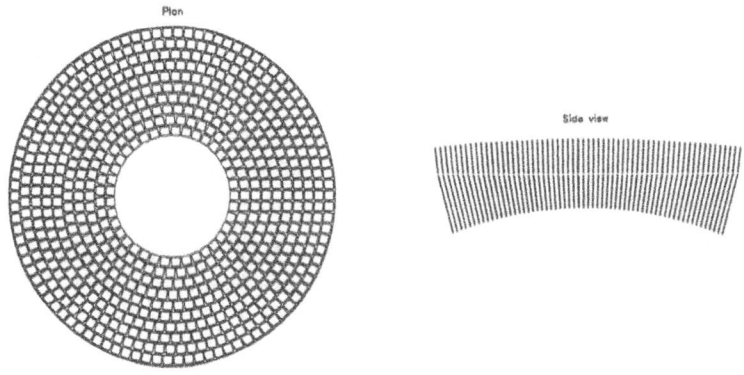

Figure 2: Left: a radially packed square pore MCP and right: the two-stage MCP geometry (see text).

THE HARD X-RAY TELESCOPE :
INSTRUMENT CONCEPT AND DETECTORS

A.D. Holland, G.W. Fraser, R. Willingale and M.J.L. Turner

X-Ray Astronomy Group, Department of Physics and Astronomy
Leicester University, Leicester LE1 7RH, U.K.

Abstract. The Hard X-ray Telescope (HXT) is a new instrument concept for the 1-60 keV band. Based on X-ray focusing using microchannel plate technology, HXT aims to combine arcminute imaging with X-ray spectroscopy and polarimetry. We describe the instrument design, its potential sensitivity and its focal plane concept - a hybrid small-pixel Si CCD / GaAs detector array. The former element provides polarimetric sensitivity and high spectral resolution (FWHM < 200 eV @ 10 keV) at the lower end of the HXT bandpass. The latter element produces good efficiency (20% @ 60 keV) and medium energy resolution (1 keV @ 60 keV) at the higher energies.

Key words: X-ray astronomy, GaAs, CCD, hard X-ray optics, X-ray polarimetry

1. Introduction

The present high energy limits on our exploration of the universe with focussing optics are represented by orbiting telescopes like the recently launched Japanese ASCA and the forthcoming Russian Spectrum X-Gamma, NASA's AXAF-I and ESA's XMM. These are limited to energies below 10 keV, while at higher energies, coded mask techniques have and will be used (as in the case of INTEGRAL) to extend sensitivity up to ~10 MeV. Coded mask devices do produce a true image but have the disadvantages of large detector area and much higher background compared with focussing optics.

We propose to extend the range of focusing optics up to 60 keV using a novel hard X-ray imager telescope concept. The Hard X-ray Telescope (HXT) [1] will have a sensitivity of 0.2 mCrab over the majority of its energy range from 1-60 keV, and a spatial resolution of 1 arc minute. With HXT extended objects and complex regions such as the Galactic centre can be resolved without decoding errors. Furthermore HXT will image the X-ray spectrum in the important 5-50 keV transition region between thermal and non-thermal emission processes with low background, high spectral and temporal resolution and a unique sensitivity to the linear polarisation of the X-rays. These capabilities will be most pertinent in the study of compact objects from Galactic X-ray binaries to quasars. The combination of imaging, spectral, timing and polarisation sensitivity in the unexploited 10-50 keV energy band will open up one of the last unexplored areas of observational astrophysics

2. Instrument Concept

Figure 1 depicts the HXT instrument concept, details of which can be found in ref. [1]. The overall instrument is formed from a set of 4 telescope modules, where each module comprises a 7x7 array of MCP "lenses" and solid state detectors, with an intermediate

L. Bassani and G. di Cocco (eds.), Imaging in High Energy Astronomy, 309–312.
© 1995 *Kluwer Academic Publishers.*

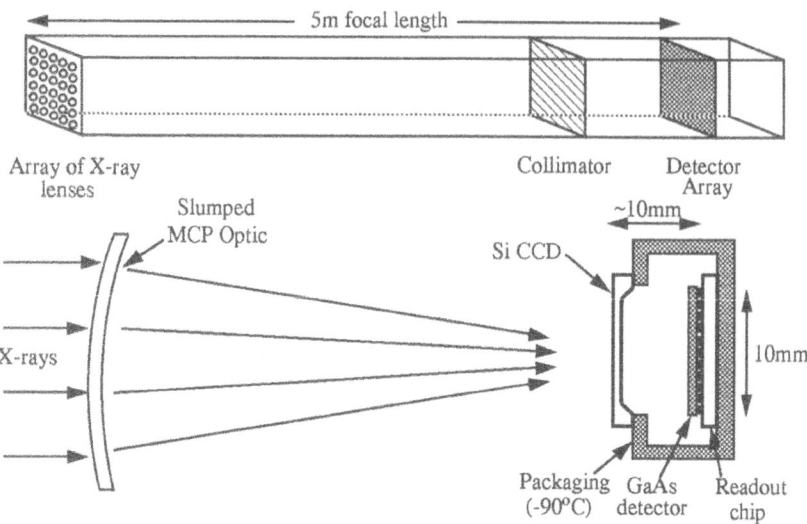

FIGURE 1. Schematic diagram of the optical design of the HXT and its hybrid detector concept

collimator to prevent scattered X-ray contamination between elements. This use of up to 200 co-aligned channels is required to produce an acceptable collecting area ($>2000cm^2$) for observation times of between 10^5-10^6s. The MCP optics are a new development for this application and are discussed in a sister paper [2].

3. X-ray Detectors

A schematic of the proposed detector arrangement is given in Figure 1, where each MCP optic is served by an imaging detector assembly. A dual-technology focal plane is used, the elements of which are both about to begin study within the ESA TRP programme. The first detection layer is formed from a CCD constructed on high resistivity silicon which is optimised for X-ray detection up to 20keV [3] and possesses small pixels (4x4μm) to enable the detection of source polarisation via the photo-electron emission direction [4]. A Monte Carlo model of CCD polarisation sensitivity has been developed and verified against existing measurements [5,6]. Some of the results are given in Figure 3, where for 4μm pixels the modulation factor of 12% at 20keV rises to 33% at 40 keV [4]. In addition to providing high resolution imaging and polarisation measurements, the CCD will also perform medium resolution spectroscopy (FWHM=250 eV @ 20 keV).

The detection efficiency of silicon is low above 20 keV. A higher-Z, pixellated, detector is therefore included behind the silicon CCD. The silicon is therefore thinned down to the active thickness to minimise unwanted absorption. Due to the depth of focus of the optics, the second detector may be up to 15 mm behind the silicon CCD. In order to determine changes in source continua above 20 keV and to resolve cyclotron lines, we require to combine high detection efficiency with an energy resolution of between 1-2 keV over the energy band 20-60 keV. We envisage the use of a hybrid array of diodes constructed on

GaAs, "flip-chip" bonded to a silicon array of charge sensitive pre-amps for the high-Z detector. Preliminary results using test structures manufactured on 250μm-thick semi-insulating GaAs have been obtained [7] where an active detection layer of 200μm thickness was observed, corresponding to a 20% detection efficiency at 60 keV.

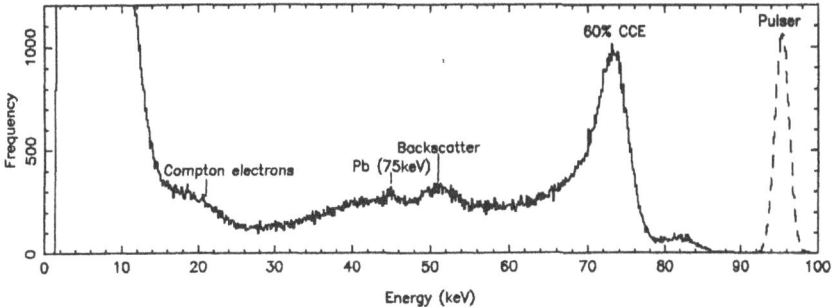

FIGURE 2. X-rays from ^{57}Co in a 300μm thick GaAs detector at a bias of 300V

Figure 2 shows the ^{57}Co spectrum obtained from a 300μm thick GaAs detector at a temperature of -10°C [7]. Using this detector an energy resolution of 4 keV was obtained which was dominated by charge loss mechanisms in the material. Work is currently underway to grow thick layers of high purity GaAs by liquid phase epitaxy which should overcome the charge loss processes and should produce Fano-limited resolution of 640eV at 60 keV [8].

4. Instrument Performance

TABLE 1. Minimum detectable polarisation, P_{min}, in the 25-35 keV band

Source Strength (mCrab)	Observation Time	
	10^5s	10^6s
10	5.3%	1.7%
1	17%	5.5%
0.25	34%	10%

The instrument collecting area for 4 telescopes each comprising a 7x7 array of MCP optics each with its focal plane detector is given as a function of energy in Figure 3. The contributions arising from detection in the Si CCD and the GaAs diode array are identified and include the edge structure of the detectors. The modelled polarisation modulation factor of the small pixel CCD is also given [4]. These data are used in Table 1 to calculate the minimum detectable polarisation (at the 99.9% confidence level) in the 25-35 keV band for a range of source strengths and integration times, assuming the continuum spectrum of the Crab with a background rate of 10^{-4}cm^{-2}s^{-1}keV^{-1}.

Figure 4 finally shows the limiting continuum sensitivity of the instrument for 5σ detection in 10^6s (with edge structure omitted), and compares the performance to that of the INTEGRAL X-ray monitor (XRM) [9] and XMM. The continuum of a 10 μCrab source is also given. This indicates that HXT will bridge an important gap between these

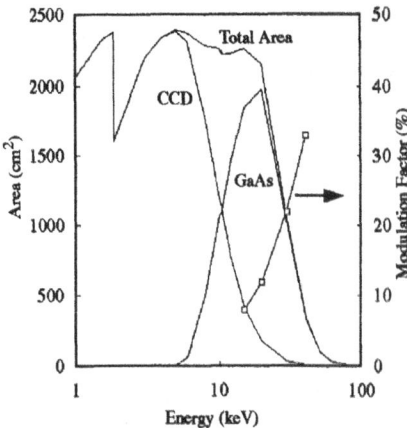

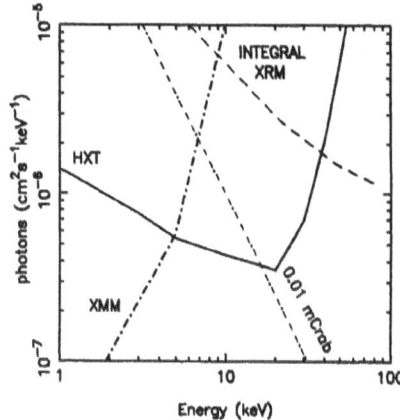

FIGURE 3. Instrument effective area and polarisation factor as a function of energy

FIGURE 4. Limiting continuum sensitivity for HXT. Detection threshold is 5σ in 10⁶s. XMM and INTEGRAL sensitivities are also shown

missions, due mainly to the large collecting area up to 30 keV. HXT will be particularly well suited to studies of continua above iron line features (>7keV) in comparison to the current planned missions in the 0.1-10 keV band and hence will enable better extraction of weak features from the continuum. The high sensitivity will also enable the first sky survey to be performed in the 10-30keV band.

5. Conclusion

The outline of an imaging instrument providing good source sensitivity and medium resolution spectroscopy over the 1-60 keV band has been presented. The instrument will particularly contribute to our knowledge in the 5-30 keV region which is not as accessible using other instruments. In addition the unique approach to polarisation detection will enable the first sky survey of polarised sources in the hard X-ray band to be performed.

References

1. "Hard X-ray Telescope (HXT) : A proposal for the ESA M3 mission opportunity", G.W. Fraser et al, University of Leicester, U.K., May 1993
2. J.E. Lees, G.W. Fraser, A.N. Brunton and R. Willingale, "Hard X-ray imaging using microchannel plate optics", This volume
3. A.D. Holland, M.J.L. Turner, D.J. Burt and P. Pool, "The MOS CCDs for the European Photon Imaging Camera", Proc. SPIE **2006** (1993) 2
4. A.D. Holland, A.D.T. Short, G.W. Fraser and M.J.L. Turner, "The X-ray polarisation sensitivity of CCDs", submitted to Nucl. Inst. Meth. A, August 1994
5. G. Buschhorn et al, "X-ray polarimetry using the photoeffect in a CCD detector", Nucl. Inst. Meth. A346 (1994) 578
6. H. Tsunemi et al, "Detection of X-ray polarisation with a charge coupled device", Nucl. Inst. Meth. A321 (1992) 629
7. A.D. Holland, A.D.T. Short and T. Cross, "X-ray detection using bulk GaAs", Nucl. Inst. Meth. A346 (1994) 366
8. J.E. Eberhardt, R.D. Ryan and A.J. Tavendale, "Evaluation of n-GaAs for nuclear radiation detectors", Nucl. Inst. Meth. 94 (1971) 463
9. Integral X-Ray Monitor : Minutes of group meeting AO4, LU-XRM-AO4, May 1994

EXPERIMENTAL RESULTS OBTAINED WITH THE POSITRON-ANNIHILATION RADIATION TELESCOPE OF THE TOULOUSE-ARGONNE COLLABORATION

J. E. NAYA, P. von BALLMOOS, F. ALBERNHE, G. VEDRENNE
Centre d'Etude Spatial des Rayonnements, 9, Av. du Colonel Roche, 31029 Toulouse, France

and

R. K. SMITHER, M. FAIZ, P. B. FERNANDEZ, T. GRABER
Advanced Photon Source, Argonne National Laboratory, 9700 South Cass Avenue, Argonne, Ill. 60439 USA

Abstract. We present laboratory measurements obtained with a ground-based prototype of a focusing positron-annihilation-radiation telescope developed by the Toulouse-Argonne collaboration. This balloon-borne telescope has been designed to collect 511-keV photons with an extremely low instrumental background. The telescope features a Laue diffraction lens and a detector module containing a small array of germanium detectors. It will provide a combination of high spatial and energy resolution (15 arc sec and 2 keV, respectively) with a sensitivity of $\sim 3 \times 10^{-5}$ photons $cm^{-2}s^{-1}$. These features will allow us to resolve a possible narrow 511-keV line both energetically and spatially within a Galactic center "microquasar" or in other broad-class annihilators.

The ground-based prototype consists of a crystal lens holding small cubes of diffracting germanium crystals and a 3x3 germanium array that detects the concentrated beam in the focal plane. Measured performances of the instrument at different line energies (511 keV and 662 keV) are presented and compared with Monte-Carlo simulations. The advantages of a 3x3 Ge-detector array with respect to a standard-monoblock detector have been confirmed.

The results obtained in the laboratory have strengthened interest in a crystal-diffraction telescope, offering new perspectives for the future of experimental gamma-ray astronomy.

1. Introduction

Recently, the Toulouse-Argonne collaboration presented a new type of gamma-ray telescope that may begin a new stage in gamma-ray astronomy. This instrument, designed to collect 511-keV photons with an extremely low instrumental background, consists of a Laue diffraction lens; a detector module with a 3x3 germanium array; and a balloon gondola stabilized to 5" pointing accuracy [1].

As a first step in the project schedule, a ground-based prototype telescope has been achieved. It consists of a diffraction lens focusing at finite distances (provided by ANL, Chicago) and a 3x3 Ge array detector (provided by the CESR, Toulouse).

In this paper, we present some of the experiments achieved and the measured performance of the system. The results obtained validate the diffraction-lens-based telescope concept and open interesting perspectives for the development of the balloon-telescope model.

2. System test at Argonne

The test was performed at Argonne National Laboratory in June/July 1994. It can be divided into three parts: the source, the lens module, and the detector module.

A characteristic of the ground-based lens is the possibility of focusing a wide range of energies without retuning the lens. To accomplish this, one adjusts the distance from the source to the lens (D_S) and the distance from the lens to the detector (D_I) to match the corresponding Bragg diffraction angle. The relationship between these distances for small

313

L. Bassani and G. di Cocco (eds.), Imaging in High Energy Astronomy, 313–317.

angles is:

$$\frac{1}{D_S} + \frac{1}{D_I} = \frac{1}{F} \tag{1}$$

where F is the focal length of the lens for a given gamma-ray energy.

2.1 THE SOURCE

The high-energy gamma-ray sources used in these experiments are described in table 1. They are enclosed in a large lead shield and are well collimated into a narrow cone of radiation just large enough to illuminate the lens. In this work, we present in detail the results from the 511-keV line, which represents the energy of interest for the proposed telescope.

Source	Energy [keV]	Activity [mCu]	Focal l. [m]	D_S [m]	D_i [m]
^{137}Cs	662	100	10.75	24.75	19.00
^{22}Na	511	39	8.30	18.48	15.07

Table 1: A description of the gamma-ray sources used in this experiment is presented. Focal l. is the focal length of the lens for the considered energy; Ds is the distance between the source and the lens; and Di is the distance between the lens and the projected image (where the detector is placed.)

2.2 LENS MODULE

The lens used in these experiments consists of 416 Ge crystals cut into small cubes of dimension 1 cm x 1cm x 1cm. These cubes are mounted in six concentric rings on a stainless steel frame. To compensate for the change in angle, each ring contains crystals cut so that a different set of crystalline planes is used for diffraction. The radius of each ring is designed such that each crystal in a ring focuses the same energy gamma-ray into a small focal spot simultaneously.

Due to the finite distance between the source and the lens, only a small percentage of the gamma-ray flux incident on the face of an individual crystal will have the correct Bragg angle and be diffracted. However, to optimize the diffraction efficiency of the crystals, they are divided in three parts by two wedged slots. This compensates for the change in the angle of the gamma rays as they impinge upon the front surface of the crystal.

The finite size of the source (3-mm diameter) produces a larger focal spot than would be expected for a point source (2.0-cm diameter instead of 1.7 cm for an ideal point source). For several of the experiments performed, we required a focal spot size smaller than the pixel surface, so a collimator consisting of a lead brick (10 cm thick) with a cylindrical hole (8-mm diameter) was used.

2.3 DETECTOR MODULE

A novel gamma-ray detector consisting of a high-purity 3x3 germanium matrix housed in a single cylindrical aluminum cryostat was used for these experiments. Each of the single Ge bars is an n-type coaxial detector with dimensions of 1.5 cm x 1.5 cm x 4 cm and an internal electrode hole of 4-mm diameter. The distance between the front surface and the electrode hole is 1 cm. Two adjacent elements are separated by 2 mm with an indium surface of 0.5 cm^2.

3. Experiment results

3.1 LENS EFFICIENCY

The efficiency of the lens for diffracting 511-keV photons is given by the formula:

$$\varepsilon_{diff} = \frac{A_D N_L D_S^2 \varepsilon_{nc}}{ns N_D (D_I + D_S)^2 \varepsilon_c} \qquad (2)$$

where N_L is the count rate of 511-keV events diffracted by the lens and seen by the detector; N_D is the count rate measured by the detector when the lens is removed from the system; D_S is the distance from the lens to the source; D_I is the distance from the lens to the detector; A_D is the area of the detector; n is the number of diffracting crystals in the lens; s is the average surface area of the front side of a single crystal, ε_{nc} is the detector efficiency for a non concentrated beam; and ε_c is the detector efficiency for a lens concentrated beam.

In order to calculate the lens efficiency, the intensity of the diffracted beam was measured with a modified Ortec HPGe detector system, normally used at ANL, instead of the array because of its less complex geometry. The ANL detector consists of an n-type coaxial germanium detector, 6 cm in diameter and 6 cm high. The diameter of the inner hole is 1 cm, and the distance from the front surface to the reentry hole has been customized to 3 cm. This modification makes the detector more efficient at these energies than the standard model, which has a distance of 1 to 1.5 cm between the front face and the reentry hole. The photo-peak efficiency for this detector is 31 % when the entire surface of the detector is illuminated (non-focused beam). When the beam is focused, the photo-peak efficiency increases since the gamma-ray beam is confined to a smaller volume within the detector. The Compton-scattered photons in this case have a higher probability of being detected since they are not created near the sides of the detector. This results in a focused photo-peak efficiency of 48%.

To calculate the lens efficiency at 511 keV, the total counting rate of the source-lens-detector system was measured. To eliminate transmission of the non-diffracted beam through the lens and into the detector, the center of the lens was blocked. After background subtraction, we found $N_L = 153$ c s^{-1} (count rate in the peak) with the lens tuned. Next, the lens was removed from the system, and the measurement was repeated. The count rate without the lens was $N_D = 77$ c s^{-1} (in the peak).

After substitution of the parameters in formula (2), we find that the diffraction efficiency for 511-keV photons is 2.8%, which corresponds to an effective diffraction surface of 17 cm^2. The low efficiency in this case is due to the narrow intrinsic rocking curve (~2 arc sec FWHM) of the crystals used in the present lens. The Na source, as seen by a single lens crystal, subtends an angle of about 25 arc sec. Because of its relatively narrow rocking curve, a typical crystal can only diffract photons that are emitted from a small 2 arc sec strip of the source. A technique for increasing the width of a germanium crystal's rocking curve has been developed and will be incorporated in any future lens. However, it should be noted that, if the source subtended a smaller angle, the efficiency of the present lens would increase significantly. For an astrophysical source with an angular radius on the order of 2 arc sec or less, the efficiency of present lens could be as much as 22%.

3.2 Ge ARRAY EFFICIENCY

The efficiency calculations were carried out by comparing the measured count rates of the detector array with the count rate observed for the ANL detector. The ratio between their count rates must be equal to the ratio between their efficiencies. The measured efficiencies for a 511-keV gamma-ray beam incident on a 0.8-cm- diameter focal spot were of 25% when the focal spot is placed on the central pixel and 18% when it is placed on the pixel of the corner.

3.3 BACKGROUND REDUCTION

The segmentation of the 3x3 Ge matrix together with the concentrated beam from the crystal lens allows application of new techniques for background reduction. The method consists of only accepting events that are compatible with the signature of a "good" 511-keV photon coming from the crystal lens. In other words, when the focal point of the lens is centered on one pixel of the array, 511-keV photons that deposit some energy in that pixel are considered as "good" events. The measured background rejection from an isotropic ambient background flux is 79.3% when the focals spot is on one pixel. This represents an improvement in the sensitivity by a factor of 2.2 in comparison with a standard detector of the same volume.

3.4 OFF AXIS SOURCE RECOGNITION

The germanium detector array allows us to take maximum advantage of a focused gamma-ray beam to spatially resolve the source. The focal spot can be easily localized looking at the 9 count rates. This also allows imaging of an off-axis source. Fig. 1 shows a series of measurements in which the source-lens-detector system has been intentionally misaligned.

The gamma-ray energy in this experiment was 662 keV. In the first drawing, the alignment is perfect, the focal spot being on the central pixel. The next three drawings show measurements with the lens shifted along the Y-axis 5 mm, 7.5 mm, and 10 mm, respectively. One can observe the movement of the focal spot as well as the decrease in the intensity of the focused beam due to the loses of diffraction efficiency because of the misalignment. Finally, the last drawing shows a measurement when the required tilt was applied to the lens in order to match the Bragg condition with a shift of 10 mm. The focal spot is on detector 4, but the initial intensity of the diffracted beam has been recovered.

4.   Discussion

The results of the tests validate the lens-telescope concept: the lens diffracts the gamma rays concentrating them into a small focal spot on the detector surface. Because the signal is associated with the collection surface and the background is associated with the detection volume, the large lens collection area combined with the small Ge array volume makes an optimal instrument for maximizing the signal-to-noise ratio.

The germanium-array-lens system also provides the following advantages:

1-   The possibility of background rejection (up to 80%) combined with a maximum detection efficiency. With the focal spot placed far from the array borders, the efficiency increase is 25% over the nonfocused beam efficiency at 511 keV.

2-   Off-axis source recognition.

3-   The possibility of simultaneous background source monitoring. The fact that the signal is localized to a small volume can be exploited by defining equivalent volumes within other detector pixels that are not receiving signal photons. This allows the use of these equivalent volumes to monitor the background; thus, eliminating the problem of background fluctuations during observations.

Focussing off-axis sources

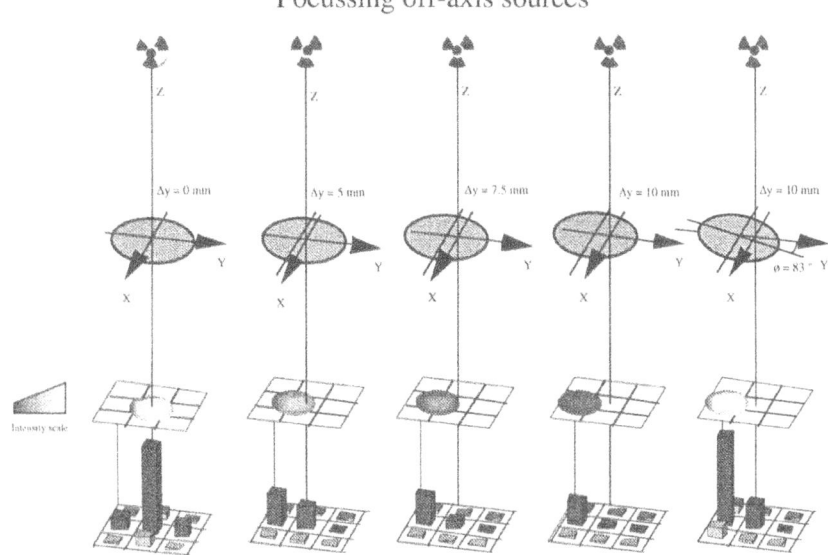

Fig. 1: The telescope response for focusing off axis sources is shown.

The extrapolation of these results to a balloon diffraction-lens telescope consisting of a 600-crystal lens, working at 511 keV, combined with the presented detector array results in an instrument with a FOV of 10"-20", an energy band of 6 keV at 511 keV, and a spatial resolution of 15" FWHM. The 3 σ sensitivity for a 20 hr balloon flight mission in Alice Springs would be 3×10^{-5} photons cm^{-2} s^{-1}. Such an instrument would be able to resolve, both energetically and spatially, point sources emitting narrow 511-keV photons. Interesting candidates to observe would be the "microquasars" at the Galactic center [3], such as 1E 1740.7-258 [4] or GRS 1758-258 [5], as well as other objectives like Cygnus X-1, X-ray binaries, and pulsars.

Acknowledgments

This work was partially supported by the U.S DoE Contract No. W-31-109-Eng-38, by the *Region Midi-Pyrénées* (Toulouse) and the French Space Agency (CNES). The authors are grateful to Francis Cotin for his contributions to this project.

References

[1] von Ballmoos, P. and Smither R..K. 1994, ApJ Sup. S., 92, 663
[2] GEANT CERN Program Library Office
[3] Ramaty, R., Leventhal, M., Chan, K. W.& Lingenfelter, R. 1992, ApJ, 392, L63
[4] Mirabel, I.F., et al. 1992, Nature, 358 (No. 6383), 215
[5] Rodríguez, L.F., et al. 1992, ApJ, 401, L15

GAMMA-RAY IMAGING WITH
GERMANIUM PLANAR ARRAYS

W. A. MAHONEY, S. C. MARTIN and L. S. VARNELL
Jet Propulsion Laboratory 169-327
4800 Oak Grove Drive
Pasadena, California 91109 USA

and

T. A. PRINCE
California Institute of Technology
220-47 Downs Laboratory
Pasadena, California 91125 USA

Abstract. A program is under way at JPL to develop position-sensitive germanium planar array detectors which will enable instrumentation capable of fine gamma-ray imaging combined with high-resolution spectroscopy. Diode arrays have been fabricated and tested and a number of junction and surface passivation techniques have been investigated with good results. Our goal is to build planar arrays with a position-resolution of 2 mm or better in two dimensions. When used with a coded aperture at a distance of several meters, arcminute imaging will be achieved. We envision employing large mosaics of these sensors in both balloon and satellite experiments.

Key words: Gamma-ray imaging – Germanium planar arrays

1. Introduction

Significant advances in gamma-ray astrophysics will require instrumentation which combines high-resolution spectroscopy, fine imaging, and very good sensitivity. To address this need, a program was initiated at JPL/Caltech to develop position-sensitive germanium planar array detectors which will allow arcminute gamma-ray imaging with instruments practical for both balloons and satellites. The experiments would be optimized near 100 keV and would be sensitive from roughly 5 to 600 keV. This energy range contains gamma-ray lines from the decay of several important radionuclei including ^{44}Ti, ^{56}Fe, ^{57}Co, and ^{60}Fe (Clayton & Leising 1994) with the upper end driven by the importance of the positron/electron annihilation line at 511 keV. The energy range also spans the important transition region between typical thermal and non-thermal astrophysical phenomena.

This new instrumentation would allow detailed investigations of several fundamental astrophysical issues including the formation of the Solar System, the life and death of stars, the composition and velocity distribution of interstellar material, and the power sources of quasars and active galaxies. For example, a particularly interesting radioisotope is ^{44}Ti which decays with the emission of gamma-ray lines at 68 and 78 keV, just the energies

L. Bassani and G. di Cocco (eds.), Imaging in High Energy Astronomy, 319–324.

where the performance of a germanium planar array is optimal. Observation of these lines should lead to the discovery of young, previously unknown Galactic supernovae (Mahoney et al. 1992), perhaps only detectable through ^{44}Ti line emission. Supernovae are believed to occur in our galaxy at the rate of 2-4 per century (Tammann et al. 1994), yet Cas A, the youngest known, is over 300 years old. A space instrument employing a large mosaic of these sensors would have the potential to detect any Galactic supernova that has occurred during the past several hundred years, and thus it could directly measure the supernova rate.

An important goal of the sensor development program is to achieve significant sensitivity through the 511 keV line which arises from the annihilation of positrons. In addition to providing a direct measurement of the positron annihilation rate, this line can also be used as a powerful diagnostic tool. Its width will help identify and characterize the different phases of interstellar gas, its centroid measures bulk velocities of interstellar material through the Doppler effect, and its profile can be used to study the geometry of compact sources of gamma-ray emission (e.g. Lingenfelter & Hua 1991). Finally, gamma-ray line and continuum emission can be used to probe the central energy sources and the emission mechanisms of active galactic nuclei (AGN) and quasars. A space mission would have the sensitivity to investigate hundreds of AGN and Galactic compact objects such as accreting black holes, white dwarfs, and neutron stars, x-ray binaries, and x-ray pulsars.

To date a number of important discoveries have been made in gamma-ray astronomy, however, this has been accomplished with instruments that achieved either high spectral resolution (i.e. *HEAO 3*: Mahoney et al. 1980) or good imaging (i.e. *GRIP*: Althouse et al. 1985 and *SIGMA*: Paul et al. 1991), but not both. A major advance in gamma-ray astronomy will occur with the launch of the International Gamma-Ray Astrophysics Laboratory (INTEGRAL: Winkler 1991). This mission will allow large steps forward in both gamma-ray imaging and spectroscopy, but it will do so with two separate instruments. Significant progress beyond INTEGRAL will require single instruments that combine (a) the fine spectroscopy needed to identify lines, to accurately measure the line shapes, and to study other sharp spectral features and (b) arcminute imaging required to avoid source confusion and to allow identification of gamma-ray sources with known objects.

2.   Concept and Technical Program

In order to enable the implementation of such an experiment, a program was recently initiated at JPL to develop position-sensitive planar germanium detectors by taking advantage of a unique mix of skills and capabilities in the JPL/Caltech community. The basic concept, shown in Figure 1, is a planar array where the prime signal is taken off one contact and pro-

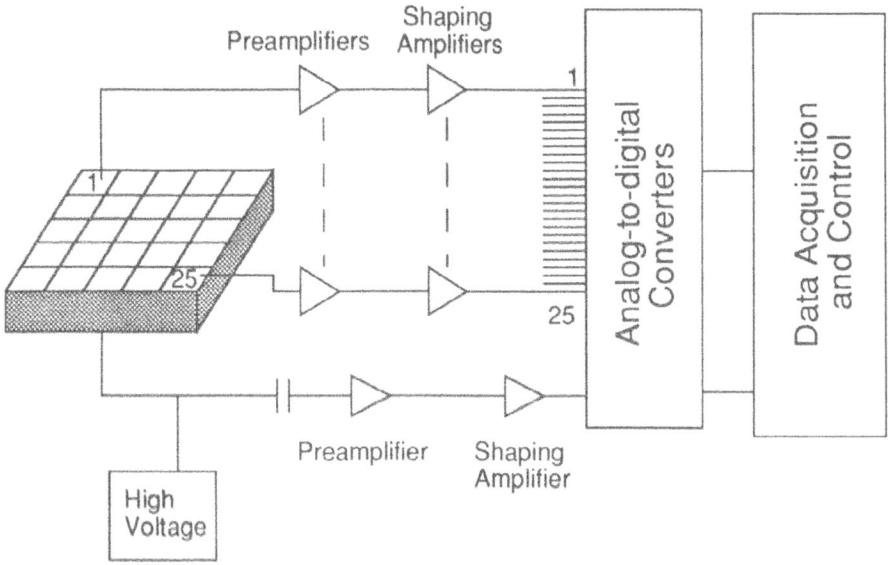

Fig. 1. Conceptual drawing of a planar germanium detector divided into a 5 × 5 array of pixels.

cessed through a high-performance signal chain. The other contact is shown divided into a 5 × 5 array of pixels, each with linear dimensions of about 1 cm, and each with a signal chain capable of determining in which pixel or combination of pixels the interaction occurred. The prime sensors are being fabricated by the JPL MicroDevices Laboratory (MDL). To date, emphasis has been on developing techniques for making high-quality junctions and on identifying surface passivation techniques which will yield arrays with low leakage current (< 1nA @ 77 K). The signal chain electronics will be a close derivative of the VLSI chips currently being built by the Caltech Space Radiation Laboratory for two experiments on the Advanced Composition Explorer (ACE). Two versions are being designed which promise high-performance with low power dissipation.

Our goal is to improve the position-resolution to 2 mm or better in two dimensions. Such sensors, when employed with a coded aperture at a distance of several meters, will allow arcminute imaging while retaining the superb energy resolution of germanium detectors. Sufficient sensitivity through 511 keV could be achieved by either fabricating planar detectors with a thickness of 1 to 1.5 cm, or by stacking thinner detectors.

3. Performance and Future Plans

The processing steps used in fabricating the detector arrays are summarized in Figure 2. Our initial approach focused on using low temperature thermal cycles ($< 400°$ C) to limit the introduction of impurities. Though this temperature is sufficient to provide dopant activation, the resulting shallow junctions gave rise to high leakage currents. To maintain the germanium purity while allowing for higher temperature processing, a quartz lined furnace that can be flushed with HCl was set up. The HCl serves to remove metal impurities present in the furnace prior to loading the detector array. In this way detector arrays were simultaneously passivated with a thermally grown germanium oxynitride (Hymes & Rosenburg 1988), and annealed using temperatures up to $600°$ C. The resulting detectors had room temperature leakage currents one sixth that of earlier devices, and 80 K values approaching usable levels. A full test of the detector performance will determine if the purity of the germanium was maintained at acceptable levels using this approach.

Future arrays will be fabricated with a field implant in the region between pixels to further reduce the surface generated portion of the dark current. In addition, if the temperature cycles are found to introduce high levels of impurities, doped low temperature molecular beam epitaxy (MBE) will be studied as an alternative to the current design which uses ion implantation as the dopant source.

Actively shielded mosaics of these germanium planar arrays would significantly advance the observational capabilities of both balloon and satellite experiments. For example, with a balloon-borne experiment containing an actively shielded mosaic of about 35 5 cm $\times$ 5 cm planar arrays, one could observe the ^{44}Ti-decay lines from Cas A, seen by COMPTEL (Iyudin *et al.* 1994), at a sensitivity of 5 - 10 σ. The first balloon-flight arrays would probably have a position-resolution of about 1 cm in two dimensions, giving an angular resolution of about $0.5°$ and a point source location capability of about $0.1°$, depending on the source strength. Cooling to approximately 80 K could be accomplished by using one of a number of small cryocoolers that are now available, all of which have been extensively characterized at JPL (Ross 1995).

A mosaic of these planar arrays having a collecting area of about 1 m^2, when used in an actively shielded satellite experiment, could achieve a narrow line sensitivity of a few times 10^{-7} photons cm^{-2} s^{-1} near 100 keV. Assuming a position-resolution of 2 mm and a coded aperture at a distance of 4 m, angular resolutions approaching 1 arcminute become possible, allowing one to map the structure of supernova remnants. The sensitivity should be sufficient to observe any supernova that has occurred in the Galaxy during the past few hundred years.

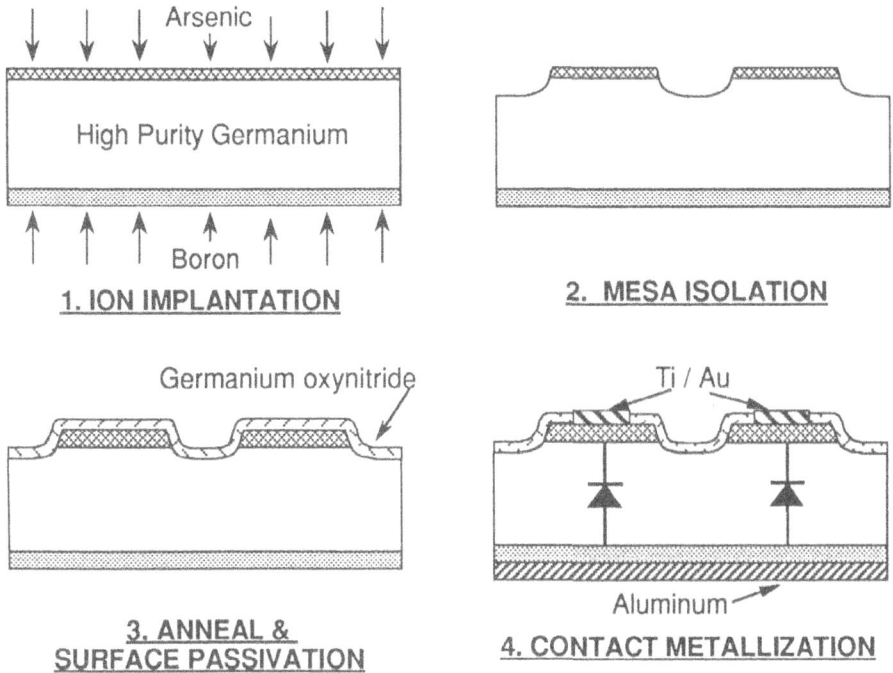

Fig 2   Outline of the processes used in the fabrication of the first diode arrays at the JPL MicroDevices Laboratory

4.   Summary

It is clear that the next generation instruments for low energy gamma-ray astronomy will require high spectral resolution, arcminute imaging, and very good sensitivity. A program is currently underway to build planar germanium array detectors with a position-resolution of about 2 mm in two dimensions. Diode arrays have been built and tested demonstrating the feasibility of the approach and fully operational detectors are anticipated within about a year. Large mosaics of array detectors on a satellite experiment should achieve narrow line sensitivities of a few $\times$ 10^{-7} photons cm^{-2} s^{-1}.

5.   Acknowledgements

The research described in this paper was performed by the Center for Space Microelectronics Technology, Jet Propulsion Laboratory, California Institute of Technology, under contract with the National Aeronautics and Space Administration.

6. References

Althouse, W. E., *et al.* 1985, Proc. 19th International Cosmic Ray Conf. (La Jolla), **3**, 299

Clayton, D. D., & Leising, M. D. 1994, in *The Second Compton Symposium*, ed. C. Fichtel, N. Gehrels, & J. Norris (New York: AIP), 137

Hymes, D. J., & Rosenburg, J. J. 1988, J. Electrochem. Soc., **135**, 961

Iyudin, A. F., *et al.* 1994, in *The Second Compton Symposium*, ed. C. Fichtel, N. Gehrels, & J. Norris (New York: AIP), 156

Lingenfelter, R. E., & Hua, X. -M. 1991, ApJ, **381**, 426

Mahoney, W. A., Ling, J. C., Jacobson, A. S., & Tapphorn, R. M. 1980, Nucl. Instr. Meth., **178**, 363

Mahoney, W. A., Ling, J. C., Wheaton, W. A., & Higdon, J. C. 1992, ApJ, **387**, 314

Paul, J. *et al.* 1991, Adv. Space Res., **11**, 289

Ross, R. G., Jr. 1995, in *Cryocoolers 8*, ed. R. Ross, Jr. (New York: Plenum), 173

Tammann, G. A., Löffler, W., & Schröder, A. 1994, ApJS, **92**, 487

Winkler, C. 1991, in *Gamma-Ray Line Astrophysics*, ed. Ph. Durouchoux & N. Prantzos (New York: AIP), 483

SPATIAL AND SPECTRAL RESOLUTION OF A
GERMANIUM STRIP DETECTOR

R.A. KROEGER, W.N. JOHNSON, R.L. KINZER and J.D. KURFESS
Naval Research Laboratory, Washington, DC 20375

S. INDERHEES and B. PHLIPS
Universities Space Research Association, Washington, DC 20024

and

N. GEHRELS
Goddard Space Flight Center, Greenbelt, MD 20771

Abstract. Germanium strip detectors combine both the excellent energy resolution typical of germanium detectors and fine spatial resolution possible in a strip detector. They are applicable to sensitive, high spectral resolution γ-ray detectors using coded-aperture or Compton telescope techniques. Our first detector is in a planar geometry with orthogonal strips on the upper and lower detector surfaces, providing 9 mm spatial resolution. The detector has 5 strips on each surface and an active volume of $45 \times 45 \times 12$ mm. Good spatial and energy resolution are demonstrated.

Key words: Ge detectors, Strip detectors, Gamma-rays

1. Introduction

Radiation detectors that combine good energy resolution with fine spatial resolution are needed to provide spectroscopy and imaging in a single instrument. Device applications in high energy astrophysics include coded-aperture and Compton scatter telescope imaging. Superior spectroscopy is needed to resolve suspected cyclotron features [1; 2], determine annihilation radiation line–width [3; 4], improve sensitivity to narrow-line features in a variety of astrophysical sources, and observe soft spectra. Superior angular resolution is needed to localize unknown sources or to resolve closely spaced sources.

Germanium strip detectors are capable of providing excellent spatial and spectral resolution. Such devices are fabricated from planar germanium by cutting [5], etching [6], and photomask [7] techniques. In a two-dimensional detector, strips on opposite faces are orthogonal (Figure 1).

Readout of the signals from each strip may be achieved in several ways: a capacitive [8] or resistive [9] charge division network may be used to to reduce the number of channels of electronics required. Readout of individual strips provides the best performance. Each strip behaves as a single-channel geramanium detector with excellent energy resolution dominated by the usual factors of detector capacitance, quality of the front-end electronics and electron-hole counting statistics. Cross-talk between the strips can be

325

L. Bassani and G. di Cocco (eds.), Imaging in High Energy Astronomy, 325–328.
© 1995 *Kluwer Academic Publishers.*

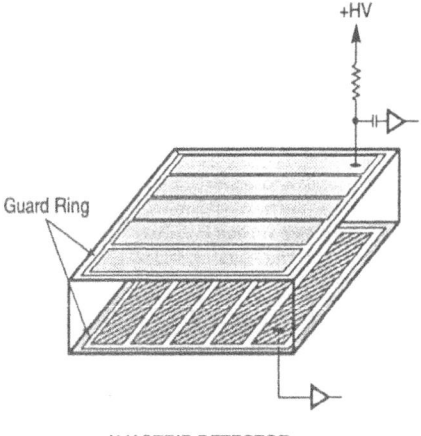

+HV

X-Y STRIP DETECTOR

Guard Ring

Fig. 1. Schematic of the germanium strip detector used in this work. Crossed electrodes provide two-dimensional position localisation of interactions. The electrodes may be read out individually or grouped in two charge-division chains, one for each side, then readout with four channels of electronics (not shown here). The guard ring provides increased immunity to leakage currents. There are 5 strips on each face of the detector, centered on a 9 mm pitch and 45 mm long. The active volume is 45 × 45 × 12 mm.

minimized and calibrated. One- and two-site gamma ray interactions can be uniquely reconstructed.

In an earlier work, we operated a two-dimensional strip detector using two capacitive charge division networks (one network for strips on each side of the detector) and four channels of electronics [10]. This approach did succeed at localizing individual interactions, but with some expected complications. First, energy resolution of the detector is degraded, getting worse with increasing numbers of strips. In our detector, resolution degraded to 5.5 keV for 5 strips in a capacitive charge division network, from 2.2 keV for a single strip. Second, the minimum energy threshold is higher than for single strip operation. Third, multiply interacting gamma rays cannot be located accurately and energy measurements are further degraded by electronic non-linearities in the system.

2. Experiment

The 5 × 5 strip detector used in this work (Figure 1) was fabricated using a photomask technique to make the lithium and boron implants. Our laboratory data acquisition system consists of 10 spectroscopy channels, each with a 13-bit ADC, permitting strips to be analyzed individually. The ADCs are read out through a CAMAC crate with a Macintosh computer. Events are stored on disk in list-mode for subsequent processing.

A collimated source of 60 keV gamma rays from ^{241}Am was used to investigate the positional response of the detector. The collimated beam is ~1.2 mm in diameter (FWHM) with a triangular intensity profile on the surface of the detector. The position of the beam is controlled by a position table under computer control.

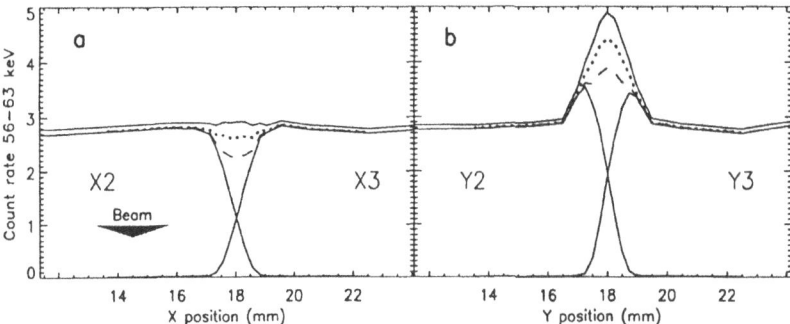

Fig. 2. Detector response *vs.* position for a 1.2 mm diameter (FWHM) collimated beam of 60 keV γ-rays. Figure 2a represents scanning the beam across the boundary between two X-strips (B side). Figure 2b represents scanning the beam across the boundary between two Y-strips (Li side). The γ-ray beam is incident on the Y-side of the detector. The uppermost solid curve shows the count rate on the Y/X strip on the opposite face of the detector (strip parallel to the scan direction). The lower solid curves show the count rates in the two strips identified in the Figures. The lower dashed line is the sum of the rates in the individual strips. The upper dotted line is the rate associated with the energy window applied to the sum of the signals from the 2 strips.

Spatial response between two X-strips and between two Y-strips is shown in Figure 2. The left panel shows a relatively flat response to 60 keV total-energy events over the surface of strip X2, then drops off rapidly to zero at the edge of the strip. Similiary, response of the X3 strip rises as X2 drops. Events that share charge between X2 and X3 are excluded from the individual strip response curves by event selection using a narrow energy window around 60 keV. Therefore, the sum of the individual strip responses (dashed curve) drops between the strips.

About half of the charge sharing events are recovered by summing the signals from adjacent strips. The upper dotted curve in Figure 2a represents the 60 keV window rate in the sum of the two strips. The remaining dip in the response curve is attributed to charge sharing events where the signal in one of the two strips is below our discriminator threshold ($\sim$20 keV), and is therefore not digitized.

Figure 2b shows a similar result from scanning between two Y-strips. The Y-strips are lithium drifted contacts on the upper surface of the detector (toward the gamma-ray source). We estimate that these contacts are currently $\sim$500 μm thick, and represent a dead absorbing layer on the surface above the active volume of the detector. The large enhancement in efficiency between the strips results from a gap between the lithium contacts where the active volume of the detector is closer to the surface.

The width of the charge-division region between strips is estimated. A model of the γ-ray beam crossing over a sharp strip edge is a good descrip-

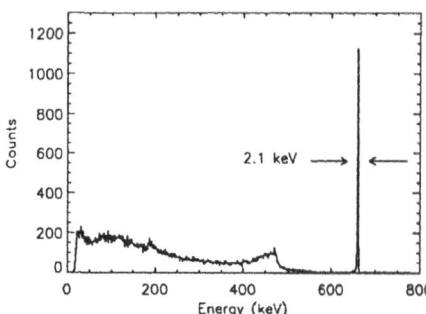

Fig. 3. Energy spectrum from all five strips using conventional room temperature electronics. Energy resolution performance is limited by the high capacitance (~30 pF) of the wiring and feed-throughs in the detector housing. A uniform illumination of 662 keV gamma-rays was used. Total-energy peak efficiencies of ~1% in a single pixel, and ~2% in an entire strip are observed, consistent with expectations for a detector of this size.

tion of the roll-off in response *vs.* position. In this model, partial energy is collected on a strip if a photon interacts beyond the edge of the strip, and partial energy events are rejected by event selection. We find the true position of the strip edge by moving an assumed position until the model fits the roll-off. This simple model provides an excellent fit to the observed roll-off. Results of fitting adjoining strip edges suggest a gap of ~0.4 mm between both the X- and Y-strips where charge division occurs between the strips. This is consistent with the size of the gap between the electrodes.

Spectroscopy of the strip detector is comparable to conventional germanium detectors as evident from Figure 3.

3. Conclusions

The device tested here demonstrates the desired properties of excellent spectroscopy with good spatial resolution. The narrow gap between strips and sharpness of the edges suggests that future devices with sub-mm imaging are possible. A device with 2 mm strips is currently being fabricated.

We thank P. Durouchoux, Saclay, France for loaning us the 5 × 5 germanium strip detector used in this work.

References

1. Makishima, 1991, *Proc. of 28th Yamada Conf. on Frontiers of X-Ray Astronomy.*
2. Kendziorra, et al., 1991, *Proc. of 28th Yamada Conf. on Frontiers of X-Ray Astron.*
3. W.N. Johnson, et al., 1972, *Astrophys. J.*, **172**, L1.
4. Leventhal, M., et al., 1978, *Astrophys. J.*, **225**, L11.
5. P.A. Schlosser, et al., 1974, *IEEE Trans. Nucl. Sci.*, NS-21, 658.
6. P.N. Luke, 1984, *IEEE Trans. Nucl. Sci.*, NS-31, 1, 312.
7. D. Gutknecht, 1990, *Nuc. Instr. and Meth.*, A228, 13.
8. D. Bloyet, et al., 1992, *IEEE Trans. Nucl. Sci.*, **39**, No. 2, 315.
9. M.S. Gerber and D.W. Miller, 1976 *Nucl Instr. and Meth.*, **138**, 445.
10. R.A. Kroeger, et al., 1993, *London Conf on Pos. Sensitive Detectors*, Brunnel Univ, Uxbridge, UK.

HARD X-RAY AND GAMMA-RAY IMAGING SYSTEMS
UTILIZING GERMANIUM STRIP DETECTORS

W.N. JOHNSON, R.A. KROEGER, R.L. KINZER and J.D. KURFESS
Naval Research Laboratory, Washington, DC

S. INDERHEES and B. PHLIPS
Universities Space Research Association, Washington, DC

and

B. GRAHAM
George Mason University, Fairfax, VA

Abstract. We investigate the characteristics of imaging systems in the 20 keV – 10 MeV energy band which incorporate the high spatial and spectral resolution of planar germanium strip detectors. A Compton scatter telescope provides sensitivity above approximately 250 keV; a coded aperture positioned above the top germanium detector plane of the Compton telescope forms a coded-aperture telescope with sensitivity in the 20 – 250 keV band. The high spectral resolution and spatial resolution of germanium strip detectors provides a Compton telescope with dramatically improved energy resolution, angular resolution, and sensitivity compared with previous Compton instruments. Such a system has excellent angular response for point source identification and spectroscopy and also provides response to high energy diffuse emissions such as the Galactic 511 keV line emission and ^{26}Al emission. Monte Carlo simulations of the concept and estimates of the sensitivity shall be presented.

Key words: Ge detectors, Strip detectors, Gamma-rays

1. Introduction

Astrophysical observations in the hard X-ray and soft gamma ray energy range have made slow but steady progress from the pioneering balloon instruments of the 1960's and the early satellite experiments of the OSO and HEAO missions. Recent satellite experiments on Gamma Ray Observatory (BATSE, OSSE and COMPTEL), GRANAT (SIGMA) and the planned INTEGRAL provide line γ-ray sensitivities approaching 10^{-5} cm^{-2} s^{-1}. High resolution γ-ray spectroscopy, as planned for INTEGRAL, is clearly important in understanding the sites and mechanisms producing the gamma ray emission, but improved sensitivity to 10^{-6} cm^{-2} s^{-1} or better is required to open up the field of γ-ray spectroscopy in astrophysics. As the sensitivity improves and the number of sources detected increases, imaging with $\sim$ arcminute resolution also becomes a requirement to avoid source confusion for both galactic and extragalactic observations. We are investigating detector technologies and instrument concepts which will provide both high resolution spectroscopy and good angular resolution imaging and can be scaled to instruments with interesting sensitivity. The objective is an instrument

329

L. Bassani and G. di Cocco (eds.), Imaging in High Energy Astronomy, 329–332.
© 1995 *Kluwer Academic Publishers.*

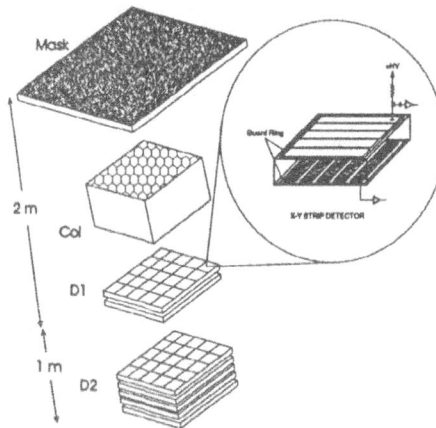

Fig. 1. Conceptual diagram of the combined coded-aperture imager and Compton telescope. The Compton telescope consists of two detector planes (D1 and D2) ~1 m apart. A coded mask is mounted ~ 2 m above the top detector plane, which forms the coded-aperture imager using the top layer of D1. A coarse collimator just above the D1 layer restricts the field of view for the imager.

which provides arc-minute imaging, a line sensitivity of 10^{-6} γ cm^{-2}s^{-1} in the 20 keV – 10 MeV region, and an energy resolution approaching that provided by germanium spectrometers.

2. Instrument Concept

Our investigations are currently centered on germanium planar strip detectors which can provide 2 – 3 keV spectral resolution and spatial resolution of $\sim$ 2 mm (see associated contribution by Kroeger et al.). These detectors, which are $\sim$ 5cm $\times$ 5cm $\times$ 1cm thick, are excellent hard x-ray detectors and can be layered to provide good sensitivity to 10 MeV. When used with a coded aperture, a single array of Ge strip detectors can provide arc-minute imaging in hard x-rays. Compton scatter telescopes using multiple layers of these detector arrays could achieve imaging resolution of a few tenths of a degree in the soft gamma ray range. We investigate the characteristics of these two configurations separately, but an intriguing instrument concept, shown in Fig. 1, uses the top detector plane of the Compton telescope as the detector plane of a coded-aperture hard x-ray imager, thus providing a system with good spectral and imaging resolution from ~20 keV to 10 MeV.

Compton scatter telescopes utilize two detector planes designed to scatter the incident radiation in the top plane and capture the scattered photon in the lower plane. Measurements of the energy losses and positions of the interactions in the two detector planes permits the reconstruction of the incident photon direction. In telescopes such as this and COMPTEL on GRO [4], it is not possible to measure the direction of the Compton electron in the top detector and the possible directions, when projected onto the sky, produce a small circle of the half-angle specified by the scatter angle and centered on the direction of the scattered photon. A point source of gamma rays is detected at the intersection of many such circles. Uncertainties in the energy loss measurements and in the interaction positions change the circle to an annulus and ultimately determine the angular resolution of Compton telescopes. Simultaneous improvements in spectral resolution and detector spatial resolution, as available in Ge planar strip detectors, are required to affect good angular resolution in these systems. We have modeled a Compton telescope system constructed from Ge planar strip detectors with 2 mm strips and 2 keV energy resolution. As discussed by Kroeger et al.[3], orthogonal strips on the two surfaces of the detector provide 2-dimensional spatial information with 2 mm resolution. The Compton telescope we have studied is shown in the bottom portion of Fig. 1. The top detector plane (D1) is formed from an array of strip detectors in two layers to provide $\sim$1 m^2 active area. The bottom plane (D2) is $\sim$1 m below D1 and comprised of five layers of strip detectors. Each layer is $\sim$ 1 m square and constructed from $\sim$ 400 detector elements of the type shown in the figure inset. The coincidence requirement for energy losses in the D1 and D2 planes produces systems with relatively low efficiency, generally 1 – 3%, but also reduces the detector background. Monte Carlo simulation of the on-axis efficiency for this configuration as a function of incident photon energy indicates $\sim$ 2% efficiency at 250 keV peaking at $\sim$ 3% in the 400 keV to 1 MeV range, and down to $\sim$ 1% by 10 MeV. For incident angles of 30° off-axis, the response is $\sim$ 80% of the on-axis response.

Below 300 keV, the coded aperture telescope using the D1 plane of detectors provides good response down to $\sim$ 20 keV. As displayed in Fig. 1, a coded mask formed from $\sim$ 1 mm-thick lead or tungsten is placed 2 m above the D1 layer. The thickness is selected to provide good modulation of hard X rays but thin enough to be reasonably transparent to higher energy photons. The figure also shows a coarse collimator which could be useful in restricting the hard x-ray field of view.

3. Discussion

The Ge-Ge Compton telescope offers significant capabilities compared to the instruments on GRO and the INTEGRAL study instruments [1]. As with COMPTEL on GRO, it has a large field of view ($\sim$ 60°) and has the ability to image diffuse emission such as the ^{26}Al emission from the Galaxy. Its excellent spectral resolution and 2 mm spatial resolution provide significant improvements over COMPTEL in spectroscopy, point source imaging and sensitivity. Improvements in sensitivity to continuum emissions by $\times$10 appear to be achievable. The Ge spectroscopy and low background of the Compton configuration will provide significant sensitivity to narrow line emissions. Fig. 2 shows the narrow line and continuum sensitivity of the Ge Compton telescope relative to current capabilities (OSSE [2], COMP-

TEL [4], INTEGRAL [1]). The sensitivity is determined by Monte Carlo response to the cosmic diffuse background as a limiting background to the detection of point sources. Other sources of background, such as local gamma ray production, spallation products and neutron interactions, which may be important have not been included at this time. The dashed line in the figure is an estimate of the Ge coded-aperture telescope sensitivity.

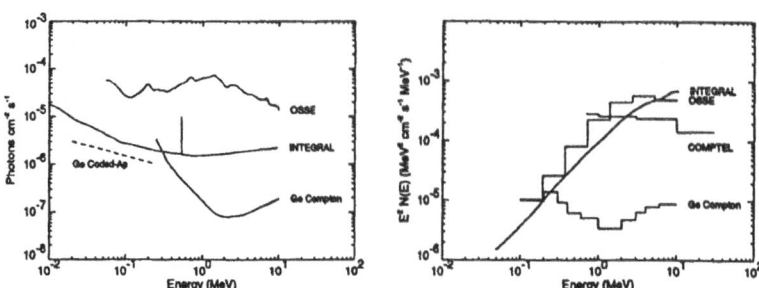

Fig. 2. Sensitivity (3σ) of the Ge-Ge Compton telescope to narrow emission lines (left panel) and continuum emissions (right panel) compared with current or planned instruments (10^6-sec observations). Sensitivity is based on diffuse background limiting sensitivity.

The studies to date indicate that imaging systems utilizing planar germanium stip detectors can provide significant improvements over the current GRO and GRANAT instruments and the planned capabilities of the INTEGRAL mission. These systems appear to be the best approach for simultaneously achieving good sensitivity to point and diffuse emissions and perhaps may also permit investigation of the diffuse γ-ray background. There are, however, many technical challenges and investigations which remain to be addressed. Some of these challenges are cryogenic support for the array of planar detectors, low-power and high-density electronics for the strip detectors, the ability to perform time-of-flight background rejection between the two detector planes, and local gamma ray production and other background issues affecting sensitivity. We will continue to investigate these issues in preparation for the next major gamma ray astronomy mission opportunity.

References

1. Bergeson-Willis, S., et al., 1993, INTEGRAL, Report on the Phase A Study
2. Johnson, W.N., et al., 1993, ApJS, 86, 693.
3. Kroeger, R.A., et al., 1994, these proceedings.
4. Schoenfelder, V., et al., 1993, ApJS, 86, 657.

THE IMAGING LIQUID XENON-CODED APERTURE

TELESCOPE (LXe-CAT)

E. APRILE, A. BOLOTNIKOV, D. CHEN, H. TAWARA, & F. XU
Physics Department and Columbia Astrophysics Laboratory
538 West 120th Street, NY, NY 10027

E. CHUPP & P. DUNPHY
Physics Department and Inst. for the Study of Earth,
Oceans and Space (EOS)
University of New Hampshire, Durham, NH 03824

T. DOKE & J. KIKUCHI
Advanced Research Center for Science & Engineering
Waseda University, Tokyo 162 Japan

K. MASUDA
Physics Laboratory, Saitama College of Health
Saitama 338 Japan

and

G. FISHMAN & G. PENDLETON
NASA/Marshall Space Flight Center
Huntsville, AL 35812

30 October 1994

Abstract. We describe a unique γ-ray imaging telescope operating in the energy range 0.3-10 MeV. The basic element of the telescope is a liquid xenon time projection chamber (LXe-TPC) as γ-ray spectrometer and 3-D imager. Location and energy analysis of multiple Compton scattering events in the chamber will permit a reconstruction of the incoming γ-ray direction, thus allowing a significant reduction of background. A 10-liter LXe-TPC prototype, with 400 cm^2 active area and 5 cm drift gap has been built and is currently being tested with γ-ray sources in the laboratory. Initial results are presented. The combination of the LXe-TPC with a coded aperture (LXe-CAT), results in a telescope with superior detection efficiency, angular and energy resolution, and excellent background rejection capability. Simulation results on the instrument's ability to image the Crab Nebula region in a balloon observation are presented, as well as to image the 1.809 MeV ^{26}Al distribution from the Vela region, in a satellite observation.

Key words: γ-ray Astronomy – Imaging – Instrumentation

1. Instrument Description

The proposed telescope (Aprile et al. 1992a) combines a liquid xenon time projection chamber (LXe-TPC) as calorimeter and position sensitive γ-ray detector with a coded aperture mask to achieve precise measurement of the energy and angular distribution of γ-ray sources in the 0.3-10 MeV energy region (Fig. 1). The coded mask is a 43×41 element URA, with elements

L. Bassani and G. di Cocco (eds.), Imaging in High Energy Astronomy, 333–338.

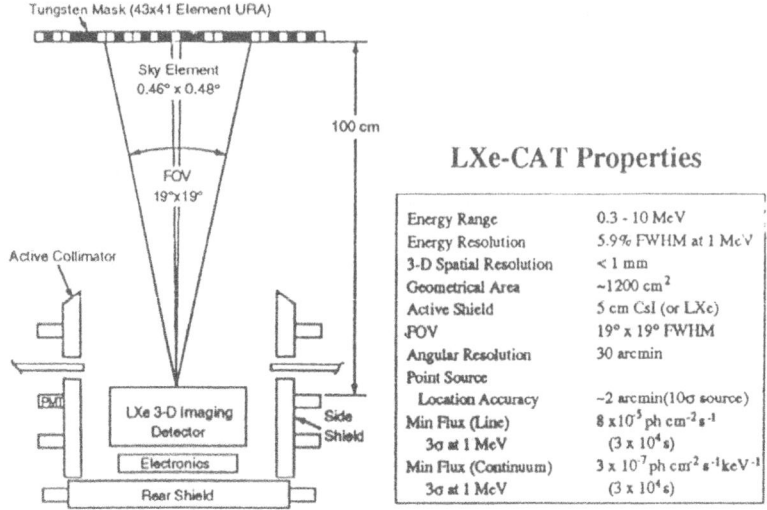

a

LXe-CAT Properties

Energy Range	0.3 - 10 MeV
Energy Resolution	5.9% FWHM at 1 MeV
3-D Spatial Resolution	< 1 mm
Geometrical Area	~1200 cm^2
Active Shield	5 cm CsI (or LXe)
FOV	19° x 19° FWHM
Angular Resolution	30 arcmin
Point Source	
Location Accuracy	~2 arcmin(10σ source)
Min Flux (Line)	8 x10^{-5} ph cm^{-2}s^{-1}
3σ at 1 MeV	(3 x 10^4 s)
Min Flux (Continuum)	3 x 10^{-7} ph cm^{-2} s^{-1}keV^{-1}
3σ at 1 MeV	(3 x 10^4 s)

Fig. 1 Schematic view of the Liquid Xenon-Coded Aperture Telescope (LXe-CAT)

$0.81 \times 0.85 \times 1.2$ cm^3 in size. With a detector-mask separation of 100 cm, the angular "pixel" is $0.46° \times 0.48°$ over a $19.3° \times 19.3°$ FOV. The telescope point source location accuracy is estimated at $2'$ for a 10 σ source, due to the excellent signal-to-noise ratio of the imaging detector. The LXe-TPC has a geometric area of 1200 cm^2 and is surrounded by an active anticoincidence shield consisting of CsI or LXe scintillator. With a 10 cm deep layer of LXe ($Z = 54$, density $= 3.06$ g cm^{-3}), the full energy peak efficiency is about 65% at 1 MeV.

The LXe-TPC works on the principle that the free ionization electrons liberated by a γ-ray interaction in the liquid can drift, under a uniform electric field, toward a signal readout structure. The ionization signals induced on the readout sensing elements provide both the spatial and the total energy information for each event. In our design, the information in the $X - Y$ plane is obtained from the signals induced on two orthogonal wire planes, while the Z information, along the direction of drift, is inferred from the drift time, measured with respect to a time zero (see Fig. 2). The fast signal (< 10 ns) from the primary scintillation light of LXe, is used for time zero mesurement. With the capability of measuring the three spatial coordinates and the energy deposited for each γ-ray interaction, the TPC is therefore ideal for event reconstruction based on Compton kinematics. The direct outcome of this event visualization is the capability of background rejection for both single- and multiple-sites energy deposition events. This, as well as the polarization sensitivity of the LXe-TPC as Compton polarimeter, has been demonstrated with Monte Carlo simulation results (Aprile et al. 1993b, 1994a). The technical feasibility of such a detector and the factors which determine its ultimate energy and spatial resolution have been stud-

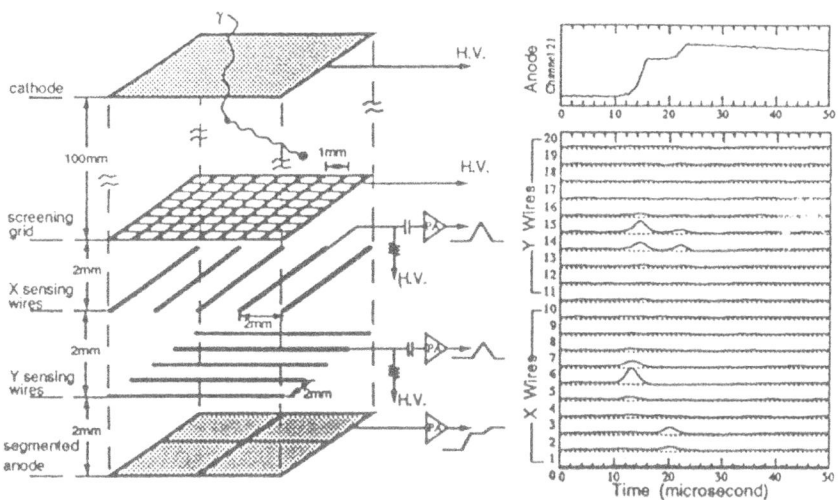

Fig. 2 Schematic view of electrodes system
for 3-D position sensitivity of LXe-TPC

Fig. 3 On-line display of a Compton scattered γ-ray
event (1 274MeV), showing the digitized waveforms
induction signals and anode signal

ied at Columbia for the past few years (Aprile et al. 1993a and references therein).

2.   Test Results with the 10-liter LXe-TPC Prototype Detector

To demonstrate the operation of a large liquid xenon detector, we have built and are currently testing a 10-liter LXe-TPC prototype, implemented with the electrode system shown in Fig. 2 (see Aprile et al. 1994b). With a wire spacing of 4 mm, the total number of sense wires for X-Y readout is 96. The prototype has half the drift gap, one third the sensitive area and a volume similar to that of the proposed flight instrument. Experiments have been carried out to test the instrument's spectroscopic and imaging performance. The detector's charge yield was as expected in high purity LXe and stable over a maximum period of 100 hr.

To simulate a parallel beam of γ-rays from a point source, a ^{60}Co or ^{22}Na source was placed 40 cm above the detector's sensitive area. Events with a single or multiple Compton scatterings, as well as photoabsorption events were accumulated. As an example, Figure 3 shows the on-line display of a 1.274 MeV γ-ray event from the ^{22}Na source experiment. The magnified view of the signal on the anode clearly shows a two-step event indicating that the γ-ray history was a Compton interaction followed by photoabsorption. The sum of the two steps pulse heights corresponds to 1.274 MeV total energy. From the amplitude and time analysis of the induction and collection signals, the coordinates and the energy for the two interaction points are inferred as well as their spatial separation. The most probable scattering

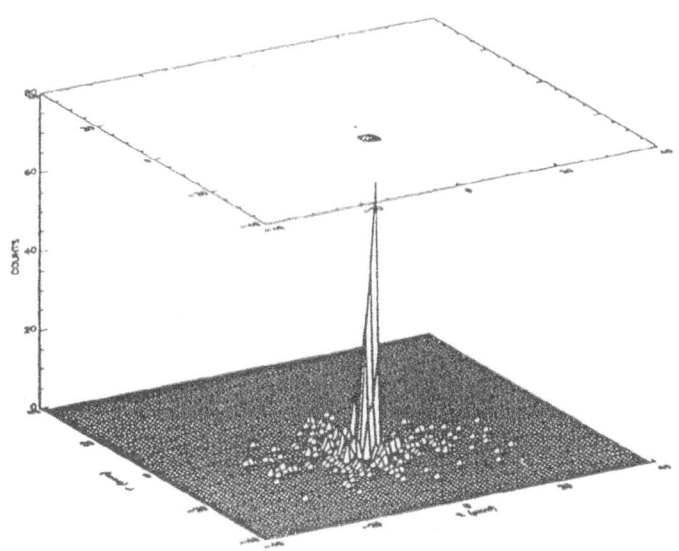

Fig 4 Distribution of 662 keV γ-ray from collimated ^{137}Cs source, detected with the
10 liter LXe-TPC

angle can then be found from Compton scattering kinematics. These initial
results directly demonstrate that the LXe-TPC can be very effective for
background rejection based on Compton event reconstruction. In addition,
they verify that the original proposal (Aprile et al. 1989) to use a LXe-
TPC as a Compton/pair telescope with large efficiency and good angular
resolution is feasible for high energy γ-ray astronomy. As an indication of
the detector's imaging performance, Fig. 4 shows the reconstructed image
of a collimated beam (2 mm diameter collimator) of 662 keV γ-rays from a
^{137}Cs source. The data are consistent with a detector RMS spatial resolution
of about 1 mm.

3. Simulation Results

To demonstrate the capability of the LXe-CAT as an imaging γ-ray telescope, we have carried out Monte Carlo calculations that simulate the response of the instrument to several source distributions. In particular, we
simulate the imaging process for a coded aperture imaging system in response
to photon continuum spectra from point sources and to a γ-ray line from
localized sources. The background rates expected in a balloon or satellite
based observation were calculated. Fig. 5 shows the total background spectrum, together with its individual components, calculated for 3 gcm^{-2} over
Palestine, TX. A 5 cm CsI thick shield was assumed for the calculation.

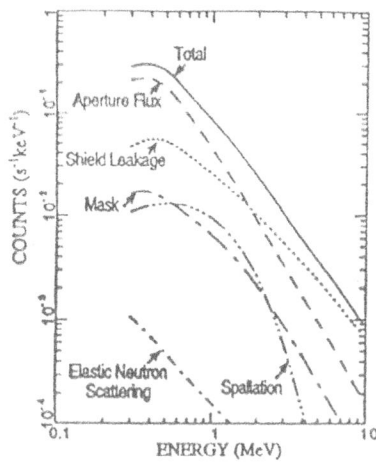

Fig. 5 Calculated background spectrum in LXe-CAT at 3 g cm^{-2} over Palestine, TX. A 5 cm thick CsI shield is assumed

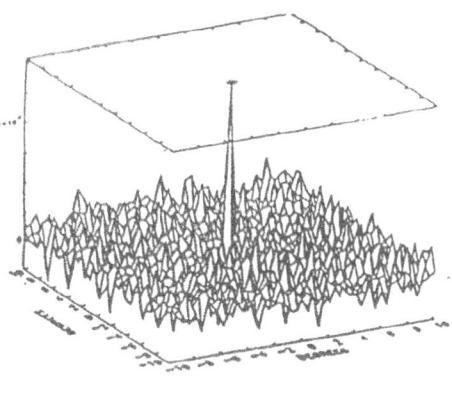

Fig. 6 Simulated image of Crab Nebula (0.5-1.0MeV) for 10 hr balloon observation

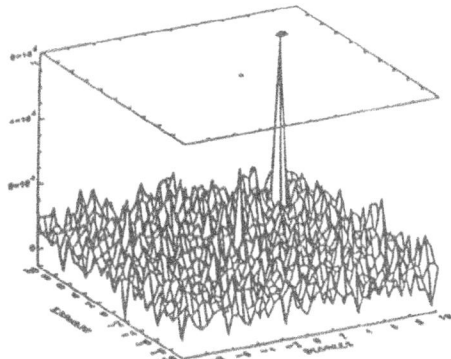

Fig. 7 Simulated image of Crab Nebula and quasar PKS 0528+134 (3-20MeV) for 336 hr satellite based observation

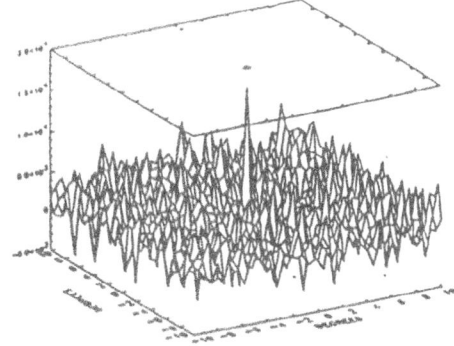

Fig. 8 Simulated image of Vela "hot spot" at 1.809MeV (^{26}Al) for an 8-week satellite-based observation

To illustrate the response of the LXe-CAT to continuum spectra of γ-rays from discrete sources, we simulate the Crab Nebula for a typical balloon observation. The Crab spectrum is taken from Penningsfeld et al. 1979. The source observation time is taken to be 10 hours. The inevitable spatial non-uniformity in the background over the position sensitive detector causes "artifacts" in the image unless corrected for. Therefore, a "background" field is observed for an equivalent time to evaluate the non-uniformity and allow for appropriate correction (McConnell et al. 1987). The background rate of Fig. 5 was reduced by a factor of ~ 2 due to event rejection based on Compton reconstruction, but was increased by a factor of 2 as required by the evaluation of background non-uniformity. The source is present at a significance level of 16σ in the range 0.5–1 MeV (Fig. 6) and 11σ in the range 1–10 MeV. Since a point source can be located with a precision of $\sigma_{loc} = \sigma_{pixel}/n_\sigma$ and the characteristic size of a pixel is about 30′, the uncertainty in the Crab Nebula location for this observation would be $\sim 2'$.

To illustrate the capability of the detector to resolve separate point sources in the same FOV, Fig. 7 shows the image for a satellite-based observation (336 hr) of the Crab Nebula and the quasar PKS 0528+134 (Collmar et al. 1993). The image of the quasar, in the energy range $3 - 20$ MeV, is seen at a significance level of 6 σ.

Finally, the measurement of the ^{26}Al distribution in the Galaxy via the 1.809 MeV γ-ray line is an important objective for γ-ray astronomy and for the LXe-CAT instrument (Aprile et al. 1994c). Based on the map of this radiation observed by *COMPTEL* (Diehl et al. 1994), we have obtained the image shown in Fig. 8 for the Vela "hot spot," with a flux of 2.2 $\times$ 10^{-5}cm^{-2}s^{-1}. The result is for an eight week satellite observation, similar to the *COMPTEL* observation time. The significance of the image is 9 σ.

This work was supported by a NASA Grant (NAGW-2013) to the Columbia Astrophysics Laboratory.

4.   References

Aprile, E., Mukherjee, R., & Suzuki, M. 1989, SPIE Proc., 1159, 295.

Aprile, E., Bolotnikov, A., Chupp, E., & Dunphy, P. 1992, NASA proposal, CAL-2015.

Aprile, E., Bolotnikov, A., Chen, D., & Mukherjee, R., 1993a, Nucl. Phys. B. (Proc. Suppl.) 32, 279.

Aprile, E., Bolotnikov, A., Chen, D., & Mukherjee, R. 1993b, Nucl. Instr. and Meth. A327, 216.

Aprile, E., Bolotnikov, A., Chen, D., Mukherjee, R., & Xu, F. 1994a, ApJS, 92, 689.

Aprile, E., Bolotnikov, A., Chen, D., Muhkerjee, R., & Xu, F. 1994b, SPIE Proc. 2305, 33.

Aprile, E., Chupp, E., Bolotnikov, A., Dunphy, P., & Xu, F. 1994c, ApJ (submitted).

Collmar, W., et al. 1993, Proc. 23rd ICRC 1, 168.

Diehl, R., et al. 1994, AIP Conf. Proc., 304, ed. C. Fichtel, (AIP: NY) 147.

McConnell, M.L., Dunphy, P.P., Forrest, D.J., Chupp, E.L., & Owens, A. 1987, ApJ, 321, 543.

Penningsfeld, F.–P., Graser, U., & Schönfelder, V. 1979, Proc. 16th ICRC 1, 101.

ISGRI: A CDTE LAYER FOR THE INTEGRAL IMAGER

F. LEBRUN, J.-P. LERAY, P. LAURENT and B. CORDIER

Service d'Astrophysique, CEA/DSM/DAPNIA, C.E. Saclay,
91191 Gif-sur-Yvette Cedex, France

and

P. MANDROU

Centre d'Etude Spatiale des Rayonnements, 9, Avenue du Colonel Roche, BP 4346,
31029 Toulouse Cedex, France

Abstract. The present paper gives a preliminary description of the INTEGRAL Soft Gamma-Ray Imager (ISGRI), a CdTe layer, replacing the first CsI layer of the INTEGRAL IMAGER model payload proposed in phase A study. It provides a much lower threshold, around 20 keV instead of 80 keV, allowing a study of the hard X-ray part of the spectrum of any source within the entire field of view. In addition it offers significantly better spatial resolution, better point source location accuracy and improved imaging sensitivity. The sensitivity to partially-coded sources is significantly enhanced. The spectroscopic performance is also strongly improved allowing the search for low-energy nucleosynthesis lines (^{44}Ti) as well as cyclotron lines.

Key words: coded mask – CdTe – INTEGRAL

1. INTRODUCTION

The INTErnational Gamma-Ray Astrophysics Laboratory is an ESA mission to be launched in 2001. The payload comprises two main instruments, the IMAGER and the SPECTROMETER and two monitors, the X-Ray Monitor (XRM) and the Optical Transient Camera. The IMAGER features a passive coded mask whose capability to produce images of the sky in the low energy part of the gamma-ray domain (30 keV - 1 MeV) has been illustrated by the SIGMA telescope. The detector of the IMAGER, as designed during the phase A study, is made up of three hexagonal detection planes of 2880 pixels each, designed to cover the energy range 70 keV - 10 MeV. A pixel is a hexagonal bar of CsI of area 1 cm^2 read out by a photodiode. With a mask distance of 3.85 m, an angular resolution of about 20' is achieved. The thickness of the top layer is 1 cm and the two others are 3 cm thick. The position and energy of each event are determined by combining the various interaction positions and energy deposits in the three planes. Events are classified according to the interactions they have experienced and selections can be applied to reduce the background. In addition, a honeycomb tantalum collimator is placed in front of the detector to reduce the diffuse gamma-ray background contribution at low energy. As thus defined, the phase A model payload does not offer a comfortable overlap between the IMAGER and the XRM, the sensitivity of the XRM at 70 keV (IMAGER lower threshold) being too low. An increase of the XRM efficiency at high energy would

339

L. Bassani and G. di Cocco (eds.), Imaging in High Energy Astronomy, 339–344.
© 1995 *Kluwer Academic Publishers.*

imply a lower sensitivity (absorption) at low energy (Fe lines). It has been proposed then to replace the first layer of the IMAGER by one of another type, offering a much lower threshold. What were the other requirements for such an imaging detector ? Above a few hundred keV, pixels have to be large enough to contain the whole series of Compton diffusions ending with a photoelectric absorption. This volume depends on the detector atomic number (Z) and density, and on the energy of the incident gamma-rays. At 511 keV, it is at least 1 cm^3 in the case of CsI. The 1 cm CsI bars foreseen for the IMAGER seem well adapted for this part of the spectrum. At low energy, the interactions are governed by photoelectric absorption and there is no need for large pixels. On the other hand the point source localization accuracy depends on the detector pixel size (see below) which should be then as small as possible. The CsI, requiring wrapping in a diffusive material, does not seem adequate to produce significantly smaller pixels. Semi-conductors operating at ambient temperature with high Z and high density seem more promising. Among them, CdTe is now produced on an industrial scale and detectors of arbitrarily small size can be produced. In this case, the limitation on the pixel size comes mainly from the number of electronic channels one can operate simultaneously: one preamplifier per pixel is necessary. To limit the volume and the power consumption of the electronics, integrated circuits should be used. ISGRI is such a CdTe gamma camera.

2. ISGRI DESIGN

Such a position sensitive device has been designed in the framework of the IMAGER model payload described in more detail in the phase A study report [ESA-SCI(93)1]. ISGRI has to replace the first CsI layer with the fewest possible changes in the overall IMAGER design. The thickness of the top CdTe layer has been fixed at 2 mm as a compromise between the spectroscopic performance and the efficiency at the lower energy threshold of the lower CsI planes (about 100 keV). Moreover, the bias voltage to apply to 2mm thick THM (Travelling Heater Method) CdTe is only 100 V. This thin CdTe layer becomes 50% transparent at 150 keV, an energy at which the 400μ thick Tantalum collimator is still efficient. The collimator structure must then still match that of the CsI planes. As a consequence the CdTe layer pixels must be adapted to this honeycomb structure. The possibility of making smaller detectors gives access to a higher spatial resolution by subdividing the hexagonal pixels. Sub-division by a factor of 6 appears to be a good compromise regarding power consumption. A CdTe pixel is therefore an equilateral triangle of 6.3 mm on a side. The 6 channels coming from one hexagon are grouped together on an ASIC circuit bonded to the 6 pixels to form a "hexacell". A breadboard model made of 4 such hexacells, but with conventional electronics, is displayed in Figure 1.

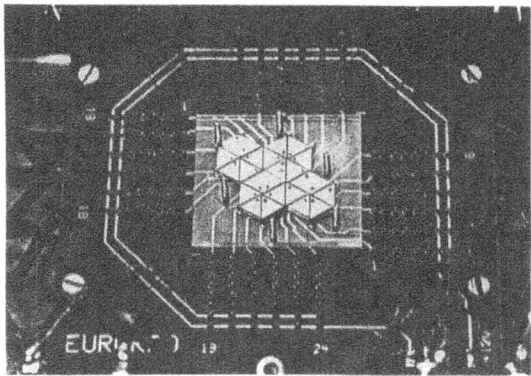

Fig. 1. A 24 pixels breadboard model of ISGRI

The modular concept of the IMAGER has been adopted with 24 triangular modules comparable to those of the CsI planes, each containing 120 hexacells. Therefore ISGRI features 17280 pixels.

3. IMAGING PERFORMANCE

The angular resolution depends only on the mask pixel size and the mask to detector distance. It is about 20' for the IMAGER model payload, a figure which is maintained with ISGRI. The finer spatial resolution of ISGRI allows a better sampling of the mask shadowgram (see Fig. 2 left) which results in a better point source localization and a higher imaging efficiency. The point source localization accuracy (PSLA) can be estimated following Stephen (1990) as:

$$PSLA(FWHM) \approx 2.36 \left(\frac{\theta}{n\sigma}\right) \sqrt{\frac{(k+1)^2(k+2)}{12k^3}}$$

where θ is the angular resolution, n the statistical significance of the source and

$$k = \sqrt{\frac{A_d}{A_m}}$$

where A_d and A_m refer to the areas of the detector and mask pixels. On this basis, the location accuracy would be improved by 50% with regard to that of the IMAGER model payload. In Figure 2 right, the imaging efficiency is given as a function of the ratio of the mask element size to the detector pixel size. per mask element (ratio of the areas). For the model payload, this efficiency was 81%. For ISGRI, with the same mask design it is 92%.

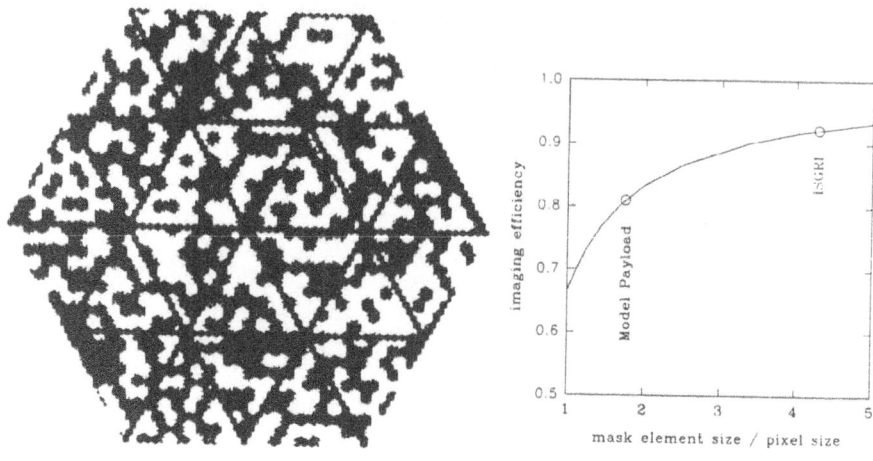

Fig. 2. Left: simulation of the mask pattern projected by a slightly off axis source
Right: imaging efficiency as a function of the ratio of the mask element size to the detector
pixel size

It is clear that the mask elements could be reduced to enhance the angular
resolution by, say. 50% without affecting dramatically the ISGRI imaging
efficiency (6% loss). The mask of the model payload offered a field of view
of 16°(FWHM), but the collimator reduced it to 6°. The collimator height
has therefore been lowered to 4 cm to take advantage of the wider mask
field of view. With this new collimator design, the field of view extends now
to 11°(FWHM). The relative sensitivity is given in Figure 3 (solid line) as
a function of the angle from the telescope axis. However, this first order
estimate may be pessimistic at large angles since the finer spatial resolution
allows a rejection of the pixels heavily shadowed by the collimator for a
given source direction. This selection permits a background reduction which
will increase the sensitivity. The dashed line in Figure 3 represents the gain
expected if this selection were continuous (infinitely small pixels). For ISGRI.
this sensitivity will be attained only by steps. With an expected background
count-rate of about 1000 s^{-1} between 20 keV and 1 MeV, the on-axis broad-
band ($\Delta E = E/2$) sensitivity at 3σ is about 2 millicrab at 100 keV for an
exposure of 10^5 s. Such a figure implies that a black hole candidate at the
galactic center, as bright as Cyg-X1 (e.g. 1E1740.7) should be detected at the
9σ level within 15 minutes. That means that a daily monitoring of most of
the galactic black holes candidates can be considered. This daily observation
of the galactic disc should also reveal not only the bright X-ray novae but
also a number of short time-scale weaker transient sources.

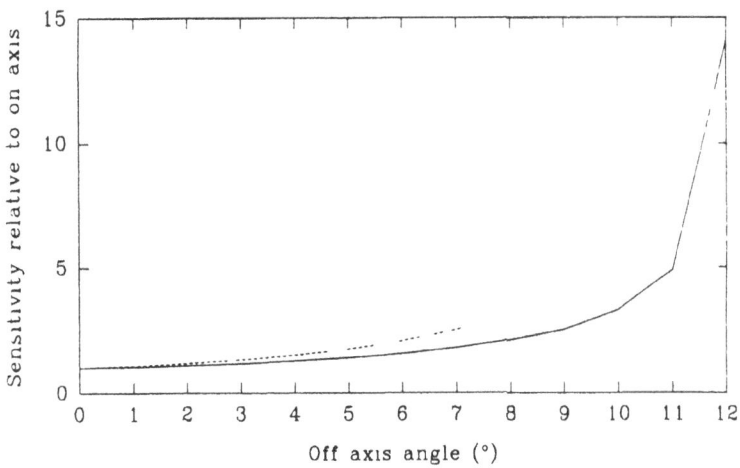

Fig. 3.   Off axis sensitivity. Dashed line: Model payload; solid line: ideal

4. SPECTROSCOPIC PERFORMANCE

With its high Z (48 for Cd and 52 for Te), CdTe has a rather high photoelectric cross section. Spectra recorded with 2 mm thick detectors should
exhibit a nice photopeak with little continuum for energies up to about 100
keV. However, as illustrated by the pulse-height spectrum shown in Figure 4
(dashed line), an important continuum is visible up to the photopeak energy.
This is due to an incomplete collection of the charges (mainly holes) created
during the gamma-ray interactions. Indeed, trapping effects are significant
in this semi-conductor. However, the charge loss is proportional to the pulse
transit time, which is roughly equal to the pulse rise-time. Under a bias of
100 V applied to a 2 mm thick THM CdTe detector the pulse rise-time varies
from a few hundred nanoseconds to five microseconds as a function of the
interaction depth. A measurement of this rise time thus allows an estimate
of the charge loss and a correction of the measured energy deposit. Such a
correction has been applied to the events used to construct the pulse-height
spectrum and the corrected spectrum is shown in Figure 4 (solid line). The
spectral resolution attained with such corrections is given in Figure 5 together with the photopeak absolute efficiency. The 3σ narrow-line sensitivity for
a 10^5 s exposure is 2-3 10^{-5} cm^{-2} s^{-1} at 80 keV and 1-2 10^{-3} cm^{-2} s^{-1}
at 511 keV. The excellent sensitivity at 80 keV should ensure the measure-
ment of the low energy lines of ^{44}Ti from the Cas A supernova remnant as
well as probably many other detections in the galactic disc. Cyclotron lines
from magnetized neutron stars (e.g. Her X-1) should be easily detected and
resolved.

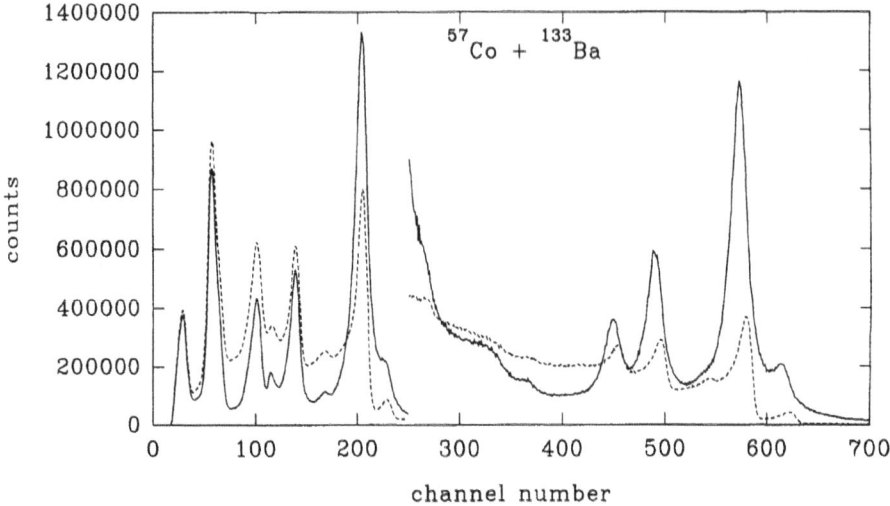

Fig. 4.   Dashed line: pulse-height spectrum; continuous line: corrected spectrum. The spectra are magnified by 25 after channel 250.

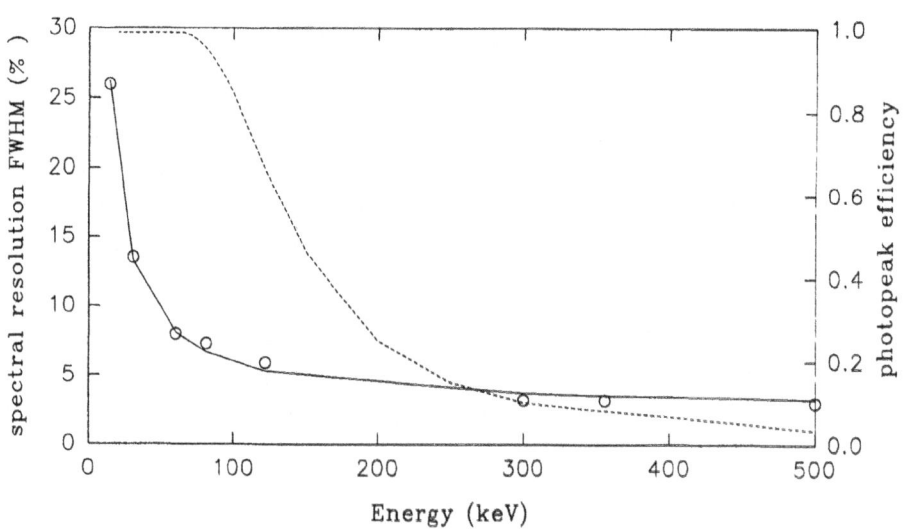

Fig. 5.   Continuous line and circles: energy resolution; dashed line: photopeak efficiency

References

Stephen, J.B.: 1991, *Adv. Space Res.* **11**, No **8**, 407

IMAGING WITH CDTE ARRAYS

E. CAROLI, J.B. STEPHEN, A. DONATI, W. DUSI and G. LANDINI

Istituto TeSRE/CNR, Via Gobetti 101, 40129 Bologna, Italy

Abstract. The next generation of high energy astronomical telescopes must provide both good imaging and spectroscopic performance together with a compact design allowing their use in small and medium size satellite missions. To fulfill these requirements the use of Cadmium Telluride (CdTe) is of particular promise. This material could offer the possibility of constructing highly compact segmented position sensitive detectors with pixel sizes down to a few square millimeters and good energy resolution ($E/\Delta E >50$ above few hundred keV). We propose a large field of view coded mask telescope operating in the range 10 keV - 1 MeV based on a detector consisting of a rectangular array of CdTe elements in conjunction with a twin-scale mask pattern in order to optimize the imaging performance and sensitivity over the whole energy range.

Key words: Solid State Detectors - Coded Aperture - X-γ ray astronomy

1. Introduction

The results from recent missions in high energy astronomy have demonstrated that in the energy range between a few tens of keV up to the MeV region future instrumentation (both for balloon and satellite payloads) must provide both high spectroscopic performance together with fine imaging over a large field of view. Spectroscopy will be required in order to obtain basic information on fundamental problems such as nucleosynthesis, supernova dynamics and compact object physics, while fine imaging is necessary in order to provide accurate source positioning, and to separate point source contributions (Diehl et al.,1994; Churazov, et al. 1994; Barret and Vedrenne, 1994; Mandrou et al., 1994). It is also clear that a strong requirement will concern the size and weight of the telescope: future instruments must be compact and (relatively) lightweight.

Nowadays an answer to all these requirements could be offered by the use of room temperature semiconductor devices. In particular Cadmium Telluride (CdTe) compounds may offer the possibility of constructing high sensitivity position sensitive detectors (PSD) based both on microstrip technology (Butler et al, 1993) and in the form of microcrystal arrays. In particular, a CdTe array detector will allow the realization of PSD's suitable for spectroscopic imaging over more than two decades of energy from $\sim$10 keV up to few MeV (Caroli, et al., 1993). In this paper we first discuss briefly the design of a compact high energy telescope based on a CdTe array PSD and a twin-scale coded mask used to optimize the imaging capability over the entire operational energy range, thereafter we present some preliminary results on the telescope's imaging performance using MonteCarlo simulations.

L. Bassani and G. di Cocco (eds.), Imaging in High Energy Astronomy, 345–349.

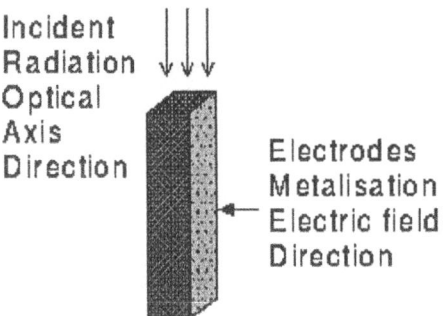

Incident
Radiation
Optical
Axis
Direction

Electrodes
Metalisation
Electric field
Direction

Fig. 1. The element of the Position Sensitive Detector: a CdTe μ-crystals of $2\times2\times10$ mm^3 used with the optical axis orthogonal to the electric collecting field (Planar Transverse Field configuration).

2. Payload Design

Cadmium Telluride II-IV is a compound semiconductor with large Z (48-52) and high density ($\sim$6 g/cm^3). Its wide band gap (1.7-2 eV) allows the material to operate as a room temperature X- and γ-ray spectrometer with good resolving power, as the energy required for generating an electron-hole pair ($\sim$5 eV) is less than twice that required for Germanium (Knoll, 1989). To date, the development of CdTe spectrometers has concentrated on devices based on Cadmium Telluride compensated either with Chlorine (Hage-Ali and Siffert, 1992) or with Zinc, the latter known as $Cd_{1-x}Zn_xTe$ (Butler et al., 1992). Due to the low charge mobility, trapping problems and noise level, the dimensions of single CdTe detectors are limited in order to maintain good spectroscopic performance: up to a few mm in the direction of the charge collection field and up to a few tens of mm in the other directions. A particularly interesting μ-crystal configuration suitable for high energy applications (up to a few MeV) has been proposed (Casali, F., et al., 1992) and is known as the Planar Transverse Field (PTF): these elements are bar-like crystals in which the collecting charge field is transverse to the photon incoming direction (Fig. 1). With PTF crystals of $2\times2\times10$ mm^3 it is possible to achieve resolving power ($E/\Delta E$) better than 50 at 500 keV, while maintaining a good efficiency up to a few MeV. Using these units it will be possible to build large area PSD's with high spatial resolution starting from linear modules of CdTe PTF units assembled on a thin (e.g. $\sim$200 μm ceramic) support with its own integrated front-end electronics (Caroli et al., 1992).

We propose a high energy telescope based on a CdTe PTF array in conjunction with a coded mask as an imaging spectrometer. In particular we suggest the construction of a $\sim$400 cm^2 PSD using PTF CdTe crystals of 2×2 mm^2 active area and 10 mm thickness. The detector will be surrounded by a BGO active shield ($\sim$ 3 cm thick) and the detector FOV will be restricted to about $12°\times12°$ (FWHM) by means of a thin (400 μ) Tantalum tube collimator efficient up to about 100 keV.

The imaging capability will be achieved by means of a tungsten coded mask at 100 cm from the PSD surface. In order to optimize the relationship between signal to noise (SNR) and angular resolution over the entire operative energy range, we foresee the use of a two scale coded mask (Skinner and Grindlay, 1993) shown in figure 2a. The mask is a two-scale design with a coarse array consisting of tungsten elements of dimensions 12×12 mm^2 by 1 cm thick arranged in a 15×17 'URA' pattern. In each open element, there is a sub-array of 3×3 elements, 4 of which are filled (randomly) with 4 mm $\times$ 4mm by 300 microns thick of tungsten. This guarantees a fine low-energy angular resolution of 14 arcmin and a corresponding high energy angular resolution of 41 arcmin. The use of such a mask design will maximize the angular resolution at low energies with only a small loss of SNR (In't Zand et al., 1994) while at higher energies the thin mask becomes transparent and only the coarse pattern is effective, thereby optimizing the imaging and sensitivity for the high energy detector PSF. The chosen mask/detector configuration gives rise to a FOV about $20° \times 20°$ (FWHM).

3. Simulation Results

We have performed numerical simulations of the payload described in the previous section in order to evaluate the imaging performance of the telescope. Some of the relevant parameters used in the MonteCarlo code are shown in Table 1. The detection efficiency and the point spread function values have been obtained from previous physical simulations (Caroli et al., 1992). The background was evaluated using $111. \times E^{-2.3}$ ph/(cm^2·s·keV·sr) as the spectrum for the cosmic diffuse component (Gehrels, 1992). For evaluating the hadronic contribution we have scaled the flat spectrum of 3.55×10^{-4} cts/(cm^2·s·keV) foreseen for the ISGRI layer on the INTEGRAL Imager (Lebrun F., et al., 1995) in the 20-1000 keV range with the volume ratio

Table 1. Values of some relevant parameters used for the numerical simulations of the telescope. E_m represents the logarithmic mean energy ($\sqrt{E_1 E_2}$); A_{eff} is the efficient area, i.e the geometric area (367.2 cm^2) times the detector efficiency; BKG is the total (cosmic plus hadronic) background; PSF is the width (FWHM) of the detector point spread function

ΔE keV	E_m keV	A_{eff} cm^2	BKG cts/(cm^2·s)	PSF cm
10-20	14.1	367.2	3.6×10^{-2}	0.
20-50	31.6	367.2	3.1×10^{-2}	0.
50-100	70.7	367.2	4.5×10^{-2}	0.1
100-300	173.2	352.3	0.21	0.2
300-1000	547.7	172.6	0.43	0.4

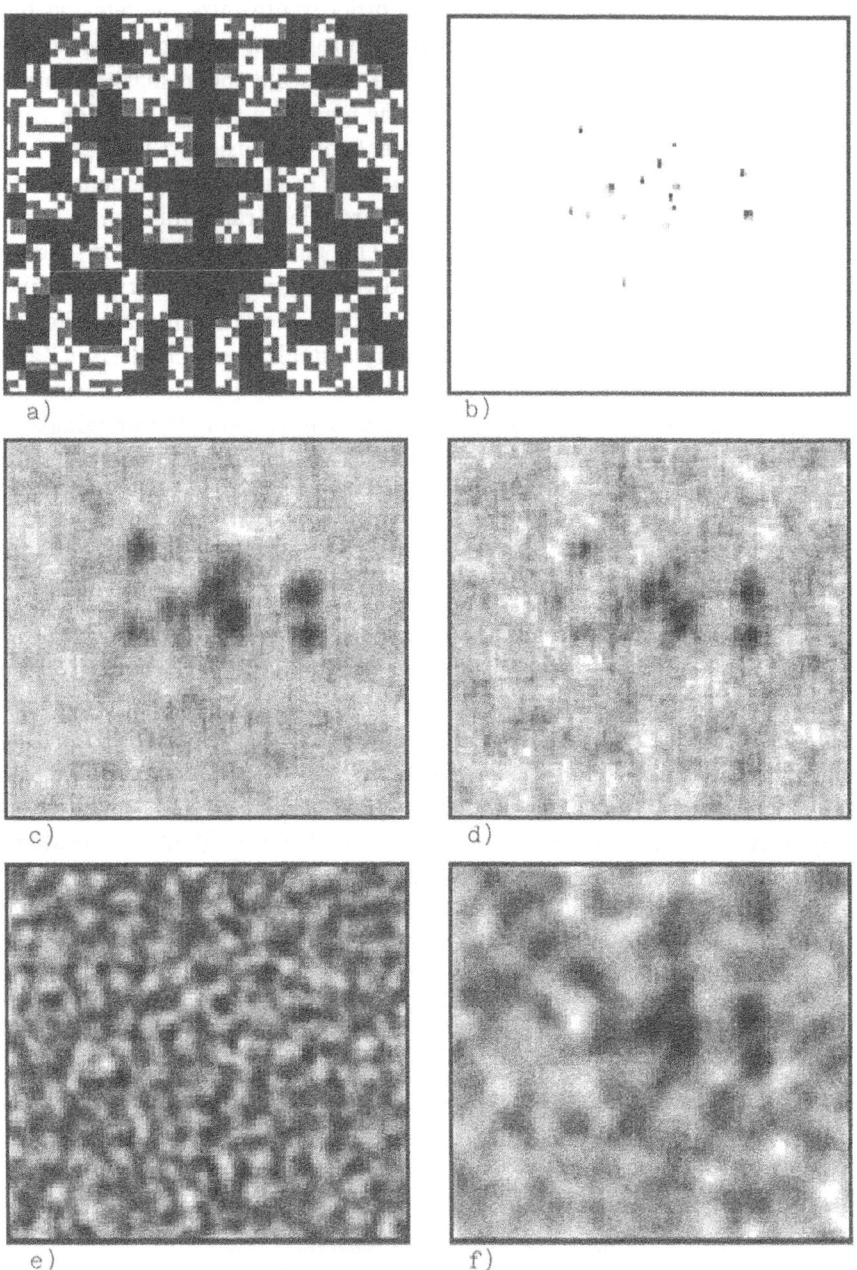

Fig. 2. The basic mask pattern employed for the simulations (a), the source field (b) and simulated images at 10-20 keV (d), and 300-1000 keV (f). Comparison images are included for the same source field observed with only the coarse mask pattern at 10-20 keV (c) and only the fine pattern at 300-1000 keV (e).

between the two detectors ($\sim$1.5). Using these data a sensitivity (3σ at 10^5s) of $\sim$3.6$\times10^{-5}$ ph/(cm$^2\cdot$s) has been evaluated in the 10-20 keV range allowing an average of about 16 sources to be detected in the FCFOV for galactic plane observations. Furthermore we have evaluated the spread of these sources at different significance level (Bassani and Stephen, 1991) between 10 and 20 keV and we have distributed them around the FOV center by means of a bidimensional gaussian distribution in order to simulate a galactic centre distribution (Figure 2b). The sources fluxes in the other energy channels have been derived supposing a common spectrum proportional to E^{-2} ph/(cm$^2\cdot$s$\cdot$keV).

We show in figure 2 the images obtained with a Monte-Carlo simulation of the instrument, for various energies. For comparison are shown the equivalent images obtained with a thick simple 15$\times$17 single scale mask at low energies and a single scale fine random thick mask at high energies. It can be seen that the two scale mask can provide a better resolved image at low energy (fig. 2d) compared with the one from a single coarse mask (fig. 2c). At the same time in the highest energy range (fig. 2f) the contribution of the strongest sources ($\sim$ 5) is well resolved while the other sources give rise to large scale emission features. On the other hand fig. 2e shows the total loss of imaging capabilities due to the bad match between mask element size and the high energy detector PSF in the case of a fine mask.

4. Conclusion

We have shown that with the use of CdTe arrays, it will be possible to construct compact lightweight spectro-imaging telescopes for use in hard X-ray and gamma-ray astronomy.

References

Barret, D., and Vedrenne, G.: 1994, *Astrophys. J. Suppl. Series* **94**, 505
Bassani, L., and Stephen, J.B.: 1991, *Adv. Space Res.* **11**, 353
Butler, J.F., et al.: 1993, in *SPIE Medical Imaging Conference* **SPIE-1896**, 30
Butler, J.F., et al.: 1992, *IEEE Trans. Nucl. Sci.* **NS-39**, 605
Caroli, E., et al.: 1993, *Astron. Astrophys. Suppl. Ser.* **97**, 393
Caroli, E., et al.: 1992, in *Photon Detector for Space Instruments* **ESA SP-356**, 27
Caroli, E., et al.: 1992, *Nucl. Instr. and Meth.* **A322**, 639
Casali, F., et al.: 1992, *IEEE Trans. Nucl. Sci.* **NS-39**, No 4, 598
Churazov, E., et al.: 1994, *Astrophys. J. Suppl. Series* **92**, 381
Diehl, R., et al.: 1994, *Astrophys. J. Suppl. Series* **92**, 429
Gehrels, N., : 1992, *Nucl. Instr. and Meth.* **A313**, 513
Hage-Ali, M., and Siffert, P.: 1992, *Nucl.Instr. and Meth.* **A322**, 313
In't Zand, J.J.M., Heise, J., and Jager, R.: 1994, *Astron. Astrophysis.* **288**, 665
Knoll, G.F.: 1989, *Radiation Detection and Measurement*, John Wiley & Sons, 466
Lebrun, F., et al.: 1995, *these proceedings* ,
Mandrou, P., et al.: 1994, *Astrophys. J. Suppl. Series* **92**, 343
Skinner, G.K., and Grindlay, J.E.: 1993, *Astron. Astrophys.* **276**, 673

A MERCURIC IODIDE X-RAY CAMERA FOR HIGH ENERGY ASTRONOMY

W. DUSI, E.CAROLI, G. DI COCCO, A.DONATI, G.RAMUNNO

Istituto TESRE/CNR ,Bologna, Italy

and

D.GRASSI, E.PERILLO

Dipartimento di Fisica, Università Federico II and Sezione INFN, Napoli, Italy.

Abstract. For fulfilling today's requirements for X-ray telescope based on concentrators focal plane detectors providing both high spatial resolution (<<1 mm) and good spectroscopic performance are nedeed. In this paper we present preliminary results on a prototype of a X- ray camera operative at room temperature for energies between few keV up to 150 keV. The camera is realised using a mercuric iodide crystal with microstrip electrodes, providing a geometrical spatial resolution of 500 µm. This detector will be suitable for use as a position sensitive spectrometer in the focal plane of X-ray telescopes based on Bragg concentrators.

Key words: Solid State Detectors, X- and γ- ray Astronomy, X- ray concentrators

1. Introduction

New hard X-ray (>10 keV) focusing techniques together with current astrophysical knowledge will put severe requirements on the development and design of focal plane detectors. In particular, for telescopes based on Bragg diffraction (Frontera and Pareschi, 1995; Christensen et al., 1995) operative between 10 keV up to ~200 keV, high spatial resolution (~100 µm) and good spectroscopic performance (E/ΔE ~20 @ 60 keV) detectors are required for use as focal plane instruments.

These capabilities could be achieved using position sensitive spectrometers made of solid state materials. Today a promising solution for this kind of application is offered by room temperature detectors based on mercuric iodide (HgI_2) crystals. A pixellated (1×1 cm^2) HgI_2 detector is already foreseen for a coded mask telescope (LEGRI) in a small satellite mission (Ballestreros et al, 1995). Indeed for focusing instruments the optimum solution will be offered by position sensitive detectors based on microstrip technology. In this paper we present preliminary results of a microstrip HgI_2 X-ray camera prototype that could be suitable for use as a focal plane detector for high energy telescopes operating between 10 and ~100 keV.

351

L. Bassani and G. di Cocco (eds.), Imaging in High Energy Astronomy, 351–355.
© 1995 *Kluwer Academic Publishers.*

2. Detector design and characteristics

Mercuric iodide (HgI_2) is a compound semiconductor with characteristics that make it an interesting material for use as a room temperature spectrometer (Iwanczyk et al. 1992, van den Berg et al. 1992). Its large bandgap of 2.14 eV ensures that thermally generated leakage current is kept at the order of a few pA at 300 K and this allows one to achieve a good energy resolution at room temperature. Because of its high Z (53-80) and density (6.3 g/cm^3) it has high photoelectric efficiency even for small thickness of material and it is particularly suitable for all those applications where a small detector with high efficiency is required (see Table 1).

Table 1. **Value** of some important parameters for X-ray absorption by 500 μm thick HgI_2 detectors: ε_{tot} and ε_p represent total and photoelectric efficiency; E_m is the average energy loss in the crystal; $\Delta E/E^{(1)}$ and $\Delta E/E^{(2)}$ are the experimentally measured energy resolution (FWHM) data respectively of: (1) a monoelectrode detector of 5x5x0.5 mm^3 (2) the X-ray camera. The errors on these values range between ~1% and 5%.

E (keV)	ε_{tot}	ε_p	E_m (keV)	$\Delta E/E^{(1)}$ %	$\Delta E/E^{(2)}$ %
5.9	1.	1.	5.9	26	30
14	1.	1.	14	13	21
18	1.	1.	18	12.2	14.5
32	0.99	0.98	31.3	9.5	10
60	0.86	0.84	44.4	6.5	7.5
81	0.59	0.55	38.3	6.	–
122	0.39	0.44	36.8	5.5	–

The microstrip detector described below is based on a Mercuric Iodide slice (Fig 1a), obtained from a high quality crystal, having a thickness of 0.5 mm and a surface area of 12x12 mm^2; the slice is mounted on a ceramic holder on which were previously deposited fan-out tracks (Dusi et al., 1994). Charge collection is realised by 20 palladium μ-strips (0.4 mm wide and 8.2 mm long with an interstrip gap of 0.1 mm) deposited, in an orthogonal manner, on the crystal surfaces so as to form a grid array. A pixel element is formed by the intersection of an upper surface strip (row) with one of the lower strips (column) having a geometrical size of 0.5x0.5 mm^2. The sensitive area is 8.2x8.2 mm^2 (256 pixels), as only 16x16 strips are used because the two external strips of each side of the surfaces are employed as a guard ring to provide a uniform electric field inside the detector. The whole detector is coated with a thin film of "GALXIL-C", a highly resistant polymer in order to provide protection from pollution and to ensure stability in time, avoiding evaporation of the crystal under vacuum.

3. Front-end electronics and experimental set-up

In preliminary tests performed on the X-ray camera a subarray of 5×5 strips was used to obtain a 25 pixel detector; each strip coupled to a hybrid charge sensitive preamplifier (CSP), model CS 507, produced by Clear Pulse (Tokyo) in a modified version with external input FET and a resistive feedback loop; CSP is characterised by an equivalent noise (r.m.s.) of about 1 keV. Pulses from CSP

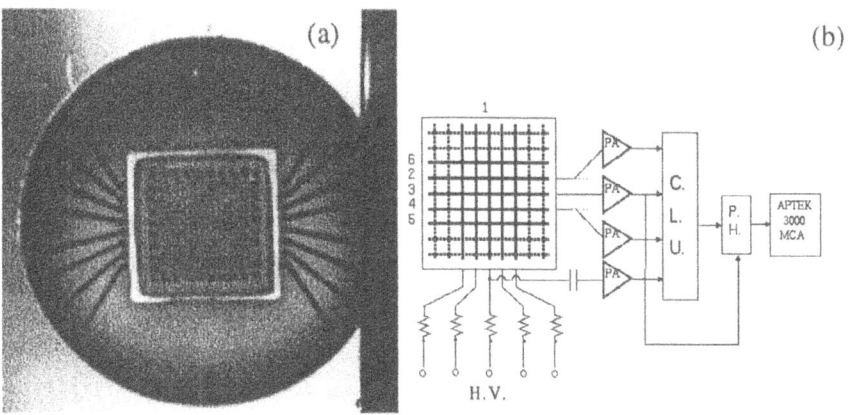

Fig. 1. (a) A back illuminated view of the HgI₂ X-ray camera integrated on the ceramic support showing the μ-strip grid; **(b)** Scheme of the electronics used for testing the central 5×5 pixels. PA's are the charge sensitive preamplifiers, C.L.U. represents the coincidence logic unit, P.H. is the Peak and Hold module.

are then amplified and shaped by main amplifiers; the scheme of the detector and the electronic chain is shown in Fig. 1b. The electronic chain includes a coincidence-anticoincidence circuit that allows the extraction of a spectrum from a pixel. The X-ray camera and the preamplifiers are mounted on a "LEXAN" board and housed in a copper box that acts as an electrostatic screen.

The tests for evaluating the spatial resolution were performed with a ^{133}Ba source placed on a two dimensional positioning system and limited by a 2.5 mm thick collimator with a pinhole of 300 μm in diameter. The source spot on the detector had an estimated diameter (~1 mm) larger than the geometrical size of a pixel. Additional tests were performed with other radioactive sources in order to evaluate the spectroscopic performance of the X-ray camera between 6 keV and ~120 keV.

4. X-Ray Camera performance

The first tests on the X-ray camera were performed in order to verify the effective size of the pixel, that is the area where the events are detected contemporarely by two orthogonal microstrips. In order to do this the detector was scanned by a collimated beam in both directions and for each position the number of the coincidence events from two orthogonal electrodes was recorded. In Fig. 2 are shown the count distributions along one direction for pixels (1,2) and (1,4). Because the diameter of the source spot is much larger than the width of the pixels the two distributions are overlapped and are not well resolved. However it is possible to estimate the size of the pixel

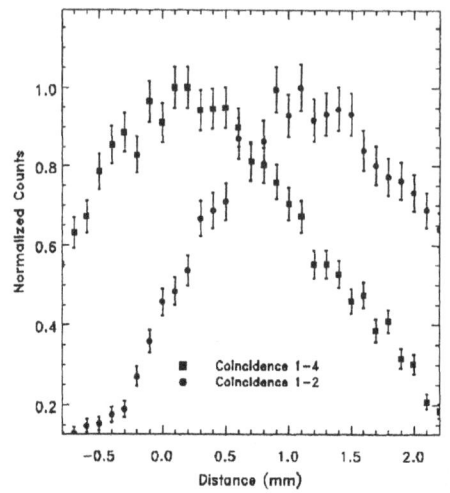

Fig. 2. Counting rate distribution from pixel 2 and 4. The counts are normalised to maximum rate of each pixel.

from the plot. Indeed the flat part of the distribution amounts to on about 700 μm without deconvolving for the width of the spot.

To evaluate the spectroscopic characteristics of the X-ray camera and the influence of multihits events (because of the charge sharing among the strips), the strip energy spectrum for the events giving a coincidence between orthogonal strips and an anticoincidence with two adjacent strips was measured. This technique allows the rejection of those events that cause a broadening of the peak spectrum, as they do not correspond to full charge collection. The spectrum obtained from the pixel (3,1) is shown in Fig. 3. The influence of multihits events on the spectroscopic

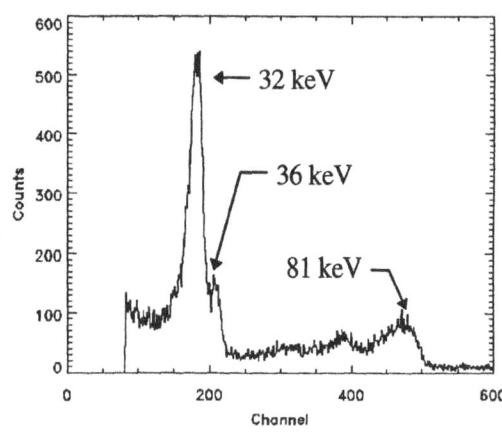

Fig.3. Spectrum of a ^{133}Ba source from strip n. 3 in coincidence with strip n. 1 and anticoincidence with strips 2 and 4.

characteristics is quite negligible up to 80 keV (Dusi et al., 1995). The energy resolution dependence on energy reported in Table 1 (last column) is compatible with a power law of index 0.63 ± 0.06 that is significantly steeper than the $1/\sqrt{E}$ law. This could be explained by the contribution of the electronic noise to the resolution that has a greater influence at lower energies. When considering only high energy data there is a much better agreement with the inverse square root law.

5. Conclusions

The preliminary results show that is possible to use our camera as a bi-dimensional PSD with digital readout of the strips from a few keV up to about 100 keV, with a spatial resolution compatible with the geometrical pixel size (500 µm), and a high spatial resolving power ($E/\Delta E \sim 13$ @ 60 keV). Other tests are in progress for improving the evaluation of the spatial resolution and its dependence on the photon energy and to study the uniformity of the response across the whole active detector surface.

References

Ballestreros, F. et al., 1995, *this proceedings*
van den Berg, L.: 1992, *Nucl. Instr. and Meth.* **A322**, 453
Christensen, F. :1995, *this proceedings*
Dusi, W., et al.: 1994, *Nucl. Instr. and Meth.* **A348**, 531
Dusi, W., et al., 1995, *Proceedings of 1994 IEEE Nuclear Science Symposium,* in press
Frontera, G., and Pareschi, G.:1995 *this proceedings*
Iwanczyk, J.S., Shnepple, W.F., and Masterson, M.J.: 1992, *Nucl. Instr.and Meth.* **A283**,421

THE NEUTRON SPECTRUM INSIDE THE SHIELDING OF
BALLOON-BORNE GE SPECTROMETERS

J.E. NAYA, P. JEAN, P. von BALLMOOS, F. ALBERNHE
Centre d'Etude Spatial des Rayonnements, UPS-CNRS, Toulouse, France

J. BOCKHOLT
Max Plank Institute für Kernphysik, Heidelberg, Germany

Abstract. Understanding the background of balloon borne germanium detectors used in high energy astrophysics is crucial for the design of future space missions. In this paper we present a method which allows to estimate the poorly understood β^- and β^+ background components inside the shield of an balloon borne instrument. The method consists of an experimental estimate of the neutron flux by determining the activation of isomers with measured gamma-ray line strengths and known cross sections. The gamma-ray line intensities of typical isomeric transitions in Germanium detectors are calculated for a neutron spectrum model. After comparing the calculated line fluxes with the measurements of balloon experiments the model neutron spectrum is changed appropriately. The procedure is repeated iteratively until an acceptable fit is achieved.
Applying this method to spectra of balloon-borne experiments results in neutron spectra that are harder than those generally used in standard background models.
The neutron spectra that we calculate for various balloon borne Ge spectrometers have made possible a re-estimation of the continuum background components due to β decays. It is shown that the enriched ^{70}Ge produces more β^- than expected. The model spectra also explain the poorly understood spectral features of ^{70}Ge detectors; e.g.; the behaviour of the neutron activation line at 198 keV and the increase of the background in the 1.5-4 MeV range, which is due to an enhanced production of isotopes that disintegrate via β^+ decays.

1. Introduction

Gamma ray spectroscopy has a unique potential to provide information on nuclear processes and high-energy particle interactions taking place in the Universe. Due to the opacity of the atmosphere, the observation of cosmic sources has to be performed either at the top of the atmosphere or in space. In space, the instruments are exposed to the radiation environment induced by high cosmic ray fluxes. The detection of the low intensity astrophysical gamma-rays against the dominant background is a major challenge in experimental gamma-ray astronomy.
A good comprehension of the background is thus crucial for the design of future space missions. Understanding the dependence of the background intensities on shield material, shield thickness, enrichment of the detector, etc., is important for the design instrument configurations that minimize the background and optimize the detection efficiency. Recent studies of the data of balloon-borne gamma-ray spectrometers have led to a multi-component background model [1] [2]. The background is understood as a four component continuum on which are superimposed a discrete nuclear gamma-ray lines. The origin of the continuum is from atmospheric and cosmic gamma rays that enter the instrument through its (1) aperture or (2) penetrate the shield, through (3) elastic neutron scattering in the detector and (4) through activation of the instrument materials by neutrons and protons induced by cosmic rays. The origin of the lines is through activation of the instrument material by atmospheric hadronic radiation and from secondary

L. Bassani and G. di Cocco (eds.), Imaging in High Energy Astronomy, 357–362.

particles produced by the cosmic ray bombardment of the instrument, as well as by natural radioactivity in the surrounding materials.

Although the overall background spectrum has been explained [3] to within 50% over most of the energy range of 30 keV to 10 MeV, a recent balloon flight [4] has raised anew the problem of the background. The detectors have shown behaviour that was not predicted by the theory, indicating that the understanding of the relative importance of the background components is still insufficient and that there is a non-negligible contribution of components that were previously not accounted for.

The analysis of the data background measured by enriched and natural germanium detectors during the ballon fligth of GRIS (1992) , have shown the following problems when compared to the existing model:

1-A ratio $I^{Enr}_{198}/I^{Nat}_{198}$ = 1 instead of the predicted value of 4 [3]. Where I^{Enr}_{198} is the intensity of the 198 keV background line from the isomeric transition ^{71m}Ge in an isotopically enriched ^{70}Ge detector and I^{Nat}_{198} is the intensity of this line in a natural germanium detector.

2-An unexpected increase of the background rate for the enriched ^{70}Ge GRIS detector in the 1.5-4 MeV range [4]. This behaviour has not been observed in HexagoneII.

These discrepancies with respect to the predictions show the necessity of more accurate calculations of the various components in order to provide a better quantitative understanding of the background.

In this paper, the gamma-ray line intensities of typical isomeric transitions in Ge detectors are used to calculate the neutron spectrum inside the shield.

In section 2, the method for calculating the neutron spectrum is explained and applied to several balloon-borne experiments. In section 3, a calculation of the continuum background components using the resulting neutron spectrum is performed. The resulting background is in good agreement with observations.

2. Neutron flux estimation from the isomeric transitions of Ge

The study of nuclear gamma-ray lines that are produced by isomeric transitions in Ge detectors offers the possibility of estimating the neutron spectrum inside the shield. This process constitutes an important source of background in gamma-ray spectrometers flown in the upper atmosphere and in earth orbit.

In the following (sections 2.1 and 2.2) we present the different isomeric transitions of Ge and our method of estimating the neutron flux is presented.

2.1 Ge ISOMERS

The nuclear reactions are described as taking place in two steps: (1) the formation of a compound nucleus in a highly excited virtual level, and (2) the dissociation of this compound nucleus leading to a state of high excitation of the product nucleus. The product nucleus, which decays promptly by electromagnetic transitions, ends up in the ground state or in some low-lying metastable level called an isomer.

Since the lifetimes of the Ge isomers are much longer than the microsecond anticoincidence times of standard instruments, the emitted lines are not vetoed by the shield. Thus the lines produced by the activation of Ge nuclei are the strongest in the background of a Ge spectrometer. These lines are the 198.4 keV, 53-66.7 keV complex and 139.7 keV. They originate in the isomeric transitions of ^{71m}Ge, ^{73m}Ge and ^{75m}Ge, respectively.

2.2 NEUTRON FLUX CALCULATION

The gamma-ray line intensities of typical isomeric transitions in Germanium detectors for a neutron spectrum are given by the equation:

$$I\,[Cs^{-1}cm^{-3}] \;=\; \frac{\langle D\rangle\rho N_A S_T}{4V\langle A\rangle}\,\varepsilon\int\limits_{E_0}^{E}\sigma_m(E)f(E)dE$$

where I is the intensity of the line (139 or 198 keV), $\langle D\rangle$ is the average distance crossed by a neutron flux impinging isotropically on the detector surface, ρ is the Ge density (5.32 g/cm^3), N_A is the Avogadro Number, S_T is the total detector surface, V is the detector volume, $\langle A\rangle$ is the Ge molecular weight, ε is the detector efficiency for the full absorption of the photon resulting from the isomeric transition, $\sigma_m(E)$ is the total cross section for the production of the isomeric state for the energy E and $f(E)$ is the calculated isotropic neutron flux in units of counts·cm^{-2}·MeV^{-1}·s^{-1}. The limits on the integral (E_0, E_1) correspond to the energy range with a significant value of the isomer cross sections which is 10^{-8}-10^2 MeV. The resulting neutron spectrum is valid within these limits.

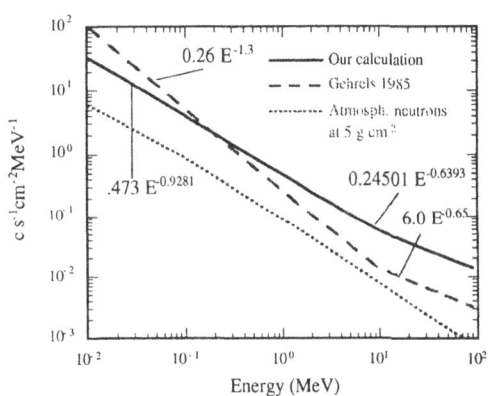

Fig. 1: Neutron spectrum inside a 15 cm NaI shield (GRIS 1992, Alice Springs). The calculated spectrum of this work is compared with a neutron spectrum previously used for background estimations [1] and the atmospheric neutron spectrum at 5 gr·cm^{-2} corresponding to Alice Springs derived from [5]

We have performed the calculations for the 1992 GRIS flight [4]. The resulting neutron spectrum is shown in fig.1 together with the neutron spectrum previously assumed for the calculation of the GRIS background. Also shown is the corresponding atmospheric neutron spectrum derived from Armstrong [5]. The GRIS neutron spectrum calculated by our method is harder than that used by Gehrels [3]: it is a factor of five higher at 100

MeV. This has implications on the continuum background component due to β decays that are calculated in the next section, i.e. the high energy neutron reactions are more dominant than previously assumed.

3.   Implications on the GRIS continuum background model

The calculated background spectra in fig 2 are based on the predictions for the GRIS experiment in fig 1 of [6]; they are plotted together with the observed spectra. Some changes have been made to account for the new calculations:

1   A calculation of the aperture flux was performed by Monte-Carlo simulation of a setup equivalent to GRIS (see point 3 of this section) using the γ-spectrum given in [1] but divided by 1.28 to account for the different γ-flux over Alice Springs [7];

2   The elastic neutron scattering considered follows the method given in [1];

3   The shield leakage component has been calculated by Monte-Carlo simulations for a setup equivalent to the GRIS experiment (15 cm NaI, 17° FOV, 80 keV of shield threshold, 500 gr of passive Al). It was irradiated by the anisotropic flux described by [1], but also divided by 1.28 to account for the lower γ-flux over Alice Springs.

4   The β⁻ decay curves are the sum of the individual decay spectra from the β⁻ unstable nuclides originiated from neutron and proton interactions with Ge. The shapes of the beta-ray pectra were calculated using the formula given in chapter 17 of [8]. The localized and non localized β⁻ components (i.e., single β⁻ decays and β⁻ decays accompanied with prompt gamma-ray emission respectively) have been treated separately.

5   The β⁺ component is produced by the decay beta-unstable nuclei induced by neutron and proton reactions. The released energy in one event consists of the kinetic energy of the positron plus the two 511 keV photons from the positron annihilation. This component was thought to be insignificant in previous models. But the harder neutron spectrum resulting from our calculations has revealed that the rate of the reactions producing this decay is not neglectible, especially for the enriched detector In order to simplify the calculation, we have assumed that only the β⁺'s that are totally absorbed in the detector are accepted (this is: not rejected by the anticoincidence system). Consequently this background component starts at an energy of 1.022 MeV.

The calculated efficiency for total absorbing both 511 keV photons of a positron annihilation in the enriched GRIS detector is 8.8 %.

In practice, a fraction of the incomplete absorbed events are not rejected by the shield. This happens when the photons leaving the detector are either absorbed by the passive material surrounding the detector or leave the instrument by the aperture. To take into account this contribution, a detailed model of the instrument should be considered in the calculations. However, this effect does not introduce a significant change in the total background of the studied instruments.

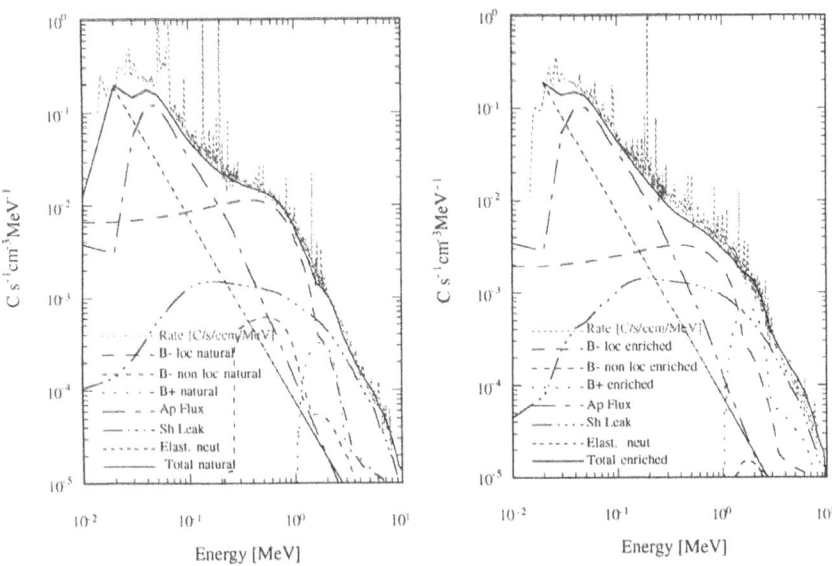

Fig 2: The calculated GRIS spectra, compared with the measured background during the 1992 flight over Alice Springs, Australia: (a) for the enriched detector; (b) for the natural detector.

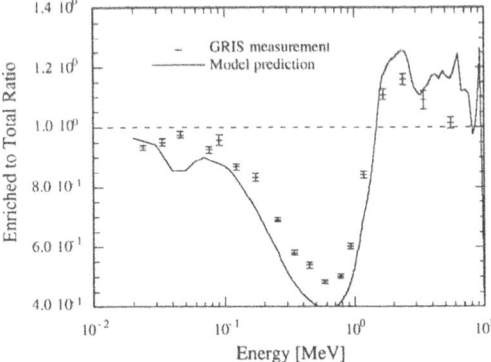

Fig. 3: Comparison of the calculated and observed background ratio between enriched and natural Ge, normalized by detector volume.

The ratio of the calculated total continuum background as a function of energy for the two isotopic abundances (enriched & normal) is diplayed in fig.3. The solid line represents our calculations the points are the GRIS measurements from the 1992 May 8 flight in Alice Springs.

The model fits well with the experimental data, at medium energies (0.3–1 MeV, the β bump) the background in the enriched is reduced by a factor of 2.2. The origin of the increase in the background for the enriched detector in the 1.5-4 MeV range can be explained by an enhanced yield of $^{68}Ga(\beta^+)^{68}Zn$ mainly produced by ^{70}Ge.

4.   Conclusion

We have presented a new method for calculating the neutron flux inside a shielded Germanium spectrometer. The principal results on the neutron-spectra and the effects on the background of the different instrument configurations are summarized below:

1   Comparing the resulting neutron spectrum with the atmospheric neutrons at this latitude, we conclude that the main part of the neutrons are induced by the cosmic-ray protons in the surrounding materials of the spectrometer.

2   We have shown that reactions induced by high energy neutrons are more important than formerly assumed because the calculated neutron spectrum is harder than previously noticed. This leads to the result that the β^- component is not mainly produced by the neutron capture reaction $^{74}Ge(n,\gamma)^{75}Ge$ induced by the atmospheric low energy neutrons, but also by reactions induced by higher energy neutrons (such as $^{76}Ge(n,2n)^{75}Ge$) that are mostly generated in the surrounding detector materials.

3   The enriched ^{70}Ge detector is a stronger β^- producer than predicted. Our estimate puts a limit on the achievable background reduction using enriched ^{70}Ge detectors: instead of the postulated 85%, the maximum achievable reduction with respect to natural Ge is about 70% at 600 keV.

4   Also, we notice that the harder neutron spectrum enhances the yield of β^+ producers (^{69}Ge, ^{68}Ga,..). Since the Ge isotope which is closest to these nuclei is ^{70}Ge it can account for a higher background spectrum for enriched ^{70}Ge detectors in the 1-4 MeV energy range forming the observed 2 MeV bump in the enriched to natural background spectra ratio. The importance of the β^+ component will be greater for a satellite instrument such as the INTEGRAL [15] spectrometer due to the higher efficiency of the detection system and the long life of the mission.

5.   References

[1] N. Gehrels, Nucl. Instr. and Meth. A239 (1985) 324

[2] A.J. Dean, F. Lei and P.J. Knight, Space Sci. Rev. 57 (1991) 109

[3] N. Gehrels, Nucl. Instr. and Meth. A292 (1990) 505

[4] S.D. Barthelmy, L.M. Bartlett, N. Gehrels, M. Leventhal, B.J. Teegarden, and J. Tueller, S. Belyaev, V. Lebedev and H.V. Klapdor-Kleingrothaus, Ap. J. 427 (1994) 519

[5] T.W. Armstrong, K.C. Chandler and J. Barish, J. Geophys. Res. 78 (1973) 2715

[15] INTEGRAL

[6] N. Gehrels, Nucl. Instr. and Meth. A313 (1992) 513

[7] J.M. Lavigne, M. Niel, G. Vedrenne, B. Arginier, E. Bonfand, B. Parlier and K.R. Rao, ApJ 261 (1982) 720

[8] R.D. Evans, The Atomic Nucleus, McGraw-Hill, New York, 1972

AN EXTENDED ENERGY RANGE CSI(TL)-PHOTODIODE
DETECTOR

A. J. BIRD, T. CARTER and A. J. DEAN

Astronomy Group, Physics Department, University of Southampton,
Southampton, SO17 1BJ ENGLAND

Abstract.
   A hybrid scintillator-photodiode detector has been developed which makes use of a pulse-shape analysis readout technique. This allows signals from gamma-ray interactions in the scintillator and from X-ray interactions in the silicon photodiode to be identified and separated. The low energy detection threshold in the silicon is below 10 keV, which can considerably extend the operational energy range. The resulting detector pixel is compact, rugged and requires no high voltage supply, and is ideal for incorporation in a position-sensitive detector array for hard X-ray and gamma-ray imaging.

Key words: Gamma-Ray Detector Pulse shape

1. Introduction

Photodiode readout of scintillation crystals is now a common radiation detection technique, with advantages over photomultiplier readout which include compactness, robustness, an unsusceptibility to magnetic fields, and no requirement for a high voltage supply.

   However, photodiode readout is inferior in terms of the low energy detection threshold which can be achieved. This is due to the lack of internal gain within the photodiode, which means that electronic readout noise will dominate the detector performance at low energies. A typical threshold for a 1cc CsI(Tl) detector will be in the range 30-50 keV. At energies below 50 keV a bare silicon photodiode becomes a useful X-ray detector due to photoelectric absorption within the bulk of the silicon.

   In this paper we investigate the use of a CsI(Tl)-photodiode detector in a hybrid mode, detecting low energy photons in the photodiode and higher energy photons in the scintillator. The different temporal characteristics of the two types of signal can then be used to distinguish them.

2. Theory

We are here attempting to distinguish between two distinct types of event with different pulse shape characteristics. The first is a scintillation event in the CsI(Tl) crystal, for which a typical yield will give ~52,000 optical photons per MeV energy deposit. When combined with the light collection and photodiode quantum efficiencies, this gives an overall conversion efficiency (OCE) of ~40 electrons/keV. The temporal decay of light from CsI

L. Bassani and G. di Cocco (eds.), Imaging in High Energy Astronomy, 363–366.

has been well reported in the literature [2,3], and is characterised by two decay components of approximately equal intensity and time constants of $\sim 0.7\mu s$ and $\sim 4\mu s$.

The second type of event is the absorption of lower energy X-ray photons within the bulk of the silicon photodiode. Here the overall conversion efficiency is determined simply by the energy required to produce an electron-hole pair, which in silicon is 3.6 eV. Thus the OCE for this type of detection is $\sim$275 electrons/keV. The signals from the silicon are much faster than those from the scintillator, and will, in practice, be limited by the risetime of the preamplifier used ($\sim$60ns). The basic principle of operation of the hybrid detector is shown in Figure 1.

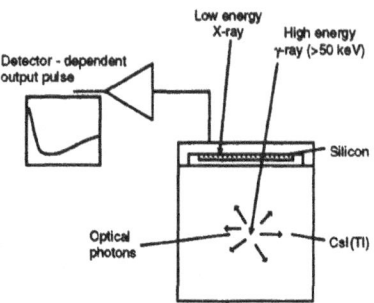

Fig. 1.   Principle of the hybrid CsI(Tl)-photodiode detector

Simulated efficiency curves for photodiodes 200 and $500\mu m$ thick have been calculated using both monte-carlo simulations of the detection efficiency and a spreadsheet model for the detection threshold (calculated using the reference data for the Hamamatsu S3590 series). As expected the $500\mu m$ diode offers better low energy detection efficiency, but at the cost of a large discontinuity in the response curve. This is a result of the higher detection threshold in the CsI(Tl) detector which is in turn caused by the higher leakage current from the thick diode. The minimum detection efficiency is calculated to be 8% and 2% for the $200\mu m$ and $500\mu m$ diodes respectively.

3.   Experimental Procedure

The tests were carried out using a typical CsI(Tl)-photodiode detector comprising a 1cm^3 scintillator read out by a 1cm^2 Hamamatsu S3590-01 photodiode, which has a thickness of $200\mu m$. The signals from the photodiode were amplified by an eV products eV-5093ULN hybrid preamplifier, and then passed directly to a Philips 3355 digital storage oscilloscope for digitisation. Each pulse was analysed to determine the 10%-90% rise-time and peak amplitude.

To evaluate the performance of the detector, 15000 events were collected in the CsI(Tl)-PD detector while it was simultaneously illuminated with ^{241}Am and ^{22}Na sources giving lines at 60, 511 and 1275 keV. At the same time a reference spectrum was collected with a standard NaI(Tl)-photomultiplier system. The pulse-shape spectrum for the raw events is shown in Figure 2. Two clear peaks are seen, easily distinguishing the fast events from the silicon and the slower scintillation events. The rise-time spectrum for the Si events is distorted as the readout system was too slow to accurately determine the rise-times of the fastest events.

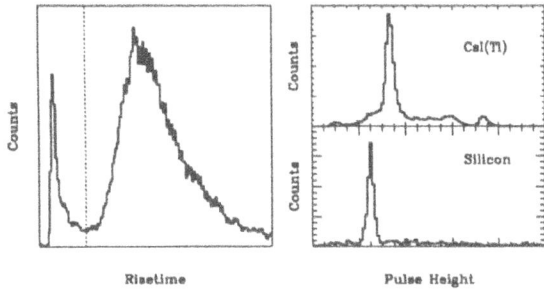

Fig. 2.   *Raw spectra from the hybrid detector showing (left) pulse shape and (right) energy*

Using this pulse-shape spectrum, the cross-over point between the two detectors was set at a rise-time of $\sim 3\mu s$, shown by the dotted line in the figure. Events with faster rise-times were analysed as coming from the photodiode and were essentially all due to 60 keV events; all others were assumed to come from the caesium iodide scintillator. The two types of event were analysed individually in order to obtain raw energy spectra from each detector (also shown in Figure 2). The peaks in the energy spectra were then used to calibrate the energy scales for the two systems.

The energy resolution obtained is poor in comparison to that which can be obtained with appropriate pulse shaping of the signals. Unfortunately this was not possible with our somewhat crude acquisition system. Figure 3. shows the predicted energy resolution for the hybrid detector, the two curves representing detection in the silicon and in the scintillator. These curves were obtained on separate test systems employing the correct pulse shaping.

Figure 3. shows a superposition of the energy spectra from the hybrid detector after they have been calibrated for energy. No attempt has been made to normalise the very different responses of the two detectors.

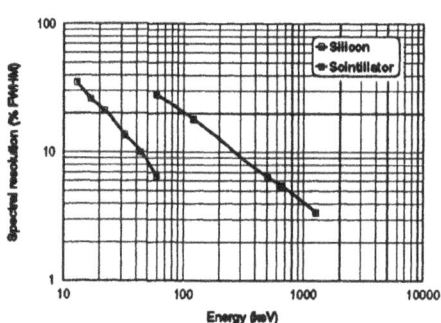

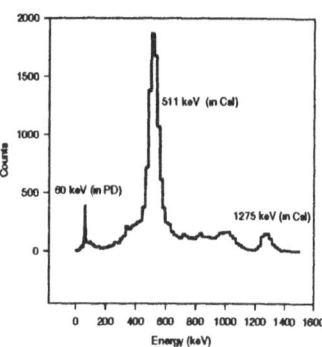

Fig. 3. *Predicted energy resolution for the hybrid detector*

Fig. 4. *Final pulse-height spectrum for the hybrid detector*

4. Conclusions

We have succesfully demonstrated the application of pulse-shape analysis to create a hybrid CsI(Tl)-silicon detector. This technique promises to lower the low energy threshold of such detectors significantly. Areas in which we aim to improve the system include:

(i) The use of more appropriate photodiodes. There is certainly more work to be done in optimising the thickness of the silicon. In addition, the thick ceramic packages on the devices used prevented their use for low-energy detection of X-rays by rear illumination - plastic packaging would be more suitable.

(ii) A suitable normalisation algorithm to enable the spectra from the two detectors to be combined more effectively. This requires very precise knowledge of the response of the system at all energies which can only be obtained by modelling and calibration.

References

[1] Friese, J. *et al.*, 1993, IEEE Trans. Nucl. Science, NS-40, no.4, p443

[2] Schotanus, P. *et al.*, 1993, IEEE Trans. Nucl. Science, NS-37, no.2, p177

[3] Valentine, J. *et al.*, 1992, Nucl. Instr. and Meth., A314, p119

The Energy Correction Algorithm for an X-ray Imaging Gas Scintillator Proportional Counter

E.Alippi,M.Biserni,P.Dalla Ricca,A.Lenti
Laben SpA, SS Padana Superiore 290, 20090 Vimodrone (Milano) - Tel. 02/250751 - Fax 02/2505515

Introduction

The Energy Correction Algorithm has been developed for the High Pressure Gas Scintillator Proportional Counter. (HP-GSPC) one of the instrument on board of the Italian Dutch X- ray Satellite , SAX.
The algorithm has been introduced in order to recover on the full instrument area the intrinsically good energy resolution of the detector.
A tabular method based on the research into a look-up table of the appropriate corrective factor has been developed. The definition of the look-up table, as well as its H/W implementation are described.
Some experimental results are reported illustrating the capability of the algorithm to correct the measured energy in function of the position.

Instrument Description

The HP-GSPC Detector Unit (fig. 1) consists of a gas cell pressurised at 5 atm. with a mixture of 90% Xe and 10% He.The electronic cloud generated inside the cell by the absorption of the X-ray radiation is drifted by an electrical field toward the region at higher electrical field in which the scintillation occurs.The UV signal produced in the scintillation region is detected by an array of seven photomultipliers (Pmts).
A guard ring region around the useful area allows the rejection of the background events.
The signal coming from the seven Pmts is at first amplified by the front end electronic mounted directly below every Pmt and then is processed by the electronic unit.
The individual signals from every Pmt , the sum signal the duration of the light burst and the arrival time are available for the successive elaboration.The individual signals of the Pmts are used to reconstruct the position of the scintillation and then to correct by means of an algorithm the Sum signal , in order to take into account the total solid angles subtended by the seven Pmts.
The position information is also used to reject the event occurring in the guard ring region.
A collimator mounted in front of the entrance window limits the field of view of the instrument.

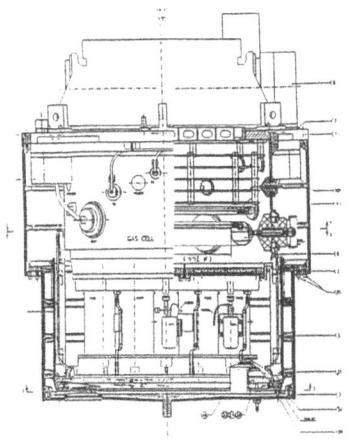

Fig 1 HP-GSPC Detector Unit

DETECTOR CHARACTERISTICS

Geometrical Area	450 cm2
Entrance Window	Be/ 1300 um
Depth of Drift Region	10 cm
Depth of Scintillatio Region	1 cm
UV Exit Window	7 Suprasil 5 mm
UV Read Out	7 pmt Emi D319
Voltage of Drift Region	10 KV
Voltage of Scintillation Region	10 KV
Filling Gas	90% Xe 10% He
Filling Pressure	5 atm
Energy Resolution	11% at 6 KeV
	3 5 % at 60 KeV
Energy Range	3 5 - 120 KeV
Field of View	1 1 °
Burst Lenght	1 - 5 usec

The scientific design of the Detector and the first prototype model have been realized by IFCAI .

L. Bassani and G. di Cocco (eds.), Imaging in High Energy Astronomy, 367–371.
© 1995 Kluwer Academic Publishers.

The Energy Correction Algorithm

Look up table definition

The good full area energy resolution of the instrument can be achieved, correcting the sum signal of the Pmts in function of the scintillation position, by means of the energy correction algorithm [1]

The need to correct can be better understood considering that the total solid angle variation along the useful radius is 20 % while the expected energy resolution of the instrument is 2 % at 120 KeV

In principle, the intrinsically good energy resolution can be recovered multiplying the measured energy for the ratio between the solid angle on axis and that subtended by the Pmts at the scintillation points

An appropriate method has been developed to define the corrective factor in function of the (x,y) position

The selection of the algorithm has been driven by a compromise between

- the degradation in the energy resolution shall be no more than 10 % respect to a perfect reconstruction
- low number of lost events during position reconstruction phase, which implies a short elaboration time

To satisfy the above requirements a tabular method based on an appropriate discretization of the image plane has been implemented

The plane has been subdivided in small areolas, at each vertex is associated the corrective factor Considering the symmetry of the detector only a 30° sector of the total surface has been analysed

The scintillation event occurring in any of the other eleven sectors is reconducted to the reference one by means of rotations and reflections

The assignment of an average corrective factor to every area and proper selection of the shape and dimensions of the areolas and the area identification parameters minimize the intrinsic errors of this method

Corrective factor

Being the number of photons picked up by the j-th Pmt(N_j), proportional to the effective solid angle subtended by the j-th Pmt($\Omega_j(x,y)$), it is possible to define the coefficient $C_j(x,y)$ as

$$S_j = N_j \propto N_{gen} \times \Omega_j(x,y)$$
$$S_{tot} = N_{tot} \propto N_{gen} \times \Omega_{tot}(x,y)$$
$$C_j(x,y) = \frac{S_j}{S_{tot}} = \frac{\Omega_j(x,y)}{\Omega_{tot}(x,y)} \qquad j = 1 \ 7$$

where

S_j is the signal coming from the j-th Pmt,

S_{tot} is the sum of the seven Pmt signals,

N_{gen} is the number of photons generated by the scintillation process,

N_{tot} is the total number of photons picked up by the seven Pmts,

$\Omega_{tot}(x,y)$ is the effective solid angle subtended by the seven Pmts at the (x,y) event position

$C_j(x,y)$ is directly obtainable by the detecting system and since at every point (x,y) is associable one and only one set of seven coefficients it is possible to come back to the event position and correct the energy

The easiest way to define the function $C(x,y)$ in the plane is by the isocoefficient line

The look up table is defined in such a way that the difference in solid angle between two adjacent areolas is less then 1% [2]

Scintillation position search

The scintillation sector identification is carried out in two steps

In the first one is searched the scintillation sector with 1/6 symmetry(each sector is centred on one lateral Pmt), this has done by comparing the signals of the lateral Pmts The highest signal is associated to the Pmt in the scintillation sector

The scintillation plane is rotated, around the detector symmetry axis, until the identified Pmt overlaps the Pmt in the reference sector

In the second step is identified the scintillation sector at 1/12 symmetry comparing the signal of the two Pmts adjacent to the reference one If the event is not in the reference sector (1/12) a reflection by π angle around the axis passing through the central and reference Pmts is performed

In this way it is possible to reconduct a scintillation that happened in any point of the scintillation plane to the a corresponding one in the reference sector

$$0 \leq \rho \leq 14 \ 5 \qquad 0° \leq \theta \leq 30°$$

Then only the coefficients relevant to two Pmts are used to find the scintillation position

- the coefficient C1 connected to central Pmt
- the coefficient C2 connected to lateral Pmt with the highest signal

The areola in which the scintillation occurred inside the reference sector is identified searching in the look up table the greatest C1 coefficient, closest to the measured one The correct identification is favoured by the fact that C1 is an increasing monotonic function, then along the C1 isocoefficient line will be searched the areolas vertex having the smallest C2 closest to the measured one

To the couple of coefficient is associated the corresponding corrective factor

Energy Correction Algorithm Implementation Description.

The correction algorithm of the HP-GSPC is implemented on a Digital Signal Processing board developed by LABEN for this particular application but with an architecture that can be easily applied in several other contests

This type of board provides the necessary interfaces with the experiment and with the upper level hardware (microprocessor boards) The choice of basing the design of the board on a Digital Signal Processor, instead of a definite not flexible hardware, has guaranteed the possibility of easy modifications in all the steps of development of the instrument On the other hand the increment of the processing time deriving from this choice does not constitute a problem for the instrument In fact, due to the "pipeline" structure of the A/D conversion/processing tasks and to the low execution time of the correction algorithm (54 4 µs max compared to the 60 µs of the A/D conversion of the 9 signals relevant to an event), the implementation of the correction algorithm is practically invisible HW design guidelines for this board have been the easy and full reprogrammability of both the data and program memory and the compatibility with the System Bus 90, a standard multimaster bus developed by LABEN for Data Processing systems

Due to these characteristics in every moment of the mission (calibration, instrument or satellite integration, mission life) both the look up table and/or any module of the correction algorithm can be modified to take into account of experimental data or Pmts degradation/failure

The board has been developed using the TMS320C25 digital signal processor, HCMOS integrated circuits for the glue logic (this technology is qualified for space applications), an FPGA (field programmable gate array) for the bus interface and on board time functions, and 2 hybrid memory modules, housing 64 Kbyte of EEPROM, developed by LABEN

In order to obtain very short processing time, the software of the algorithm has been tailored on the existing hardware and written in the assembly language of the TMS320C25 This digital processor has two internal blocks of RAM, one for data and one for program (in fact, due to its Hardware architecture, the TMS320C25 has internally separated program and data busses) During the initialisation the main module of the program is loaded in the dedicated RAM

The constants and the variables of frequent use are loaded in the data RAM

Due to the limited dimensions of this block all the routines that are not part of the main loop or the routines to handle low frequency interrupts, are located on the external EEPROM

Also all the look up tables are located on the EEPROM This type of memory is slow compared to the DSP, and introduces one wait state for each read operation To avoid excessive time wasting during the consultation of the look up tables, a search algorithm has been defined that reduces drastically the number of read operations

Table 1 shows the processing time estimated by simulation and successively verified experimentally

CONDITION	N OF MACHINE CYCLES	ELABORATION TIME (us)
Max time for pr a single event	272	54 4
Max time for pr a double event	524	104 8
Max time for rej a single event for E < Emin	206	41 2
Max time for rej a double event for E < Emin	415	83 0
Max time for rej a single event for Rad > Rad_max	212	42 4
Max time for rej a double event for Rad > Rad_max	421	84 2
Max time for rej a single event for FIFO full	54	10 8
Max time for rej a double event for FIFO full	58	11 6

Tab 1 Elaboration Times

Experimental results

Some spectra coming from the detector are reported in fig.2 and fig.3.

The measurements have been performed positioning a collimated source at 2 meters from the entrance window in order to simulate a full area illumination . A filter on the radius at 8 cm. has been used, since for this measurement the calculated map has been used, and not the measured one that takes into account also the boarder effect.

The spectrum of cadmium is shown in fig. 2.a for the uncorrected energy and 2.b for the corrected energy.

The two lines of cadmium 22 KeV and 25 KeV undistinguished in fig. 2.a, are instead resolved in fig 2.b, after correction .

Fig. 3 shows the spectrum of americium ; even in this case it can be seen after the reconstruction the good resolution of 60 KeV line and of the 26 and 29.8 escape peaks and of the 33.4 peak due the fluorescence of Xenon.

No sensible degradation in the energy resolution respect to a collimated source can be detected .

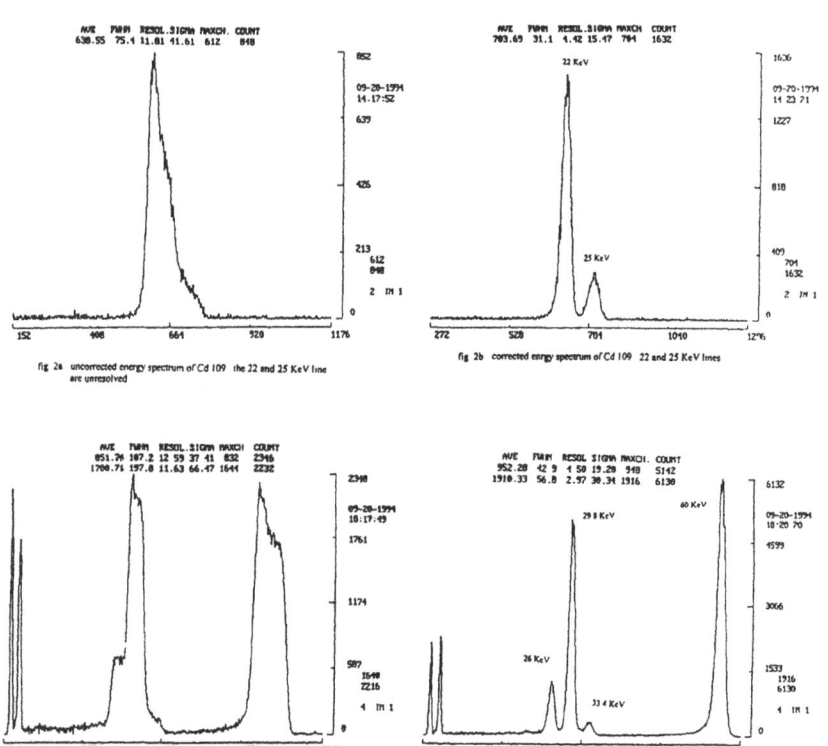

fig 2a uncorrected energy spectrum of Cd 109 the 22 and 25 KeV line are unresolved

fig 2b corrected energy spectrum of Cd 109 22 and 25 KeV lines

fig 3a uncorrected energy spectrum of Am 241 the 26 and 29 8 Kev escape peaks and the Xe fluorescent peak at 33 4 are unresolved

fig 3b corrected energy spectrum of Am 241 line at 60 KeV, escape peaks at 26 and 29 8 KeV, Xe fluorescent peak at 33 4 KeV

Conclusions

The Position Reconstruction Algorithm developed for the SAX HP-GSPC instrument represents not only a performing solution for the on-board spectra data acquisition, but also a promising method for all the position sensitive detectors, especially for those detectors presenting a geometrical symmetry in the scintillation-photo-detection volume.

A further application of this algorithm could be thought in the X-ray Imaging detector field whenever a coded mask is placed on the entrance window of a position sensitive device.

References

[1]   S.Giarrusso at. al. " Energy and Position Resolution of High Pressure Gas Scintillation Proportional Counter
       on board of the Italian- Dutch X-ray Astronomy Satellite SAX" , SPIE Vol. 982,1988
[2]   S.Giarrusso et al. " Reconstruction Algorithm and Full Area Energy Resolution of the Scientific Prototype
       of the High Pressure Gas Scintillation Proportional Counter on Board the SAX Satellite, to be published
       on Nuclear Instruments & Methods

THE PHOTON STATISTICS OF POINT SOURCE

CORRELATION IMAGES

IN CODED APERTURE IMAGING

MARK. H. FINGER

Compton Observatory Science Support Center, USRA

and

THOMAS A. PRINCE

California Institute of Technology

Abstract.
We discuss continuous source images for coded aperture gamma-ray telescopes that employ a mask-antimask pair in conjunction with a continuous position sensitive detector. The images discussed are constructed by correlation of the measured mask-antimask differenced count rates with the expected point source response. Results are presented on flux and point source location errors for background dominated observations. These show the impact of the detector's position resolution on the telescope's flux sensitivity and source location accuracy. We also discuss the expected frequency of noise peaks of a given significance in the image.

Key words: Gamma-Rays: Imaging – Imaging: Coded Aperture

1. Introduction

Coded aperture imaging [1, 2] is a technique for imaging sources of photons that can be used with x-rays and gamma-rays because no mirrors and lenses are required. The technique employs a mask composed of transparent and opaque regions which is interposed between the photon sources and a position sensitive detector. The flux from any given source is spatially modulated by the mask forming a shadow pattern on the detector. In a gamma-ray telescope this shadow pattern must be seperated from a large background of photon and particle interactions which vary in rate with position. This problem is solved by employing an antimask, a mask based on the same pattern but with transparent and opaque regions interchanged. The difference between the fluxes detected with the mask and antimask is independent of the position dependent background. For some mask patterns the antimask can be implemented by rotation of the mask [3, 4] or by a simple mechanical transformation of the mask [5].

After introducing the correlation image we discuss several problems associated with the detection and location of sources using this image. For concreteness we will assume that the mask-antimask pair have the imaging properties of Hexagonal Uniformly Redundant Arrays (HURAs)[3, 6], but much of the analysis has wider application.

373

L. Bassani and G. di Cocco (eds.), Imaging in High Energy Astronomy, 373–378.
© *1995 Kluwer Academic Publishers.*

2. Experiment Model

The difference $\Delta I(\mathbf{x})$ of the incident flux at detector position $\mathbf{x}$ between the mask and antimask measurements may be described as

$$\Delta I(\mathbf{x}) \; = \; \sum_j \Delta M(\mathbf{x} + \mathbf{z}_j) F_j \tag{1}$$

where j indexes sources in the field of view, and $\mathbf{z}_j$ is the plan of sky position of source j which has flux F_j. Due to the finite resolution of the detector the expected difference in counting rates will will be blurred compared to to the incident flux. If $p(\Delta \mathbf{x})$ is the point spread function of the detector, then the expected difference $D(\mathbf{x})$ in counting rates is

$$D(\mathbf{x}) \; = \; \int p(\Delta \mathbf{x}) I(\mathbf{x} - \Delta \mathbf{x}) d^2 \Delta x \; = \; \sum_j \Delta M^{eff}(\mathbf{x} + \mathbf{z}_j) F_j \tag{2}$$

where the effective mask difference function is given by

$$\Delta M^{eff}(\mathbf{y}) \; = \; \int p(\Delta \mathbf{x}) \Delta M(\mathbf{y} - \Delta \mathbf{x}) d^2 \Delta x \; . \tag{3}$$

3. The Correlation Image

The correlation image $C(\mathbf{z})$ is constructed by correlating the detected difference in count rates with the difference expected for a point source at a given sky position $\mathbf{z}$:

$$C(\mathbf{z}) \; = \; \sum_k \alpha_k \Delta M^{eff}(\mathbf{x}_k + \mathbf{z}) \tag{4}$$

where k indexes events, $\mathbf{x}_k$ is the detected location of an event, and α_k is one or minus one depending on whether the event occurred during mask or antimask observations. For an observation of duration τ evenly divided between mask and antimask data collection the expected value of the correlation image is

$$< C(\mathbf{z}) > = \frac{A\tau}{2} \sum_j f(\mathbf{z}, \mathbf{z}_j) F_j \; . \tag{5}$$

where A is the detector area from which photons are collected and the image function

$$f(\mathbf{z}, \mathbf{z}') \; \equiv \; \frac{1}{A} \int_A \Delta M^{eff}(\mathbf{x} + \mathbf{z}) \Delta M^{eff}(\mathbf{x} + \mathbf{z}') d^2 x \; . \tag{6}$$

The image is correlated from point to point. The covariance is given by

$$\begin{aligned}
\rho(\mathbf{z}, \mathbf{z}') &\equiv \; < C(\mathbf{z}) C(\mathbf{z}') > \; - \; < C(\mathbf{z}) > < C(\mathbf{z}') > \\
&= \; \frac{\tau}{2} \int_A R(\mathbf{x}) \Delta M^{eff}(\mathbf{x} + \mathbf{z}) \Delta M^{eff}(\mathbf{x} + \mathbf{z}') d^2 x \\
&\approx \; A\tau R f(\mathbf{z}, \mathbf{z}') \; ,
\end{aligned} \tag{7}$$

assuming that the observations are background dominated. Here $R(\mathbf{x})$ is the mean total count rate per unit detector area, which for this calculation has been approximated by the constant R.

Because the image function $f(\mathbf{z}, \mathbf{z}')$ determines both the expected value and the covariance of the correlation image, it is important to understand its form. When the masks are based on a HURA of order v, and photons are collected from a detector area A of the same size and shape as the basis mask pattern (which contains v cells), the image function takes on a very simple form in the fully coded field of view [7]:

$$f(\mathbf{z}, \mathbf{z}') = f(\mathbf{z} - \mathbf{z}') = \frac{v}{A} \int_A h^{eff}(\mathbf{x}) h^{eff}(\mathbf{x} + \mathbf{z} - \mathbf{z}') d^2 x \ - \ \frac{1}{v} \qquad (8)$$

where

$$h^{eff}(\mathbf{x}) = \int p(\Delta \mathbf{x}) h(\mathbf{x} - \Delta \mathbf{x}) d^2 \Delta x \qquad (9)$$

with $h(\mathbf{y})$ being a zero-one function describing a single hexagonal mask cell. Note that the correlation image for an HURA is the same (but with better statistics) as that obtained with a mask with a single hexagonal hole.

4. Source Flux and Location

For a single point source the flux is estimated from the peak value of the correlation image, and the source location from the location of the peak. Using equations 6 and 7 the flux and its error can be shown to be:

$$F = \frac{2C(\mathbf{z}_{max})}{A\tau f(\mathbf{0})} \pm \sqrt{\frac{4R}{A\tau f(\mathbf{0})}} \ . \qquad (10)$$

The statistical significance of a source detection is given by

$$\kappa_F = \frac{F}{\sigma_F} = \kappa_0 \sqrt{f(\mathbf{0})} \text{ where } \kappa_0 = \sqrt{\frac{A\tau F^2}{4R}} \ . \qquad (11)$$

Here, κ_0 is the statistical significance that would be obtained for a maskless source-region minus back-region measurement with an instrument of the same effective area and the same observation time. For suitably designed coded aperture systems $f(\mathbf{0})$ is close to unity, yielding close to optimum sensitivity.

For an isolated source the source direction is estimated from the location of the peak in the correlation image. The error in the source direction along any axis is then given by [7]:

$$\sigma_z = \frac{1}{\kappa_0} \sqrt{\frac{2}{|\nabla^2 f(\mathbf{0})|}} \ . \qquad (12)$$

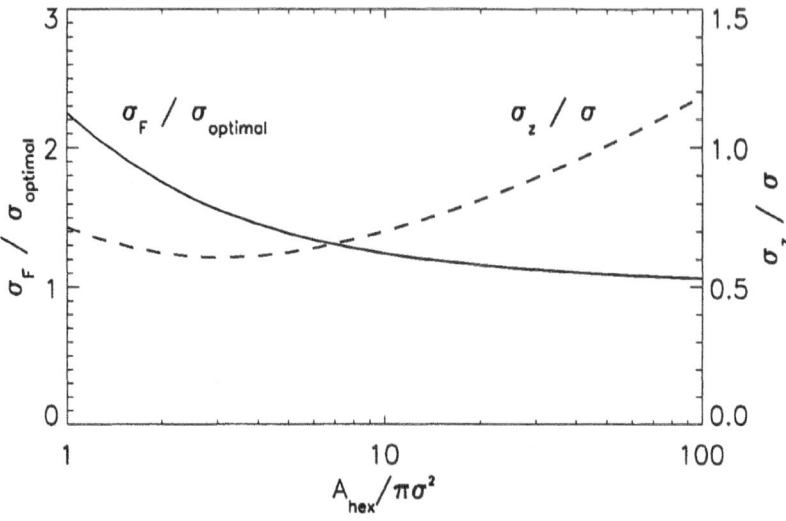

Fig. 1.   The dependence of source flux error σ_F (solid curve) and source location error σ_z (dashed curve) on the choice of mask cell area A_{hex}. The point spread function of the continuous position sensitive detector is taken to be gaussian with variance σ^2. For the assumed conditions a maskless source-region minus back-region measurement would have a flux error of $\sigma_{optimal}$ with a flux significance of $\kappa_0 = 5.0$.

In Figure 1 we show the dependence of the flux error σ_F and the source location error σ_z on the choice of the hexagonal cell size for fixed detector resolution. We assume that the point spread function is gaussian with standard deviation σ. The ratio of the flux error to $\sigma_{optimal}$, the error obtained from a maskless source-region minus background-region measurement, and the ratio of the location error to the position resolution σ are plotted versus the ratio of the hexagonal cell area, $A_{hex} = A/v$, to $\pi\sigma^2$. For the computation of the source location error it is assumed that $\kappa_0 = 5.0$. As can be seen from Figure 1, the source location error is optimized at small cell size, while the flux sensitivity is optimized in the limit of large cell size, requiring a trade-off between flux sensitivity and source localization ability.

5.   Noise Peaks in the Image

The correlation image will contain statistical noise peaks and the process of searching the image for sources will enhance the chance that noise peaks of a given significance will be encountered. To be confident that a peak in the correlation image is due to a source, we must therefore know the probability of such a peak occurring purely because of the statistical nature of the image. It can be shown that the probability per unit area of a peak (local maximum) of significance κ occurring in an image when no sources are present is given by [7]:

$$P(\kappa) = \frac{F}{A_0}e^{-B\kappa^2/2}\left[Ge^{G^2\kappa^2/2}Z(G\kappa)\right.$$
$$\left. + r^2(\kappa + [\kappa^2 - 1]He^{H^2\kappa^2/2}Z(H\kappa))\right] \tag{13}$$

where

$$\beta = \frac{-1}{2f(0)}\nabla^2 f(0) \ , \ \gamma = \frac{1}{8f(0)}\nabla^4 f(0) \ , \ r = \frac{\beta}{\sqrt{2\gamma}} \ ,$$

$$Z(x) = \int_{-\infty}^{x} e^{-y^2/2}dy \ , \ A_0 = 2\pi\sqrt{3}\frac{\beta}{\gamma} \ , \ B = \frac{1}{1-r^2} \ , \tag{14}$$

$$F = \frac{\sqrt{3}}{\pi}\frac{\sqrt{1-r^2}}{r} \ , \ G = \frac{r}{\sqrt{(3-2r^2)(1-r^2)}} \ , \ H = \frac{r}{\sqrt{1-r^2}} \ .$$

The area A_0 contains on average one peak.

In Figure 2 we show the integral of this distribution, for several choices of the ratio of the mask cell size A_{hex} to $\pi\sigma^2$. The integral distribution is defined by

$$P(> \kappa) = \int_{\kappa}^{\infty} P(\kappa)d\kappa \ . \tag{15}$$

Each curve is the expected number of peaks greater then κ in significance expected in the area $\pi\sigma^2$. These curves are useful in determining the confidence with which we can assume a peak in the image is due to a source. If we take as our confidence level the probability that an image has no noise peak of significance greater then κ then we have

$$Conf \geq 1 - A_I P(> \kappa) \ , \tag{16}$$

where A_I is the total area of the image. The equality will hold if the probability vanishes that two or more peaks of significance greater than κ can occur in the image due to noise.

6. Conclusion

We have described the construction of continuous images from the data of a coded aperture instrument with mask minus anti-mask background subtraction and a continuous detector. We have assumed that the mask used is an HURA, but simular results should hold for all masks with good imaging properties. The error analysis presented shows the the importance of correctly matching mask cell size to detector position resolution, and reveals that a trade-off must be made between flux sensitivity and source localization ability. Our results on the distribution of noise peaks in images can be used in setting realistic thresholds for source detection.

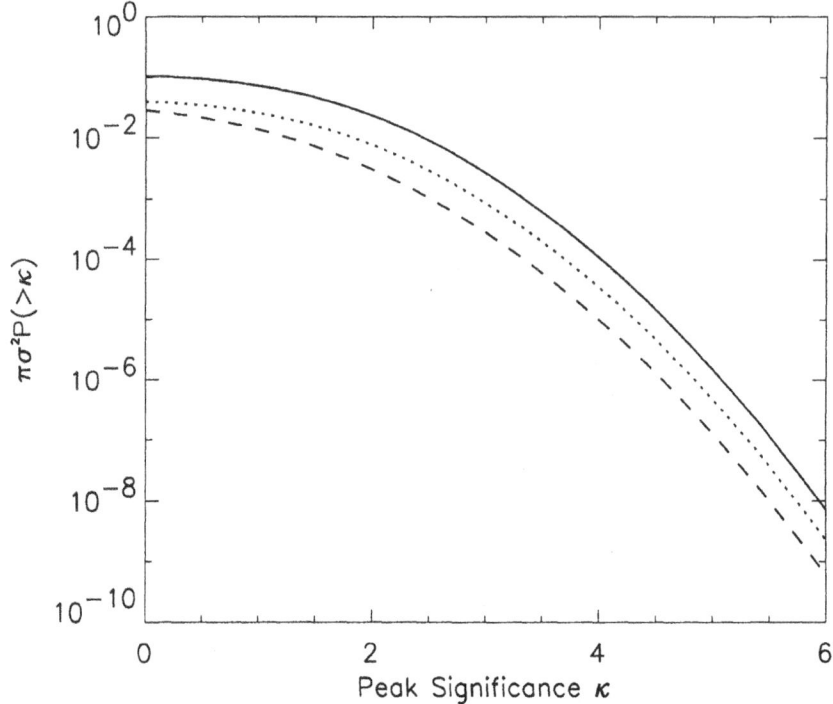

Fig. 2. The probability of finding a peak of significance greater than κ in an image area of $\pi\sigma^2$. The curves are parameterized by the ratio of the area A_{hex} of the mask cell hexagon to the position resolution area $\pi\sigma^2$, with $A_{hex}/\pi\sigma^2$ equal to 1, 10, and 100 for the solid, dotted, and dashed curve respectively.

References

1. E. Caroli, B. Stephen, G. Di Coco, L. Natalucci, and A. Spinzzichino, 1987, Sp. Sci. Rev., 45, 349
2. H. H. Barrett and W. Swindell, 1981, "Radiological Imaging, Vol. 1 and 2", Academic Press Inc., Orlando, Florida
3. W. R. Cook, M. Finger, T. A. Prince and E. C. Stone, 1984, IEEE Tran. Nuclear Science, NS-31, 771
4. K. Byard,1992, Experimental Astronomy, 2, 227
5. J. Braga, T. Villela, U. B. Jayanthi, F. D'Amico and J. A. Neri, 1991, Experimental Astronomy,2, 101
6. M. H. Finger and T. A. Prince, 1985, Proc. 19th International Cosmic Ray Conference,3, 295
7. M. H. Finger, 1988, Ph. D. Thesis, Caltech

Author Index

The manufacturer's authorised representative in the EU is Springer
Nature Customer Service Centre GmbH, Europaplatz 3, 69115 Heidelberg,
Germany. If you have any concerns regarding our products, please
contact ProductSafety@springernature.com

Printed and bound by CPI Group (UK) Ltd, Croydon, CR0 4YY
23/04/2026
02095628-0004